United States Nuclear Regulatory Commission

Protecting People and the Environment

NUREG-1910
Supplement 4, Vol. 2

I0493579

Environmental Impact Statement for the Dewey-Burdock Project in Custer and Fall River Counties, South Dakota

Supplement to the Generic Environmental Impact Statement for *In-Situ* Leach Uranium Milling Facilities

Draft Report for Comment

Chapter 5 and Appendices

Office of Federal and State Materials and Environmental Management Programs

AVAILABILITY OF REFERENCE MATERIALS
IN NRC PUBLICATIONS

United States Nuclear Regulatory Commission

Protecting People and the Environment

NUREG-1910
Supplement 4, Vol. 2

Environmental Impact Statement for the Dewey-Burdock Project in Custer and Fall River Counties, South Dakota

Supplement to the Generic Environmental Impact Statement for *In-Situ* Leach Uranium Milling Facilities

Draft Report for Comment

Chapters 5 and Appendices

Manuscript Completed: October 2012
Date Published: November 2012

Office of Federal and State Materials and
    Environmental Management Programs

COMMENTS ON DRAFT REPORT

Any interested party may submit comments on this report for consideration by the NRC staff. Comments may be accompanied by additional relevant information or supporting data. Please specify the report number NUREG–1910, Supplement 4, in your comments, and send them by the end comment period specified in the *Federal Register* notice announcing the availability of this report to the following address:

Cindy Bladey, Chief
Rules, Announcements, and Directives Branch
Division of Administrative Services
Office of Administration
Mail Stop: TWB-05-B01M
U.S. Nuclear Regulatory Commission
Washington, DC 20555-0001

For any questions about the material in this report, please contact:

Haimanot Yilma
Mail Stop T8F5
U.S. Nuclear Regulatory Commission
Washington, DC 20555-0001
Phone: 301-415-8029
E-mail: Haimanot.Yilma@nrc.gov

Please be aware that any comments that you submit to the NRC will be considered a public record and entered into the Agencywide Documents Access and Management System (ADAMS). Do not provide information you would not want to be publicly available.

NUREG-1910, Supp. 4, Vol. 2 has been reproduced from the best available copy.

ABSTRACT

The U.S. Nuclear Regulatory Commission (NRC) issues licenses for the possession and use of source material provided that proposed facilities meet NRC regulatory requirements and will be operated in a manner that is protective of public health and safety and the environment. Under the NRC environmental protection regulations in 10 CFR Part 51, which implement the National Environmental Policy Act of 1969 (NEPA), issuance of a license to possess and use source material for uranium milling, as defined in 10 CFR Part 40, requires an environmental impact statement (EIS) or a supplement to an EIS.

In May 2009, NRC issued NUREG–1910, the Generic Environmental Impact Statement for *In-Situ* Leach Uranium Facilities (GEIS) (NRC, 2009). In the GEIS, NRC assessed the potential environmental impacts from the construction, operation, aquifer restoration, and decommissioning of an *in-situ* leach uranium recovery facility [also known as an *in-situ* recovery (ISR) facility] located in four specified geographic regions of the western United States. As part of this assessment, NRC determined which potential impacts will be essentially the same for all ISR facilities and which will result in varying levels of impact for different facilities, thus requiring further site-specific information to determine potential impacts. The GEIS provides a starting point for NRC NEPA analyses for site-specific license applications for new ISR facilities, as well as for applications to amend or renew existing ISR licenses.

By letter dated August 10, 2009, Powertech (USA), Inc. (Powertech, referred to herein as the applicant) submitted a license application to NRC for a new source and byproduct material license for the Dewey-Burdock ISR Project. The proposed Dewey-Burdock ISR Project will be located in Fall River and Custer Counties, South Dakota, which is in the Nebraska-South Dakota-Wyoming Uranium Milling Region identified in the GEIS. The NRC staff prepared this draft Supplemental Environmental Impact Statement (SEIS) to evaluate the potential environmental impacts from the applicant proposal to construct, operate, conduct aquifer restoration, and decommission an ISR uranium facility at the proposed Dewey-Burdock ISR Project. This draft SEIS describes the environment potentially affected by the proposed site activities, presents the potential environmental impacts resulting from reasonable alternatives to the proposed action, and describes the applicant environmental monitoring program and proposed mitigation measures. In conducting its analysis in this draft SEIS, the NRC staff evaluated site-specific data and information to determine whether the applicant's proposed activities and site characteristics were consistent with those evaluated in the GEIS. NRC staff then determined relevant sections, findings, and conclusions in the GEIS that could be incorporated by reference and areas that required additional analysis. Based on its environmental review, the preliminary NRC staff recommendation is that a source and byproduct material license for the proposed action be issued as requested, unless safety issues mandate otherwise.

Paperwork Reduction Act Statement

This NUREG contains and references information collection requirements that are subject to the Paperwork Reduction Act of 1995 (44 U.S.C. 3501 et seq.). These information collections were approved by the Office of Management and Budget (OMB), approval numbers 3150-0014, 3150-0020, 3150-0021, and 3150-0008.

Public Protection Notification

NRC may not conduct or sponsor, and a person is not required to respond to, a request for information or an information collection requirement unless the requesting document displays a currently valid OMB control number.

References

10 CFR Part 40. Code of Federal Regulations, Title 10, *Energy*, Part 40. *"Domestic Licensing of Source Material."* Washington, DC: U.S. Government Printing Office.

10 CFR Part 51. Code of Federal Regulations, Title 10, *Energy*, Part 51. *"Environmental Protection Regulations for Domestic Licensing and Related Regulatory Functions."* Washington, DC: U.S. Government Printing Office.

NRC. NUREG–1910, "Generic Environmental Impact Statement for *In-Situ* Leach Uranium Milling Facilities." ML091480244, ML091480188. Washington, DC: NRC. May 2009.

CONTENTS

CONTENTS (continued)

CONTENTS (continued)

CONTENTS (continued)

CONTENTS (continued)

CONTENTS (continued)

CONTENTS (continued)

xi

CONTENTS (continued)

CONTENTS (continued)

CONTENTS (continued)

CONTENTS (continued)

CONTENTS (continued)

CONTENTS (continued)

FIGURES

FIGURES (continued)

FIGURES (continued)

TABLES

TABLES (continued)

EXECUTIVE SUMMARY

BACKGROUND

By letter dated August 10, 2009, Powertech (USA), Inc. (Powertech) submitted an application to the U.S. Nuclear Regulatory Commission (NRC) for a new source and byproduct material license for the Dewey-Burdock *In-Situ* Uranium Recovery Project, located in Fall River and Custer Counties, South Dakota. The applicant is proposing to recover uranium using the *in-situ* leach (ISL) [also known as *in-situ* recovery (ISR)] process. The proposed Dewey-Burdock ISR Project would include processing facilities and sequentially developed wellfields sited in two contiguous areas, the Burdock area and the Dewey area. Proposed facilities include a central processing plant in the Burdock area, a satellite facility in the Dewey area, wellfields, Class V deep injection wells and/or land application areas for disposal of liquid wastes, and the attendant infrastructure (e.g., pipelines and surface impoundments).

The Atomic Energy Act of 1954 (AEA), as amended by the Uranium Mill Tailings Radiation Control Act of 1978, authorizes NRC to issue licenses for the possession and use of source material and byproduct material. These statutes require NRC to license facilities, including ISR operations, in accordance with NRC regulatory requirements to protect public health and safety from radiological hazards. Under the NRC environmental protection regulations in 10 CFR Part 51, which implement the National Environmental Policy Act of 1969 (NEPA), preparation of an environmental impact statement (EIS) or supplement to an EIS is required for issuance of a license to possess and use source material for uranium milling [10 CFR 51.20(b)(8)].

In May 2009, the NRC staff issued NUREG–1910, the Generic Environmental Impact Statement for *In-Situ* Leach Uranium Milling Facilities (herein referred to as the GEIS) (NRC, 2009). In the GEIS, NRC assessed the potential environmental impacts from the construction, operation, aquifer restoration, and decommissioning of an ISR facility located in four specified geographic regions of the western United States. The proposed Dewey-Burdock ISR Project is located within the Nebraska-South Dakota-Wyoming Uranium Milling Region identified in the GEIS. The GEIS provides a starting point for NRC NEPA analyses for site-specific license applications for new ISR facilities, as well as for applications that amend or renew existing ISR licenses. This Supplemental EIS (SEIS) incorporates by reference information from the GEIS and also uses information from the applicant's license application and other independent sources to fulfill the requirements set forth in 10 CFR 51.20(b)(8).

This draft SEIS includes the NRC staff analysis that considers and weighs the environmental effects of the proposed action, the environmental impacts of alternatives to the proposed action, and mitigation measures to either reduce or avoid adverse effects. It also includes the NRC staff's preliminary recommendation regarding the proposed action.

This draft SEIS was prepared in cooperation with the U.S. Bureau of Land Management (BLM). BLM has requested to be and is acting as a cooperating agency with NRC to evaluate the impacts of Powertech's Plan of Operations (POO) in accordance with the National Memorandum of Understanding with NRC. BLM manages 97 ha [240 ac] of land within the proposed Dewey-Burdock ISR Project area. Under 43 CFR Part 3809, BLM is required to review the environmental impacts of federal actions on surface lands to assure that there is no "unnecessary or undue degradation of public lands." To fulfill this requirement, the applicant submitted a POO to BLM for the Dewey-Burdock ISR Project on August 26, 2009. Powertech modified the POO and resubmitted it to BLM on January 28, 2011.

1 **PURPOSE AND NEED FOR THE PROPOSED ACTION**
2
3 NRC regulates uranium milling, as defined in 10 CFR 40.4, including the ISR process, under
4 10 CFR Part 40, "Domestic Licensing of Source Material." The applicant is seeking an NRC
5 source and byproduct material license to authorize commercial-scale ISR uranium recovery at
6 the proposed Dewey-Burdock ISR Project. The purpose and need for the proposed federal
7 action is to either grant or deny the applicant a license to use ISR technology to recover
8 uranium and produce yellowcake at the proposed project. Yellowcake is the uranium oxide
9 product of the ISR milling process used to produce various products including fuel for
10 commercially operated nuclear power reactors.
11
12 This definition of purpose and need reflects the Commission's recognition that, unless there are
13 findings in either the AEA-required safety review or in the NEPA environmental analysis that
14 would lead NRC to reject a license application, NRC has no role in a company's business
15 decision to submit a license application to operate an ISR facility at a particular location.
16
17 The BLM purpose and need for the proposed action is to provide for orderly, efficient, and
18 environmentally responsible mining of the uranium resource. The uranium resource is needed
19 to fulfill market demands for this product for power generation and other needs. These public
20 lands are open to mineral entry, and the applicant has filed mining claims on them. Within the
21 proposed project area, Powertech maintains the mining claims associated with 1,708 ha
22 [4,220 ac] of federal minerals that the U.S. Government reserved under the Stock-Raising
23 Homestead Act. The BLM federal decision is to either approve the Powertech-modified POO
24 subject to mitigation included in the license application and this draft SEIS, or deny approval of
25 the POO. BLM's responsibility to respond to the POO establishes the need for the action. The
26 mining claimant has the right to mine and develop the mining claims as long as it can be done
27 without causing unnecessary or undue degradation of the public lands and follows pertinent
28 laws and regulations under 43 CFR Part 3800.
29
30 **THE PROJECT AREA**
31
32 The proposed Dewey-Burdock ISR Project is located in Custer and Fall River Counties,
33 South Dakota, within the Great Plains physiographic province on the edge of the Black Hills
34 uplift. The proposed site is located approximately 21 km [13 mi] north-northwest of the city
35 of Edgemont, approximately 64 km [40 mi] west of the city of Hot Springs, and approximately
36 80 km [50 mi] southwest of the city of Custer. The total land area of the proposed Dewey-
37 Burdock Project is 4,282 ha [10,580 ac]. Sections within the proposed project area are split
38 estate, in which two or more parties own the surface and subsurface mineral rights. The
39 surface rights are both publicly and privately owned. Approximately 4,185 ha [10,340 ac] of
40 land is privately owned, and the remaining 97 ha [240 ac] of surface rights are owned by the
41 U.S. Government and administered by BLM. The subsurface mineral rights are owned by
42 various private entities and federally reserved by the U.S. Government.
43
44 The proposed Dewey-Burdock ISR Project will consist of processing facilities and sequentially
45 developed wellfields in two contiguous areas: the Burdock area and the Dewey area. Planned
46 facilities associated with the proposed project include buildings associated with a central
47 processing plant in the Burdock area and a satellite facility in the Dewey area; surface
48 impoundments; wellfields and their associated infrastructure (e.g., wells, header houses, and
49 pipelines); Class V deep injection wells and/or land application areas for disposal of liquid
50 wastes; and access roads. The applicant estimated that the land surface area that would be

1 affected by proposed ISR operations would be approximately 98 ha [243 ac] if Class V deep
2 injection wells alone are used to dispose of process-related liquid wastes and approximately
3 566 ha [1,398 ac] if land application alone is used to dispose of liquid wastes.
4
5 ***IN-SITU* RECOVERY PROCESS**
6
7 During the ISR process, an oxidant-charged solution, called a lixiviant, is injected into the
8 production zone aquifer (uranium ore body) through injection wells. Typically, a lixiviant
9 uses native groundwater (from the production zone aquifer), carbon dioxide, and sodium
10 carbonate/bicarbonate, with an oxygen or hydrogen peroxide oxidant. As the lixiviant circulates
11 through the production zone, it oxidizes and dissolves the mineralized uranium, which is present
12 in a reduced chemical state. The resulting uranium-rich solution is drawn to recovery wells by
13 pumping and then transferred to a processing facility via a network of pipelines, which may be
14 buried just below the ground surface. At the processing facility, the uranium is removed from
15 solution (typically via ion exchange). The resulting barren solution is then recharged with the
16 oxidant and reinjected to recover more uranium.
17
18 During production, the uranium recovery solution continually moves through the aquifer from
19 injection wells to recovery wells. These wells can be arranged in a variety of geometric patterns
20 depending on the location and orientation of the ore body, aquifer permeability, and operator
21 preference. Wellfields are typically designed in a five-spot or seven-spot pattern, with each
22 recovery (i.e., production) well located inside a ring of injection wells. Monitoring wells are
23 installed in the production zone aquifer and surround the wellfield pattern area. Monitoring
24 wells are screened (i.e., open to allow water to enter) in the appropriate stratigraphic horizon
25 to detect the potential migration of lixiviant away from the production zone. Monitor wells are
26 also installed in the overlying and underlying aquifers to detect the potential vertical
27 migration of lixiviant outside the production zone. The uranium that is recovered from the
28 solution is processed, dried into yellowcake, packaged into NRC- and U.S. Department of
29 Transportation (USDOT)-approved 208-L [55-gal] steel drums, and trucked offsite to a licensed
30 conversion facility.
31
32 Once production is complete, the production zone groundwater is restored to NRC-approved
33 groundwater protection standards, which are protective of the surrounding groundwater. The
34 site is decommissioned according to an NRC-approved decommissioning plan and in
35 accordance with NRC-approved standards. Once decommissioning is approved, the site may
36 be released for public use.
37
38 **ALTERNATIVES**
39
40 The NRC environmental review regulations that implement NEPA in 10 CFR Part 51 require
41 NRC to consider reasonable alternatives, including the No-Action alternative, to a proposed
42 action. The NRC staff considered a range of alternatives that would fulfill the underlying
43 purpose and need for the proposed action. From this analysis, a set of reasonable alternatives
44 was developed, and the impacts of the proposed action were compared with the impacts that
45 would result if a given alternative was implemented. This SEIS evaluates the potential
46 environmental impacts of the proposed action and the No-Action alternative and also considers
47 alternative wastewater disposal options to the proposed action. Under the No-Action
48 alternative, the applicant would not construct and operate ISR facilities at the proposed site.
49 Other alternatives considered at the proposed Dewey-Burdock ISR Project site but eliminated
50 from detailed analysis include conventional mining and milling, conventional mining and heap

1   leach processing, alternative lixiviants, alternative site locations, and alternative well completion
2   methods. These alternatives were eliminated from detailed study because they either would not
3   meet the purpose and need of the proposed project or would cause greater environmental
4   impacts than the proposed action. This SEIS also discusses alternative wastewater disposal
5   options (evaporation ponds and surface water discharge) that were not included in the
6   proposed action.
7
8   **SUMMARY OF ENVIRONMENTAL IMPACTS**
9
10 This draft SEIS includes the NRC staff analysis that considers and weighs the environmental
11 impacts from the construction, operation, aquifer restoration, and decommissioning of ISR
12 operations at the proposed Dewey-Burdock ISR Project site and the No-Action alternative. This
13 draft SEIS also describes mitigation measures for the reduction or avoidance of potential
14 adverse impacts that (i) the applicant has committed to in its NRC license application, (ii) will be
15 required under other federal and state permits or processes, or (iii) are additional measures
16 NRC staff identified as having the potential to reduce environmental impacts but that the
17 applicant did not commit to in its application. The draft SEIS uses the assessments and
18 conclusions reached in the GEIS in combination with site-specific information to assess and
19 categorize impacts.
20
21 As discussed in the GEIS and consistent with NUREG–1748 (NRC, 2003), the significance of
22 potential environmental impacts is categorized as follows:
23
24 SMALL: The environmental effects are not detectable or are so minor that they will
25   neither destabilize nor noticeably alter any important attribute of the resource.
26
27 MODERATE: The environmental effects are sufficient to alter noticeably, but not
28   destabilize, important attributes of the resource.
29
30 LARGE: The environmental effects are clearly noticeable and are sufficient to
31   destabilize important attributes of the resource.
32
33 Chapter 4 of this draft SEIS provides the NRC evaluation of the potential environmental impacts
34 from the construction, operation, aquifer restoration, and decommissioning of the proposed
35 Dewey-Burdock ISR Project. The significance of impacts from the ISR facility lifecycle is listed
36 next, followed by a summary of impacts by environmental resource area and ISR phase for the
37 proposed action.
38
39 **Impacts by Resource Area and ISR Facility Phase**
40
41 **Land Use**
42
43 Construction: Impacts will be SMALL. If deep well disposal via Class V injection wells alone is
44 used to dispose of liquid wastes, approximately 98 ha [243 ac] or 2.3 percent of the proposed
45 project area will be disturbed by the construction phase. If land application alone is used to
46 dispose of liquid wastes, the construction phase will disturb approximately 566 ha [1,398 ac] or
47 13.2 percent of the proposed project area. Topsoil will be stripped and stockpiled to build
48 surface facilities, develop the initial wellfields and the attendant infrastructure, and construct
49 access roads. Livestock grazing and recreational activities will be excluded from fenced areas
50 surrounding the central plant, satellite facility, surface impoundments, and wellfields.

1   Operation: Impacts will be SMALL. Land use impacts during the operations phase will be
2   limited to the wellfields and will be similar to, or less than, those during the construction phase.
3   Wellfields will be sequentially developed resulting in the disturbance of approximately 57 ha
4   [140 ac]. Land disturbance and access restrictions will result from drilling new wells and
5   constructing additional header houses and pipelines. Livestock grazing and recreational
6   activities will continue to be restricted from the central plant, satellite facility, surface
7   impoundments, and wellfields. Potential land application areas may also be fenced to control
8   livestock access.
9
10   Aquifer Restoration: Impacts will be SMALL. Land use impacts will be similar to, or less than
11   those described for the operations phase. Land use impacts will decrease as fewer wells and
12   pump houses are used and overall equipment traffic and use diminish. Access to wellfields
13   and surface facilities will continue to be restricted. No additional land will be disturbed to
14   construct facilities.
15
16   Decommissioning: Impacts will be SMALL to MODERATE. Land use impacts during the
17   decommissioning phase will be similar to those experienced during the construction phase.
18   Decommissioning the buildings, wellfields, storage ponds, and access roads and removing
19   potentially contaminated soil will result in a temporary, short-term increase in land-disturbing
20   activities. Upon completion of the plugging and abandonment of wells, the soil will be
21   returned to areas in the wellfield where it had been removed and reseeded. At the end of
22   decommissioning, because the reclaimed land will be released for other uses and no longer
23   restricted, the land use impact in disturbed areas will be MODERATE until vegetation becomes
24   reestablished. After vegetation is reestablished in reclaimed areas, the land will be returned to
25   a condition that can support a variety of land uses; therefore, the impact will be SMALL.
26
27   **Transportation**
28
29   Construction: Impacts will be SMALL to MODERATE. Dewey Road, the road nearest the
30   proposed site, will experience a sixteenfold increase in daily vehicle traffic during the ISR
31   construction phase. This increase in traffic will accelerate degradation of road surfaces,
32   increase the generation of dust, and increase the potential for traffic accidents and wildlife or
33   livestock kills. The well-traveled regional roads will not be significantly impacted by the
34   construction traffic.
35
36   Operation: Impacts will be SMALL to MODERATE. Dewey Road, the road nearest the
37   proposed site, will experience a fivefold increase in daily vehicle traffic during the ISR
38   operations phase. This increase in traffic will accelerate degradation of road surfaces, increase
39   the generation of dust, and increase the potential for traffic accidents and wildlife or livestock
40   kills. Additionally, the transport of yellowcake product, hazardous materials, uranium-loaded
41   resins from the Dewey Unit to the Burdock Unit, and wastes could result in spills or leakage if an
42   accident occurred; however, this risk was determined to be low and will be further limited by
43   compliance with existing NRC and USDOT transportation regulations and the implementation of
44   best management practices (BMPs) for containing leakage and spills.
45
46   Aquifer Restoration: Impacts will be SMALL. Transportation impacts will be less than those
47   estimated for the construction and operation phases because the need to transport yellowcake
48   product, hazardous materials, and uranium-loaded resins between units will decrease as aquifer
49   restoration progressed. The decrease in the supply shipments, waste shipments, and employee

1   commuting (because fewer workers will be involved) will reduce the potential for spills or
2   leakage from accidents.
3
4   Decommissioning: Impacts will be SMALL. Transportation impacts will be less than those
5   during the construction and operation phases because the transport of yellowcake product and
6   processing chemicals will end during decommissioning. Access roads will either be reclaimed
7   or left in place for future use. Waste shipments will increase temporarily, but will still represent a
8   small contribution to daily traffic. Fewer workers will be employed, further reducing the potential
9   transportation impact during this phase.
10
11   **Geology and Soils**
12
13   Construction: Impacts will be SMALL. Earthmoving activities associated with construction of
14   the Burdock central plant and Dewey satellite plant facilities, access roads, wellfields, pipelines,
15   and surface impoundments will include topsoil clearing and land grading. Topsoil removed
16   during these activities will be stored and reused later to restore disturbed areas. The limited
17   areal extent of the construction area, the soil stockpiling procedures, the implementation of
18   BMPs, the short duration of the construction phase, and mitigative measures such as
19   reestablishment of native vegetation will further minimize the potential impact on soils.
20
21   Operation: Impacts will be SMALL. The operation phase will not remove rock matrix or
22   structure and will not dewater production zone aquifers. Therefore, no significant matrix
23   compression or ground subsidence is expected. The occurrence of potential spills during
24   transfer of uranium-bearing lixiviant to and from the Burdock central plant and Dewey satellite
25   facility will be mitigated by implementing onsite standard procedures and by complying with
26   NRC requirements for spill response and reporting of surface releases and cleanup of any
27   contaminated soils. The U.S. Environmental Protection Agency (EPA) will determine the
28   suitability of deep geologic formations for deep Class V disposal of liquid waste before issuing a
29   underground injection control (UIC) permit for Class V injection wells. Treated wastewater
30   disposed of in Class V injection wells will be required to meet release standards as referenced
31   in 10 CFR Part 20, Subparts D and K and Appendix B. Potential soil contamination in
32   proposed land application areas will be mitigated by implementing soil collection and monitoring
33   procedures. Treated wastewater applied to land application areas will be required to meet NRC
34   release limit criteria, as referenced in 10 CFR Part 20, Appendix B, and applicable state
35   groundwater quality standards under a Groundwater Discharge Permit (GDP) issued by South
36   Dakota Department of Environmental and Natural Resources (SDDENR).
37
38   Aquifer Restoration: Impacts will be SMALL. During aquifer restoration, the processes of
39   groundwater sweep and groundwater transfer will not remove rock matrix or structure. The
40   formation groundwater pressure within the extraction zone will be decreased during restoration
41   as groundwater is removed to ensure the direction of groundwater flow is into the wellfields to
42   reduce the potential for lateral migration of constituents. However, the change in groundwater
43   pressure will not result in collapse of overlying rock strata as it is supported by the rock matrix of
44   the formation. The potential impact to soils from spills, leaks, and land application of treated
45   wastewater will be comparable to that described for the operations phase. The NRC
46   requirements for spill response and recovery and routine monitoring programs will also apply.
47
48   Decommissioning: Impacts will be SMALL. Disruption or displacement of soils will occur during
49   dismantling of the facilities and reclamation of the land; however, the disturbed lands will be

1   restored to their preextraction land use. Topsoil will be reclaimed and the surface regraded to
2   the original topography.
3
4   **Surface Waters and Wetlands**
5
6   Construction: Impacts will be SMALL. The occurrence of surface water at the proposed
7   Dewey-Burdock site is limited, and surface water flow in channels is intermittent. The applicant
8   will construct ISR processing and support facilities on level areas and outside the 100-year
9   floodplain. National Pollutant Discharge Elimination System (NPDES) permits issued by
10   SDDENR will set limits to control the amount of pollutants that can enter surface water bodies.
11   Implementation of a storm water pollution management plan (SWMP) will control storm water
12   runoff during construction and ensure that surface water runoff from disturbed areas meets
13   NPDES permit limits. U.S. Army Corps of Engineers permits under Section 404 of the Clean
14   Water Act will be required before conducting work in jurisdictional wetlands identified in the
15   project area.
16
17   Operation: Impacts will be SMALL. The applicant's SDDENR-approved NPDES permit and
18   SWMP will be in place to mitigate impacts to surface water from erosion, runoff, and
19   sedimentation. The applicant will implement an emergency response plan to identify and clean
20   up accidental spills and leaks. Processing facilities and chemical and fuel storage tanks will
21   have secondary containment to contain potential spills. Operations will create liquid wastes that
22   will be contained in radium-settling and storage ponds for eventual Class V injection well
23   disposal and/or land application. Radium settling and storage ponds will be constructed with
24   liners, underdrains, and leak detection systems. Liquid waste applied to land application areas
25   will be required to meet NRC release limit criteria for radiological contaminants, as referenced in
26   10 CFR Part 20, Appendix B. SDDENR will require liquid waste applied to land application
27   areas to meet applicable state discharge requirements under a GDP.
28
29   Aquifer Restoration: Impacts will be SMALL. Impacts will be similar to those during the
30   operations phase because the same infrastructure will be used and the same activities will be
31   conducted. The applicant's SDDENR-approved NPDES permit and SWMP will be in place to
32   mitigate impacts to surface water from erosion, runoff, and sedimentation. Restoration of
33   groundwater aquifers will create wastewater that will be contained in radium settling and storage
34   ponds for eventual Class V injection well disposal and/or land application. Radium settling and
35   storage ponds will be constructed with liners, underdrains, and leak detection systems. Treated
36   wastewater applied to land application areas will be required to meet NRC release limit criteria
37   for radiological contaminants, as referenced in 10 CFR Part 20, Appendix B. SDDENR will
38   require wastewater applied to land application areas to meet applicable state discharge
39   requirements under a GDP.
40
41   Decommissioning: Impacts will be SMALL. The impacts will be similar to those during the
42   construction phase. Activities to cleanup, recontour, and reclaim the land surface during
43   decommissioning will mitigate long-term impacts to surface water. The applicant's SDDENR-
44   approved NPDES permit and SWMP will be in place to mitigate impacts to surface water from
45   erosion, runoff, and sedimentation.
46
47   **Groundwater**
48
49   Construction: Impacts will be SMALL. The primary impact to groundwater during the
50   construction phase will be from the consumptive use of groundwater, introduction of drilling

1   fluids into the environment during well installation, and from surface spills of fuels and
2   lubricants. The applicant is required to obtain water appropriation use permits from SDDENR
3   prior to withdrawing water from aquifers. During well installation, drilling fluids (mud) will have
4   the potential to impact surficial aquifers; however, all wells will undergo mechanical integrity
5   tests of the casing and therefore ensure against well leakage prior to entering service. Impacts
6   to groundwater from surface spills of fuels and lubricants will be mitigated by the applicant's
7   implementation of BMPs and by following a spill prevention program that will require an
8   immediate cleanup response to prevent soil contamination or infiltration to groundwater.
9
10 Operation: Impacts will be SMALL. The operations phase may impact near-surface (alluvial)
11 aquifers, production zone aquifers containing the orebodies and surrounding aquifers, and deep
12 aquifers below the ore production zone used for the disposal of liquid wastes.
13
14 Alluvial aquifers are separated from production zone and surrounding aquifers by thick aquitards
15 (confining units) and, therefore, are not hydraulically connected to production zone and
16 surrounding aquifers. In addition, alluvial aquifers do not serve as a water supply for domestic
17 use or livestock. The impacts from spills and leaks will be SMALL. The applicant's leak
18 detection and cleanup program will include rapid response and remediation to minimize impacts
19 to soils and groundwater. Liquid waste applied to land application areas will be required to meet
20 NRC release limit criteria for radiological contaminants, as referenced in 10 CFR Part 20,
21 Appendix B and applicable state discharge requirements under a GDP issued by SDDENR.
22
23 The applicant has committed to removing and replacing existing domestic wells drawing water
24 from production zone aquifers within the project area from private use prior to ISR operations.
25 In addition, the applicant will monitor all domestic wells within 2 km [1.2 mi] of the project
26 boundary during operations and replace these wells in the event of significant drawdown or
27 degradation of water quality. Water levels in affected wells will recover with time after ISR
28 operations and aquifer restoration activities are complete.
29
30 The establishment of an inward hydraulic gradient during wellfield operations along with the
31 applicant-installed groundwater monitoring network to detect potential vertical and horizontal
32 excursions will limit the potential for undetected lixiviant excursions that could degrade
33 groundwater quality. Because the ore production zones are overlain and underlain by
34 impermeable shale layers, this further ensures the hydraulic isolation of the ore production
35 zones, which helps to limit potential groundwater contamination in surrounding aquifers.
36
37 Liquid wastes generated from operation of the proposed Dewey-Burdock ISR Project will be
38 disposed of via Class V deep well injection, land application, or a combination of Class V deep
39 well injection and land application. The groundwater in deep formations targeted for Class V
40 deep well injection must not be a potential underground source of drinking water. Class V
41 injection wells will be permitted in accordance with the EPA Underground Injection Control
42 Program. Liquid wastes injected into Class V injection wells may not be classified as hazardous
43 under the Resource Conservation and Recovery Act. NRC will require the liquid waste pumped
44 into Class V injection wells to be treated and monitored to verify it meets NRC release
45 standards in 10 CFR Part 20, Subparts D and K and Appendix B.
46
47 Aquifer Restoration: Impacts will be SMALL to MODERATE. Groundwater restoration will be
48 initiated once a wellfield is no longer being used to produce uranium. Larger withdrawals will
49 produce larger drawdowns in production aquifers during aquifer restoration, resulting in a
50 greater impact on yields of nearby wells. As with operations, the applicant will monitor all

1   domestic wells within 2 km [1.2 mi] of the project boundary during aquifer restoration and
2   replace these wells in the event of significant drawdown or degradation of water quality. Water
3   levels in affected wells will recover with time after ISR operations and aquifer restoration
4   activities are complete. Natural recovery and the well monitoring measures established by the
5   applicant will reduce impacts to nearby wells, ensuring the long-term environmental impact from
6   consumptive use will be SMALL.
7
8   During aquifer restoration, hydraulic control for the former production zone will be maintained;
9   this will be accomplished by maintaining an inward hydraulic gradient through a production
10 bleed. During aquifer restoration activities, water will be pumped from the wellfield (without
11 reinjection), resulting in an influx of "fresh" groundwater into the affected (mined) portion of the
12 aquifer. Hydraulic connection (leakage) between production aquifers (Fall River and Chilson
13 aquifers) through the intervening confining unit (Fuson Shale) in the Burdock area may impact
14 aquifer restoration. The Fall River aquifer is hydraulically connected to abandoned open pit
15 mines in the Burdock area. Water in the abandoned open pit mines has elevated dissolved
16 uranium and gross alpha concentrations exceeding EPA-regulated maximum concentration
17 levels. If contaminants are drawn into production zones within the Chilson aquifer from
18 abandoned open pit mines through the hydraulically connected Fall River aquifer during aquifer
19 restoration, the impacts will be MODERATE.
20
21 During the aquifer restoration phase, disposal of liquid wastes via Class V injection wells, land
22 application, or a combination of Class V injection wells and land application will occur as
23 described for ISR operations. The goal of aquifer restoration will be to restore groundwater
24 quality in the ore production zone to Commission-approved background conditions under
25 10 CFR Part 40, Appendix A, Criterion 5B(5). If the aquifer cannot be restored to background
26 conditions, then NRC will require that either the production zone be returned to maximum
27 contaminant levels in 10 CFR Part 40, Appendix A, Table 5C or to NRC-approved alternate
28 concentration limits. Postrestoration groundwater quality will be protective of public health and
29 the environment.
30
31 Decommissioning: Impacts will be SMALL. The potential impact to groundwater quality during
32 decommissioning and reclamation is comparable to that described in the construction phase.
33 Groundwater consumptive use will be less than that of the operation and restoration phases. All
34 monitoring, injection, and production wells will be plugged and abandoned in accordance with
35 UIC program requirements. Wells will be filled with cement and clay to ensure groundwater
36 does not flow through the abandoned wells. Abandoned wells will be properly isolated from the
37 flow domain. NRC will review and approve the wellfield restoration efforts to ensure that
38 restoration standards were followed and public health and safety is protected.
39
40 **Ecological Resources**
41
42 Construction: Impacts will be SMALL to MODERATE. Construction disturbance under current
43 development plans, which require vegetative removal, will affect approximately 98 ha [243 ac] if
44 deep well injection is used to dispose of treated wastewater or approximately 566 ha [1,398 ac]
45 if land application or a combination of deep well injection and land application is used to dispose
46 of treated wastewater. Some habitat loss or alteration, displacement of wildlife, and mortality
47 due to encounters with vehicles or heavy equipment will occur, though wildlife species will likely
48 disperse from the area once construction commences. Following recommended fencing and
49 power line construction designs will minimize impediments to game and avian movement.
50 Mitigation will control the introduction and spread of undesirable and invasive, nonnative plants;

1 reduce the likelihood of injury or mortality to wildlife; and ensure no loss of aquatic habitat.
2 Impacts to wildlife and habitat will be minimized with mitigation measures and the timely
3 reseeding of disturbed areas following construction. Any trees with raptor nests will not be
4 removed, and following U.S. Fish and Wildlife Service (FWS) and South Dakota Game Fish and
5 Parks (SDGFP) seasonal noise, vehicular traffic, and human proximity guidelines will help to
6 ensure the continued nesting success of area raptors. No federally threatened or endangered
7 species are known to occur within the proposed project area. Impacts to state-protected
8 species will not noticeably affect species' populations within the vicinity of the proposed
9 project site.
10
11 Operation: Impacts will be SMALL to MODERATE. Ecological impacts due to noise, vehicles,
12 structures, and the presence of humans will be similar to, but less than, those experienced
13 during construction for either disposal option because fewer earthmoving activities will occur.
14 However, larger areas of habitat will be converted to crops and animals will be disturbed with
15 irrigation activities during the land application disposal option. The applicant will reseed
16 disturbed areas with SDDENR- or BLM-approved seed mixtures to restore habitat. Spill
17 detection and response plans will reduce the potential impact to terrestrial and aquatic species.
18 Fencing and netting will limit wildlife access to liquid waste holding ponds. Potential conflicts
19 between active raptor nest sites and project-related activities will continue to be mitigated by
20 annual raptor monitoring and mitigation plans.
21
22 Aquifer Restoration: Impacts will be SMALL to MODERATE. Impacts will be similar to those
23 experienced during the operations phase with no major differences in type or degree of impact.
24 The existing infrastructure will be used during this phase, and mitigation measures will continue
25 to apply from the construction and operations phases.
26
27 Decommissioning: Impacts will be SMALL to MODERATE. Temporary disturbances to land
28 and soils during decommissioning could displace vegetation and wildlife species that had
29 recolonized the proposed project area since initiation of ISR activities. Shrubland vegetative
30 communities will be more difficult to reestablish and achieve full site recovery. The applicant
31 commits to vegetation reestablishment efforts to be ongoing throughout the ISR facility life
32 cycle. However, new vegetative growth could be affected by future grazing, droughts, or
33 intense winters, thus reducing the rate of plant productivity and delaying full recovery,
34 Revegetation and recontouring will restore habitat previously altered during construction
35 and operations.
36
37 **Air Quality**
38
39 Construction: Impacts will be SMALL to MODERATE. The proposed Dewey-Burdock ISR
40 Project is located in the Black Hills-Rapid City Intrastate Air Quality Control Region, which is
41 classified as being in attainment for all National Ambient Air Quality Standards (NAAQS)
42 primary pollutants. Air emissions during the construction phase of the proposed project will
43 consist primarily of combustion emissions from drill rigs and fugitive road dust. The magnitude
44 of the pollutant concentrations around the proposed project site from the construction phase
45 combustion emissions are below NAAQS and Prevention of Significant Deterioration (PSD)
46 Class II regulatory thresholds. This also holds true for the peak year pollutant emission levels.
47 The peak year accounts for when all four phases occur simultaneously and represents the
48 highest amount of emissions the proposed action will generate in any one project year. The
49 construction phase and peak year fugitive dust concentrations are also below NAAQS and PSD
50 Class II thresholds. However, the mass of particulate matter generated from fugitive emissions

1   is much greater than that generated from combustion emissions. In addition, these fugitive dust
2   emission sources are spread out over a large area and tend to generate emissions sporadically.
3   Due to the level and nature of these fugitive emissions, there is potential for short-term,
4   intermittent impacts to localized areas in and around the site particularly when vehicles travel on
5   unpaved roads. Wind Cave National Park, a Class I area located about 47 km [29 mi] northeast
6   of the proposed project area, has experienced visibility impacts from air pollution. The initial air
7   dispersion modeling the applicant conducted only considered the area in and around the
8   proposed site. The applicant committed to perform additional air dispersion modeling before the
9   final SEIS is prepared (Powertech, 2012). Meanwhile, based on the modeling results from a
10   similar project, the Dewey-Burdock ISR Project will contribute to visibility impacts at Wind Cave
11   National Park but the impact magnitude will be minimal.
12
13   The deep Class V injection well disposal option has more combustion emissions than the land
14   application option due to the contribution of the deep well drill rig. The land application option
15   has more fugitive emissions due to the greater amount of land disturbed. However, these
16   differences are relatively small and NRC staff do not expect to see any appreciable difference in
17   the overall air emission levels between the two disposal options. Therefore, the impact
18   magnitudes are expected to be the same.
19
20   Operation: Impacts will be SMALL to MODERATE. Combustion emission and fugitive dust
21   emission pollutant levels will be less than those experienced during construction. ISR facilities
22   are not major point source emitters of regulated pollutants. Combustion emissions in this phase
23   are basically evenly divided between light duty vehicles and construction and field equipment.
24   The combustion and fugitive dust emissions around the proposed site will be below NAAQS and
25   PSD Class II regulatory thresholds. However, due to the level and nature of the fugitive
26   emissions, there is potential for short-term, intermittent impacts to localized areas in and around
27   the site particularly when vehicles travel on unpaved roads. The Dewey-Burdock ISR Project
28   will contribute to visibility impacts at Wind Cave National Park but the impact magnitude will
29   be minimal.
30
31   The land application disposal option has more fugitive emissions than the Class V injection well
32   option due to the greater amount of land disturbed. However, this difference is relatively small
33   and NRC staff do not expect to see any appreciable difference in the overall air emission
34   levels between the two disposal options. Therefore, the impact magnitudes are expected to
35   be the same.
36
37   Aquifer Restoration: Impacts will be SMALL to MODERATE. Combustion emission and fugitive
38   emission levels for the aquifer restoration phases are the lowest relative to the other three
39   phases. For the aquifer restoration phase, combustion emissions are primarily from light duty
40   vehicles and wind erosion can generate more fugitive emissions than travel on unpaved roads.
41   Fugitive emissions can result in short-term, intermittent impacts to localized areas. The
42   proposed project can contribute to visibility impacts at Wind Cave National Park, but the impact
43   magnitude will be minimal.
44
45   The land application disposal option can generate up to about twice the amount of fugitive
46   emissions compared to the Class V injection well disposal option. Although there is some
47   difference in the overall fugitive dust emissions levels between the two disposal options, the
48   impact magnitude is expected to be similar.
49

1 Decommissioning: Impacts will be SMALL to MODERATE. The decommissioning phase
2 pollutant sources and emission levels closely match those from the operation phase. Therefore,
3 the decommissioning phase will produce the same impact magnitude as the operation phase.
4 As in the operation phase described previously, NRC staff do not expect to see any appreciable
5 difference in the overall decommissioning phase air emission levels between the Class V
6 injection well and land application disposal options.
7
8 **Noise**
9
10 Construction: Impacts will be SMALL. Increased traffic, as well as use of drill rigs, heavy
11 trucks, bulldozers, and other equipment to construct and operate the wellfields, drill wells,
12 access roads, and build the central plant and satellite facility, will generate noise audible above
13 ambient (background) levels. The sound from construction activities will return to background
14 levels at a distance of approximately 305 m [1,000 ft]. Two onsite dwellings will be impacted by
15 noise above background levels from heavy equipment use. The Daniels residence is within
16 305 m [1,000 ft] of wellfields B-WF6 and B-WF7 in the Burdock area, and the Beaver Creek
17 Ranch Headquarters is within 305 m [1,000 ft] of land application areas in the Dewey area.
18 Increased noise levels at these residences during construction will be short term (1 to 2 years)
19 and mitigated by using sound abatement controls on operating equipment. Administrative and
20 engineering controls will be expected to maintain noise levels in work areas below Occupational
21 Health and Safety Administration (OSHA) regulatory limits and be mitigated by use of personal
22 hearing protection. Noise impacts to raptors will be mitigated by adhering to FWS and SDGFP
23 seasonal noise guidelines, locating all planned facilities outside of BLM-recommended buffer
24 zones of all raptor nests, and following an FWS-approved raptor monitoring and mitigation plan.
25
26 Operation: Impacts will be SMALL. Impacts from traffic-related noise will be similar to those
27 during construction. Because wellfields will be developed and operated sequentially, potential
28 noise impacts at the Daniels residence will be short term (1 to 2 years each for wellfields B-WF6
29 and B-WF-7). In addition, the Daniels residence will not be occupied year round. Residents at
30 the Beaver Creek Ranch Headquarters will only be exposed to noise from nearby land
31 application areas during the growing season (May 11 to September 24). Noise impacts will be
32 mitigated by using sound abatement controls on operating equipment. The central plant and
33 satellite facility will generate indoor noise audible to workers. OSHA regulatory limits will be
34 maintained and mitigated by use of personal hearing protection. Potential noise-related impacts
35 to active raptor nest sites will continue to be mitigated by adherence to timing and spatial
36 restrictions within specified distances of active raptor nests as determined by appropriate
37 regulatory agencies (e.g., FWS, SDGFP, and BLM).
38
39 Aquifer Restoration: Impacts will be SMALL. Noise impacts will be similar to, or less than,
40 those experienced during the operations phase. Pumps and other wellfield equipment
41 contained in buildings would reduce the potential sound impact to an offsite individual. Because
42 the aquifers in wellfields will be restored sequentially, potential noise impacts at the Daniels
43 residence will be short term (1 to 2 years each for wellfields B-WF6 and B-WF7). In addition,
44 the Daniels residence will not be occupied year round. During aquifer restoration, residents at
45 the Beaver Creek Ranch Headquarters will only be exposed to noise from nearby land
46 application areas during the growing season (May 11 to September 24). Noise impacts will be
47 mitigated by using sound abatement controls on operating equipment. Noise impacts from
48 traffic will be SMALL because there will be fewer vehicular trips than during the operations
49 phase. Potential noise-related impacts to active raptor nest sites will continue to be mitigated by

1   adherence to timing and spatial restrictions within specified distances of active raptor nests as
2   determined by appropriate regulatory agencies (e.g., FWS, SDGFP, and BLM).
3
4   Decommissioning: Impacts will be SMALL. Noise impacts will either be similar to, or less than,
5   those experienced during the construction phase. Noise during this phase will be temporary,
6   and when decommissioning and reclamation activities are complete, the noise levels will return
7   to baseline. Noise impacts from traffic will be SMALL because there will be fewer shipments to
8   and from the proposed site as decommissioning progressed. Potential noise-related impacts to
9   active raptor nest sites will continue to be mitigated by adherence to timing and spatial
10 restrictions within specified distances of active raptor nests as determined by appropriate
11 regulatory agencies (e.g., FWS, SDGFP, and BLM).
12
13 **Historic and Cultural Resources**
14
15 Construction: Impacts will be SMALL to LARGE. Archaeological and historic sites may
16 potentially be disturbed during construction. Within the area of potential effect at the proposed
17 Dewey-Burdock site, 18 historic sites are either listed in the National Register of Historic Places
18 (NRHP) or eligible for listing in the NRHP. Based on the proposed location of ISR facilities and
19 infrastructure, avoidance of 12 of these sites is possible during the construction phase and,
20 therefore, no impacts are anticipated. Avoidance and mitigation, such as fencing and data
21 recovery excavations, are recommended for the remaining six NRHP-eligible sites. In addition,
22 avoidance is recommended for two unevaluated historic burial sites located in proximity to
23 proposed construction activities until their NRHP eligibility is determined. Avoidance and
24 mitigation is also recommended for 4 unevaluated site located within 76 m [250 ft] of proposed
25 wellfields or land application areas.
26
27 Prior to construction, an agreement between NRC, South Dakota State Historic Preservation
28 Office (SD SHPO), BLM, interested Native American tribes, the applicant, and other interested
29 parties will be established outlining the mitigation process for each affected resource. Prior to
30 construction, the applicant will also develop an Unexpected Discovery Plan that will outline the
31 steps required if unexpected historical and cultural resources are encountered.
32
33 Consultation efforts to identify properties of religious and cultural significance to Native
34 American tribes have not been completed. Thus, NRC cannot determine effects to these
35 properties at this time. Section 106 consultation between NRC, SD SHPO, BLM, tribal
36 representatives, and the applicant regarding potential impacts to these sites is ongoing.
37
38 Operation: Impacts will be SMALL. Minimal impacts will result during the operations phase
39 because impacts to cultural resources will be mitigated before facility construction and identified
40 resources will be avoided. If historical or cultural resources are encountered during operations,
41 the Unexpected Discovery Plan will be implemented. Work would stop in the immediate area,
42 and appropriate agencies would be notified.
43
44 Aquifer Restoration: Impacts will be SMALL. Impacts to historical and cultural resources
45 during the aquifer restoration phase will be similar to operations. Minimal impacts will
46 result because impacts to cultural resources will be mitigated before facility construction, and
47 identified resources will be avoided. If historical or cultural resources are encountered during
48 operations, the Unexpected Discovery Plan will be implemented. Work would stop in the
49 immediate area, and appropriate agencies would be notified.
50

1   Decommissioning: Impacts will be SMALL. Minimal impacts will result during the
2   decommissioning phase because impacts to cultural resources will be mitigated prior to facility
3   construction. If historical or cultural resources are encountered during operations, the
4   Unexpected Discovery Plan will be implemented. Work would stop in the immediate area, and
5   appropriate agencies would be notified.
6
7   **Visual/Scenic Resources**
8
9   Construction: Impacts will be SMALL. During facilities construction, short-term (1 to 2 years)
10 visual and scenic impacts will result from construction equipment and fugitive dust emissions.
11 Temporary and short-term visual impacts during the construction period in each wellfield
12 will result from header house construction, well drilling, and construction of access roads
13 and electrical distribution lines. Dust suppression and selecting building materials and paint that
14 complement the natural environment will reduce overall visual and scenic impacts of
15 project construction. Center pivot irrigation systems in proposed land application areas in the
16 Dewey area will be visible to travelers on Dewey Road; however, Dewey Road is a lightly
17 traveled county road with few residences. Proposed activities at the project will be consistent
18 with the BLM visual classification of this area.
19
20 Operation: Impacts will be SMALL. Visual impacts will be similar to, or less than, those
21 experienced during construction. Less heavy machinery will be used, and standard dust control
22 measures (e.g., water application and speed limits) will be implemented to reduce visual
23 impacts from fugitive dust. Wellfields will be developed sequentially, and there will be no large
24 expanse of land undergoing development at one time. Buildings and other structures will be
25 painted so they blend in to the natural landscape, and power lines and pipelines will be buried
26 where appropriate. Center pivot irrigation systems in proposed land application areas in the
27 Dewey area will be visible to travelers on Dewey Road; however, Dewey Road is a lightly
28 traveled county road with few residences. Proposed activities at the project will be consistent
29 with the BLM visual classification of this area.
30
31 Aquifer Restoration: Impacts will be SMALL. Visual impacts will be similar to, or less than,
32 those experienced during the operations phase. Aquifer restoration activities will use in-place
33 infrastructure; therefore, no modifications to either scenery or topography will occur. There will
34 be less vehicular traffic, creating less of a visual impact. The applicant identified mitigation
35 measures, such as dust suppression, which will be used to further reduce visual impacts.
36
37 Decommissioning: Impacts will be SMALL. Temporary impacts to the visual landscape will be
38 comparable to those during the construction phase. Reclamation will return the visual
39 landscape to baseline contours and will reduce the visual impact by removing buildings and the
40 associated infrastructure. Implementation of mitigation measures (e.g., dust suppression) will
41 further reduce the visual impacts from decommissioning.
42
43 **Socioeconomics**
44
45 Construction: Impacts will be SMALL. Because of the small size of the construction workforce
46 (86 workers) and because of the short duration of the ISR construction phase (1 to 2 years), the
47 overall potential socioeconomic impact, including the effects of ISR facility construction on
48 demographic conditions, income, housing, employment rate, local finance, education, and
49 health and social services, will be SMALL.
50

1   <u>Operation</u>: Impacts will be SMALL. Because of the small size of the operations workforce (84
2   workers), the migration of workers and their families to nearby towns will have a SMALL impact
3   on demographics. Although wage rates will be higher for Dewey-Burdock employees than for
4   workers in similar skilled positions in Fall River, Custer, and Weston Counties, the operations
5   workforce will be small in comparison to the combined labor force in the counties; therefore,
6   income impacts will be SMALL. The impact on housing will be SMALL because of available
7   housing in the immediate area surrounding the proposed ISR facility. Operation of the proposed
8   Dewey-Burdock ISR Project will create new jobs, but because of the small workforce size and
9   because most skilled workers will be drawn from areas outside of the region of influence,
10 impacts on employment will not be noticeable. The local economy will experience a SMALL
11 beneficial impact from the purchasing of local goods and services and an increase in sales and
12 income tax revenues. An increased demand for schools will have a SMALL impact on
13 education because the current school systems are not at full capacity and can accommodate
14 more students. Increased demand for health and social services will have a SMALL impact.
15
16 <u>Aquifer Restoration</u>: Impacts will be SMALL. Impacts will be less than those experienced
17 during the operations phase. Fewer workers will be required, which will reduce pressure on
18 housing, education, and health and social services.
19
20 <u>Decommissioning</u>: Impacts will be SMALL. Impacts will be less than those during the
21 construction and operations phases because fewer workers will be required. Demand for
22 housing, education, and health and social services will also be reduced.
23
24 **Environmental Justice**
25
26 <u>All Phases</u>: The percentage of minority populations living in affected block groups in the vicinity
27 of the proposed Dewey-Burdock ISR Project site in Custer and Fall River Counties in South
28 Dakota and Weston County in Wyoming does not significantly exceed the percentage of
29 minority populations recorded at the state and county level and is well below the national level.
30 Furthermore, the percentage of low-income populations living in affected census tracts in the
31 vicinity of the proposed project site in Custer, Fall River, and Weston Counties does not
32 significantly exceed the percentage of low-income populations recorded at the state or county
33 level. Therefore, there will be no disproportionately high and adverse impacts to minority and
34 low-income populations from the construction, operation, aquifer restoration, and
35 decommissioning of the proposed Dewey-Burdock ISR facility.
36
37 The closest population to the proposed Dewey-Burdock ISR Project that could be impacted by
38 environmental justice concerns is the Pine Ridge Indian Reservation located approximately 80
39 km [50 mi] east in Shannon County, South Dakota. Based on 2010 United States Census
40 Bureau data, this reservation has both minority {greater than 95 percent Native American
41 (Oglala Sioux Tribe)} and low-income populations. Environmental justice impacts to Native
42 American tribes living in the vicinity of the proposed project will be no different than those
43 experienced by other populations. The proposed action may potentially affect certain sites of
44 religious and cultural significance to Native American tribes; however, the impacts to such sites
45 could be reduced through mitigation strategies developed through the National Historic
46 Preservation Act Section 106 consultation process.
47
48
49
50

1   **Public and Occupational Health**
2
3   Construction: Impacts will be SMALL. Construction activities, including the use of construction
4   equipment and vehicles, will disturb the topsoil and create fugitive dust emissions. Fugitive dust
5   generated from construction activities will be short term (1 to 2 years), and the levels of
6   radioactivity in soils at the proposed project site are low; therefore direct exposure, inhalation,
7   and ingestion of fugitive dust will not result in a radiological dose to workers and the public.
8   Construction equipment will be diesel powered and will exhaust particulate diesel emissions.
9   The potential impacts and potential human exposures from these emissions will be SMALL,
10   because of the short duration of the release and because the emissions will be readily
11   dispersed into the atmosphere.
12
13   Operation: The radiological impacts from normal operations will be SMALL. Public and
14   occupational exposure rates at ISR facilities during normal operations have historically been
15   well below regulatory limits. Dose assessments using the MILDOS computer code indicate that
16   the 10 CFR Part 20 public dose limit of 1 mSv/yr [100 mrem/yr] will not be exceeded at any
17   property boundary. The remote location of the proposed Dewey-Burdock site and the use of the
18   proposed ISR technology coupled with the applicant procedures to minimize exposure
19   demonstrate that the potential impact on public and occupational health and safety from facility
20   operation will be SMALL. The radiological impacts from accidents will be SMALL for workers (if
21   the applicant's radiation safety and incident response procedures in an NRC-approved radiation
22   protection plan are followed) and SMALL for the public because of the facility's remote location.
23   The nonradiological public and occupational health and safety impacts from normal operations
24   and accidents, due primarily to risk of chemical exposure, will be SMALL if handling and storage
25   procedures are followed.
26
27   Aquifer Restoration: Impacts will be SMALL. Impacts will be similar to, but less than, those
28   during the operations phase. The reduction or elimination of some operational activities will
29   further reduce the magnitude of potential worker and public health impacts and safety hazards.
30
31   Decommissioning: Impacts will be SMALL. Impacts will be similar to those experienced during
32   construction. Soil and facility structures will be decontaminated, and lands will be restored to
33   preoperational conditions.
34
35   **Waste Management**
36
37   Construction: Impacts will be SMALL. Small-scale and incremental wellfield development will
38   generate small volumes of construction waste. Waste will primarily consist of building materials,
39   piping, and other solid wastes. No byproduct material will be generated during construction.
40   Nonhazardous solid waste will be disposed of at a nearby municipal solid waste landfill with
41   available capacity to accommodate estimated construction-phase waste volumes.
42
43   Operation: Impacts will be SMALL. Liquid byproduct material, including production bleed,
44   waste brine streams from elution and precipitation, resin transfer wash, laundry water, plant
45   wash-down water, and laboratory chemicals will be treated and disposed using Class V injection
46   wells. If a permit cannot be obtained from EPA for Class V injection, the applicant would pursue
47   land application of treated liquid effluent. If the capacity of either method is limited, the applicant
48   will pursue a combination of both Class V injection and land application. Deep well injection in a
49   Class V well requires an EPA permit, and wastes will have to meet EPA permit conditions and
50   NRC effluent discharge limits in 10 CFR Part 20, Appendix B (both would limit potential

1   impacts). Land application will require SDDENR-permitting of discharge water, and the land
2   application area would be monitored to assess compliance with NRC and SDDENR
3   requirements that would limit impacts. Solids classified as byproduct material will be sent to a
4   licensed facility for disposal. A preoperational agreement with a licensed facility to accept
5   wastes the proposed action generates will avoid capacity impacts. Capacity is available for
6   disposal of nonradiological, nonhazardous wastes at regional municipal landfills. Capacity will
7   be sufficient for disposal of low volumes of generated hazardous wastes.
8
9   Aquifer Restoration: Impacts will be SMALL based on the type and quantity of waste expected
10 to be generated and the available capacity for disposal. Waste disposal procedures will be the
11 same as those during the operations phase, resulting in similar impacts. One exception is the
12 addition of reverse osmosis treatment of aquifer restoration water if a Class V deep disposal
13 well is used. The applicant proposal includes adequate disposal capacity, and the applicant is
14 required to comply with EPA Class V disposal permit conditions, NRC effluent limits, and other
15 NRC safety regulations. Although the wastewater volume could increase during aquifer
16 restoration activities, this will be offset by the reduction in production capacity from completion
17 of wellfield production and removal from service.
18
19 Decommissioning: Impacts will be SMALL to MODERATE. Safe handling, storage, and
20 disposal of decommissioning wastes will be described in a required decommissioning plan for
21 NRC review before decommissioning activities begin. A preoperational agreement with a
22 licensed disposal facility to accept solid byproduct material will ensure that sufficient disposal
23 capacity will be available at the time of decommissioning. Equipment and building materials
24 that meet release criteria will be reused, recycled, or disposed as construction waste at a
25 landfill. The available local landfill capacity may be insufficient to accommodate all
26 decommissioning nonhazardous solid waste from the proposed Dewey Burdock ISR Project.
27 The potential impacts on waste management resources will depend on the long-term status of
28 the existing local landfill resources. If the capacity of the Newcastle or Custer-Fall River landfills
29 is expanded prior to project decommissioning, the impacts to local landfills will be SMALL. If
30 capacity at either landfill is not expanded prior to the Dewey-Burdock decommissioning, the
31 NRC staff conclude the Newcastle landfill will have no disposal capacity at the time of
32 decommissioning. Impacts to the Custer-Fall River landfill are expected to be MODERATE
33 because the increase in solid waste disposal will more rapidly consume storage capacity during
34 the last years of the landfill's projected operational life. The disposal of any waste from the
35 Dewey-Burdock facility in the Rapid City landfill will have a SMALL impact due to the projected
36 operational life and available capacity of that landfill.
37
38 **CUMULATIVE IMPACTS**
39
40 Chapter 5 of this SEIS provides the NRC evaluation of potential cumulative impacts from
41 the construction, operations, aquifer restoration, and decommissioning of the proposed
42 Dewey-Burdock ISR Project considering other past, present, and reasonably foreseeable future
43 actions. Cumulative impacts from past, present, and reasonably foreseeable future actions
44 were considered and evaluated in this draft SEIS, regardless of what agency (federal or
45 nonfederal) or person undertook the action. The NRC staff determined that the SMALL to
46 MODERATE impacts from the proposed Dewey-Burdock ISR Project are not expected to
47 contribute perceptible increases to the SMALL to LARGE cumulative impacts, due primarily to
48 ongoing uranium and oil and gas exploration activities, potential wind energy projects, and
49 proposed infrastructure and transportation projects.
50

1   **SUMMARY OF COSTS AND BENEFITS OF THE PROPOSED ACTION**
2
3   The implementation of the proposed action would generate primarily regional and local costs
4   and benefits. The regional benefits of building the proposed project would be increased
5   employment, economic activity, and tax revenues in the region around the proposed site. Costs
6   associated with the proposed Dewey-Burdock ISR Project are, for the most part, limited to the
7   immediate area surrounding the site. The NRC staff determined the benefit from constructing
8   and operating the facility would outweigh the economic, environmental, and social costs.
9
10   **COMPARISON OF ALTERNATIVES**
11
12   For the No-Action alternative, the applicant would not construct or operate ISR facilities at the
13   proposed Dewey-Burdock ISR Project site. As a result, no uranium ore would be recovered
14   from the proposed site. This alternative would result in neither positive nor negative impacts to
15   any resource area.
16
17   **PRELIMINARY RECOMMENDATION**
18
19   After weighing the impacts of the proposed action and comparing the alternatives, the NRC
20   staff, in accordance with 10 CFR 51.71(f), set forth its preliminary NEPA recommendation
21   regarding the proposed action (issuing a source material license for the proposed Dewey-
22   Burdock ISR Project). Unless safety issues mandate otherwise, the preliminary NRC staff
23   recommendation to the Commission related to the environmental aspects of the proposed
24   action is that a source and byproduct material license for the proposed action be issued as
25   requested.
26
27   The NRC staff conclude that the overall benefits of the proposed action outweigh the
28   environmental disadvantages and costs based on the following:
29
30   •   Potential adverse impacts to all environmental resource areas are expected to be
31   SMALL, with the exception of
32
33   1.   Land use resources during decommissioning. Land disturbance during
34   decommissioning will be MODERATE until vegetation is reestablished in seeded
35   areas (see SEIS Sections 4.2.1.1.4, 4.2.1.2.4, and 4.2.1.3).
36
37   2.   Transportation resources during construction and operation. Increases in
38   traffic during construction and operations will have a MODERATE impact on
39   Dewey Road, the road nearest the proposed site (see SEIS Sections 4.3.1.1.1,
40   4.3.1.2.1, 4.3.1.1.2, 4.3.1.2.2, and 4.3.1.3).
41
42   3.   Groundwater resources during aquifer restoration. During aquifer restoration in
43   the Burdock area, drawdown-induced migration of contaminants into the
44   production zone (i.e., the Chilson aquifer) from abandoned open pit mines could
45   adversely affect restoration goals and have a MODERATE impact (see SEIS
46   Sections 4.5.2.1.1.3, 4.5.2.1.2.3, and 4.5.2.1.3).
47
48   4.   Ecological resources during construction, operations, aquifer restoration, and
49   decommissioning. Under the land application and combined Class V deep well
50   disposal and land application options, construction, operations, and aquifer

restoration activities would have a MODERATE impact on vegetation, small- to medium-sized mammals, raptors, nongame and migratory birds, and reptiles (see SEIS Sections 4.6.1.2.1, 4.6.1.2.2, 4.6.1.2.3, and 4.6.1.3). Under all disposal options, land-disturbing activities during decommissioning would have a MODERATE impact on vegetation until it is reestablished (see SEIS Sections 4.6.1.1.4, 4.6.1.2.4, and 4.6.1.3).

5. Air quality during construction, operations, aquifer restoration, and decommissioning. During all phases of the ISR lifecycle, there will be the potential for MODERATE air impacts from short-term, intermittent fugitive dust emissions (see SEIS Sections 4.7.1.1.1 through 4.7.1.1.4, 4.7.1.2.1 through 4.7.1.2.4, and 4.7.1.3).

6. Historical and cultural resources during construction. Construction could have a MODERATE or LARGE impact on 18 historic properties—those sites currently listed or eligible for listing on the NRHP—and other unevaluated historic, cultural, and religious properties in the project area (see SEIS Sections 4.9.1.1.1, 4.9.1.2.1, and 4.9.1.3).

7. Waste management resources during decommissioning. Impacts from disposal of nonhazardous solid waste may be MODERATE depending on the long-term status of existing local landfill resources (see SEIS Sections 4.14.1.1.4 and 4.14.1.2.4).

- Regarding groundwater, the portion of the aquifer(s) designated for uranium recovery must be exempted as underground sources of drinking water prior to the start of ISR operations. Additionally, the applicant will be required to monitor for excursions of lixiviant from the production zones and to take corrective actions in the event of an excursion. Prior to operations, the applicant will be required to provide detailed hydrologic pump test data packages and operational plans for each wellfield at the proposed project. The applicant will also be required to restore groundwater parameters affected by ISR operations to levels that are protective of human health and safety.

- The costs associated with the proposed project are, for the most part, limited to the area surrounding the site.

- The regional benefits of building the proposed project will be increased employment, economic activity, and tax revenues in the region around the proposed site.

This preliminary recommendation is based on NRC staff's independent review of (i) the license application the applicant submitted; (ii) applicant responses to NRC staff requests for additional information; (iii) consultation with federal, state, tribal, and local agencies; and (iv) the assessments summarized in this draft SEIS, including the potential mitigation measures identified in the license application and this draft SEIS.

References

10 CFR Part 40. Code of Federal Regulations, Title 10, *Energy*, Part 40. "*Domestic Licensing of Source Material*." Washington, DC: U.S. Government Printing Office.

1   10 CFR Part 51. Code of Federal Regulations, Title 10, *Energy*, Part 51. *"Environmental*
2   *Protection Regulations for Domestic Licensing and Related Regulatory Functions."*
3   Washington, DC: U.S. Government Printing Office.
4
5   43 CFR Part 3800. Code of Federal Regulations, Title 43, *Public Lands: Interior*, Part 3800.
6   *"Mining Claims Under the General Mining Laws."* Washington, DC: U.S. Government
7   Printing Office.
8
9   43 CFR Subpart 3809. Code of Federal Regulations, Title 43, *Public Lands: Interior*, Subpart
10 3809. *"Subsurface Management."* Washington, DC: U.S. Government Printing Office.
11
12 NRC. NUREG–1910, "Generic Environmental Impact Statement for *In-Situ* Leach Uranium
13 Milling Facilities." ML091480244, ML091480188. Washington, DC: NRC. May 2009.
14
15 NRC. NUREG–1748, "Environmental Review Guidance for Licensing Actions Associated With
16 NMSS Programs." Washington, DC: NRC. August 2003.

ABBREVIATIONS/ACRONYMS

ACHP   Advisory Council on Historic Preservation
ACL   alternate concentration limit
ADAMS   Agencywide Documents Access and Management System
AEA   Atomic Energy Act
AET, Inc.   American Engineering Testing, Inc.
ALAC   Archaeology Laboratory Augustana College
ALARA   as low as reasonably achievable
AUM   animal unit month
APE   area of potential effect
ARC   Archaeological Research Center
ARPA   Archaeological Resources Protection Act
ARSD   Administrative Rules of South Dakota
ASLB   Atomic Safety and Licensing Board
AWEA   American Wind Energy Association

BGEPA   Bald and Golden Eagle Protection Act
bgs   below ground surface
BHNF   Black Hills National Forest
BLM   U.S. Bureau of Land Management
BMP   best management practice
BNSF   Burlington Northern Santa Fe

CAB   Commission-approved background
CCSDWPC   Custer County, South Dakota, Weed and Pest Control
CFR   *U.S. Code of Federal Regulations*
CEQ   Council on Environmental Quality
CERCLA   Comprehensive Environmental Response, Compensation, and Liability Act
CESQC   conditionally exempt small quantity generator
CNWRA   Center for Nuclear Waste Regulatory Analyses
cpm   counts per minute
CPP   central processing plant

dBA   decibels
DM&E   Dakota Minnesota and Eastern (Railroad)
DOE   U.S. Department of Energy

EFRC   Energy Fuels Resources Corporation
EIA   Energy Information Administration
EIS   environmental impact statement
E.O.   Executive Order
EPA   U.S. Environmental Protection Agency
ESA   Endangered Species Act
ESRI   Environmental Systems Research Institute

FACU   facultative upland
FACW   facultative wet
FHWA   Federal Highway Administration
FR   *Federal Register*

ABBREVIATIONS/ACRONYMS (continued)

FRA	Federal Railroad Administration
FWS	U.S. Fish and Wildlife Service
GCRP	U.S. Global Change Research Program
GDP	Groundwater Discharge Permit
GEIS	generic environmental impact statement
GHG	greenhouse gas
GPS	global-positioning-system
HABS	Historic American Buildings Survey
HDPE	high-density polyethylene
ID	well identification
IQR	interquartile range
ISL	*in-situ* leach
ISR	*in-situ* recovery
IX	ion exchange
MBTA	Migratory Bird Treaty Act
MCL	maximum contaminant level
MILDOS	computer code
MIT	mechanical integrity test
MOA	Memorandum of Agreement
mya	million years ago
NAAQS	National Ambient Air Quality Standards
NAGPRA	Native American Graves Protection and Repatriation Act
NAU	Rapid City Campus of the National American University
NCRP	National Council on Radiation Protection and Measurements
NEPA	National Environmental Policy Act
NESHAP	National Emission Standards for Hazardous Air Pollutants
NHPA	National Historic Preservation Act of 1966, as amended
NOGCC	Nebraska Oil and Gas Conservation Commission
NPDES	national pollutant discharge elimination system
NPWRC	Northern Prairie Wildlife Research Center
NRC	U.S. Nuclear Regulatory Commission
NRCS	National Resource Conservation Service
NRHP	National Register of Historic Places
OBL	obligate
OMB	Office of Management and Budget
OSHA	Occupational Safety and Health Administration
OTGR	Office of Tribal Government Relations
OW	Open Water
PABJh	Palustrine Aquatic Bed Intermittently Flooded Diked
PEM	Palustrine Emergent
PEMC	Seasonally Flooded
POO	Plan of Operations

POP	Perimeter of Operational Pollution
Powertech	Powertech (USA) Inc.
PRB	Powder River Basin
PSD	Prevention of Significant Deterioration
PUB	Palustrine Unconsolidated Bottom
PUS	Palustrine Unconsolidated Shore
PUSA	Palustrine Unconsolidated Shore Temporarily Flooded
R2EM	Riverine Lower Perennial Emergent
R4SB7	Riverine Intermittent Streambed Vegetated
R4US	Riverine Intermittent Unconsolidated Streambed
RCRA	Resource Conservation and Recovery Act
RMP	regional management plan
RO	reverse osmosis
ROI	region of influence
ROW	right of way
SDCL	South Dakota Codified Law
SDDENR	South Dakota Department of Environment and Natural Resources
SDDOA	South Dakota Department of Agriculture
SDDOE	South Dakota Department of Education
SDDOH	South Dakota Department of Health
SDDOL	South Dakota Department of Labor
SDDOT	South Dakota Department of Transportation
SDDLR	South Dakota Department of Labor and Regulation
SDDRR	South Dakota Department of Revenue and Regulation
SDGFP	South Dakota Game, Fish, and Parks
SDGS	South Dakota Geological Survey
SDNHP	South Dakota Natural Heritage Program
SDRMP	South Dakota Resource Management Plan
SD SHPO	South Dakota State Historic Preservation Office
SDSMT	South Dakota School of Mines and Technology
SDSU	South Dakota State University
SDWA	Safe Drinking Water Act
SEA	U.S. Department of Transportation Section of Environmental Analysis
SEIS	supplemental environmental impact statement
SER	safety evaluation report
SERP	safety and environmental review panel
SF	satellite facility
SMCL	secondary maximum concentration limit
SNAP	Supplemental Nutrition Assistance Program
SOW	statement of work
SPAW	soil-plant-atmosphere-water
SQR	scenic quality rating
SRI	SRI Foundation
STB	Surface Transportation Board
SUNSI	sensitive unclassified non-safeguards information
SWMP	storm water pollution management plan

ABBREVIATIONS/ACRONYMS (continued)

TANF	Temporary Assistance for Needy Families
TCP	traditional cultural property
TDS	total dissolved solids
TEDE	total effective dose equivalent
THPO	Tribal Historic Preservation Office
TLD	thermoluminescent dosimeter
TVA	Tennessee Valley Authority
UCL	upper control limit
UDEQ	Utah Department of Environmental Quality
UMTRCA	Uranium Mill Tailings Radiation Control Act
UIC	underground injection control
UPL	upland
USACE	U.S. Army Corps of Engineers
USCB	U.S. Census Bureau
USDA	U.S. Department of Agriculture
USDOT	U.S. Department of Transportation
USDW	underground source of drinking water
USFS	U.S. Forest Service
USGS	U.S. Geological Survey
UXC	The Ux Consulting Company
VRM	Visual Resource Management
WDAI	Wyoming Department of Administration and Information
WDEQ	Wyoming Department of Environmental Quality
WDTI	Western Dakota Technical Institute
WDWS	Wyoming Department of Workforce Services
WGFD	Wyoming Game and Fish Department
WIA	walk-in hunting area
WSDOT	Washington State Department of Transportation
WUS	waters of the United States
WYOGCC	Wyoming Oil and Gas Conservation Commission

5 CUMULATIVE IMPACTS

5.1   Introduction

The Council on Environmental Quality (CEQ) regulations implementing the procedural provisions of the National Environmental Policy Act of 1969, as amended (NEPA), are found in 40 CFR Parts 1500–1508. Cumulative effects are defined in 40 CFR 1508.7 as

> "the impact on the environment that results from the incremental impact of the action when added to other past, present, and reasonably foreseeable future actions regardless of what agency (Federal or non-Federal) or person undertakes such other actions."

Cumulative effects or impacts[1] can result from individually minor but collectively significant actions taking place over a period of time. This SEIS considers the cumulative impacts of past, present, and future actions in the proposed project area. These actions include oil and gas production; coal mining and coal bed methane operations; gold, sand, gravel, and limestone mining; ISR operations; conventional uranium mining; wind farms; and livestock grazing.

The identification of cumulative impacts of the proposed action resulted from an analysis of an extensive body of publicly available information on ongoing and proposed federal projects, information presented in the Generic Environmental Impact Statement (GEIS) (NRC, 2009a), and review of the literature of the environmental and socio-economic conditions in South Dakota and in the nearby communities.

A number of uranium exploration and oil and gas operations are underway within 16 km [10 mi] of the proposed Dewey-Burdock ISR Project. Several ISR uranium projects within the broader region of the proposed Dewey-Burdock ISR Project are in the operation, licensing, or prelicensing stages. Oil and gas operations are underway throughout the area. There is potential for wind energy generation within and in the vicinity of the proposed project area. The Nuclear Regulatory Commission (NRC) anticipates growth in extraction of coal, coal bed methane, and limestone, as well as government support for and industry interest in developing transmission and transportation infrastructure at distances beyond 16 km [10 mi] from the Dewey-Burdock site.

The GEIS (NRC, 2009a) provides a methodology for conducting a cumulative impacts assessment following CEQ guidance (CEQ, 1997). SEIS Section 5.1.1 describes past, present, and reasonably foreseeable future actions identified and analyzed in the cumulative impacts analysis. The methodology NRC staff used in conducting the cumulative impact analysis in this SEIS is described in Section 5.1.2.

5.1.1   Other Past, Present, and Reasonably Foreseeable Future Actions

The proposed Dewey-Burdock ISR Project is located within the Nebraska-South Dakota-Wyoming Uranium Milling Region defined in the GEIS (NRC, 2009a). This region

---

[1] In this SEIS, "cumulative impacts" is deemed synonymous with "cumulative effects."

1   encompasses parts of Sioux and Dawes Counties in Nebraska; Fall River, Custer, Pennington,
2   and Lawrence Counties in South Dakota; and Niobrara, Weston, and Crook Counties in
3   Wyoming (Figure 5.1-1). The Nebraska-South Dakota-Wyoming Uranium Milling Region holds
4   significant reserves of uranium and has a history of conventional uranium surface mining (NRC,
5   2009a). Other natural resources that are currently being exploited within the milling region and
6   in surrounding counties include oil and gas, wind, coal, coal bed methane, limestone, and gold.
7   Federal agencies have completed several environmental impact statements (EISs) related to
8   activities within the Nebraska-South Dakota-Wyoming Uranium Milling Region. Most of these
9   EISs are related to resource management actions on federal lands administered by the
10 U.S. Forest Service (USFS) or U.S. Bureau of Land Management (BLM) and are focused
11 on improving natural resources conditions and reducing adverse impacts from various
12 human-related activities.
13
14 The various past, present, and reasonably foreseeable future actions in the vicinity of the
15 proposed Dewey-Burdock ISR Project are discussed next.
16
17 **5.1.1.1   Uranium Recovery Sites**
18
19 Uranium milling operations within the Nebraska-South Dakota-Wyoming Uranium Milling Region
20 exist in the Crow Butte Uranium District located in northwestern Nebraska, in the Southern
21 Black Hills Uranium District in southwestern South Dakota and east-central Wyoming, and in the
22 Northern Black Hills Uranium District in northeastern Wyoming (Figure 5.1-2). Existing and
23 potential uranium recovery sites in the Nebraska-South Dakota-Wyoming Uranium Milling
24 Region are listed in Table 5.1-1.
25
26 Seven ISR facilities and one uranium recovery and mill tailings facility licensed under Uranium
27 Mill Tailings Radiation Control Act (UMTRCA) Title II are in the region. The only operating
28 ISR facility is at Crow Butte in Dawes County, Nebraska, approximately 105 km [65 mi]
29 south-southeast of the proposed Dewey-Burdock ISR Project. Three satellite facilities or ISR
30 expansions are planned for the Crow Butte site: North Trend, Three Crow, and Marsland.
31 License applications for North Trend and Marsland were submitted to NRC in June 2007 and
32 May 2012, respectively, and are under review. A license application for Three Crow was
33 submitted in August 2010 and withdrawn and has not yet been resubmitted.
34
35 In addition to the proposed Dewey-Burdock ISR Project, the applicant has identified other
36 potential uranium orebodies in the region at Dewey Terrace in Niobrara and Weston Counties,
37 Wyoming, and at Aladdin in Crook County, Wyoming (Powertech, 2009b). Dewey Terrace is
38 just west of the proposed Dewey-Burdock ISR Project in Weston and Niobrara Counties,
39 Wyoming (Figure 5.1-3). The uranium orebodies at Dewey Terrace are a continuation of the
40 mapped orebodies at the Dewey-Burdock site (Powertech, 2009b). To date, the applicant has
41 not submitted a letter of intent to NRC for either Dewey Terrace or Aladdin. NRC therefore has
42 no specific information that the applicant plans to go forward with these projects. It is also
43 uncertain whether, if either project went forward, the applicant would seek to operate these
44 projects as satellite facilities and ship uranium-loaded resins from Dewey Terrace or Aladdin to
45 the proposed Dewey-Burdock site for processing into yellowcake. NRC staff and other local
46 government agencies will monitor these potential projects, which will be discussed within the
47 context of cumulative impacts in this SEIS based on the available information.
48

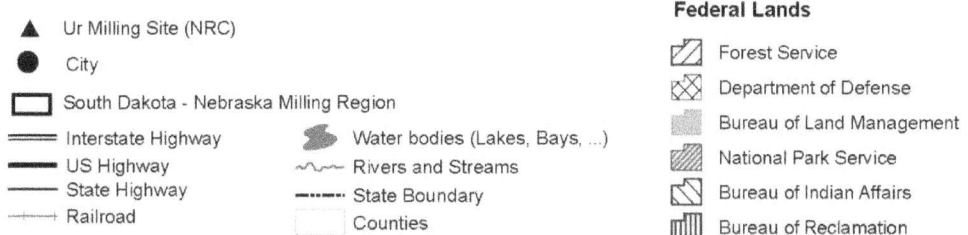

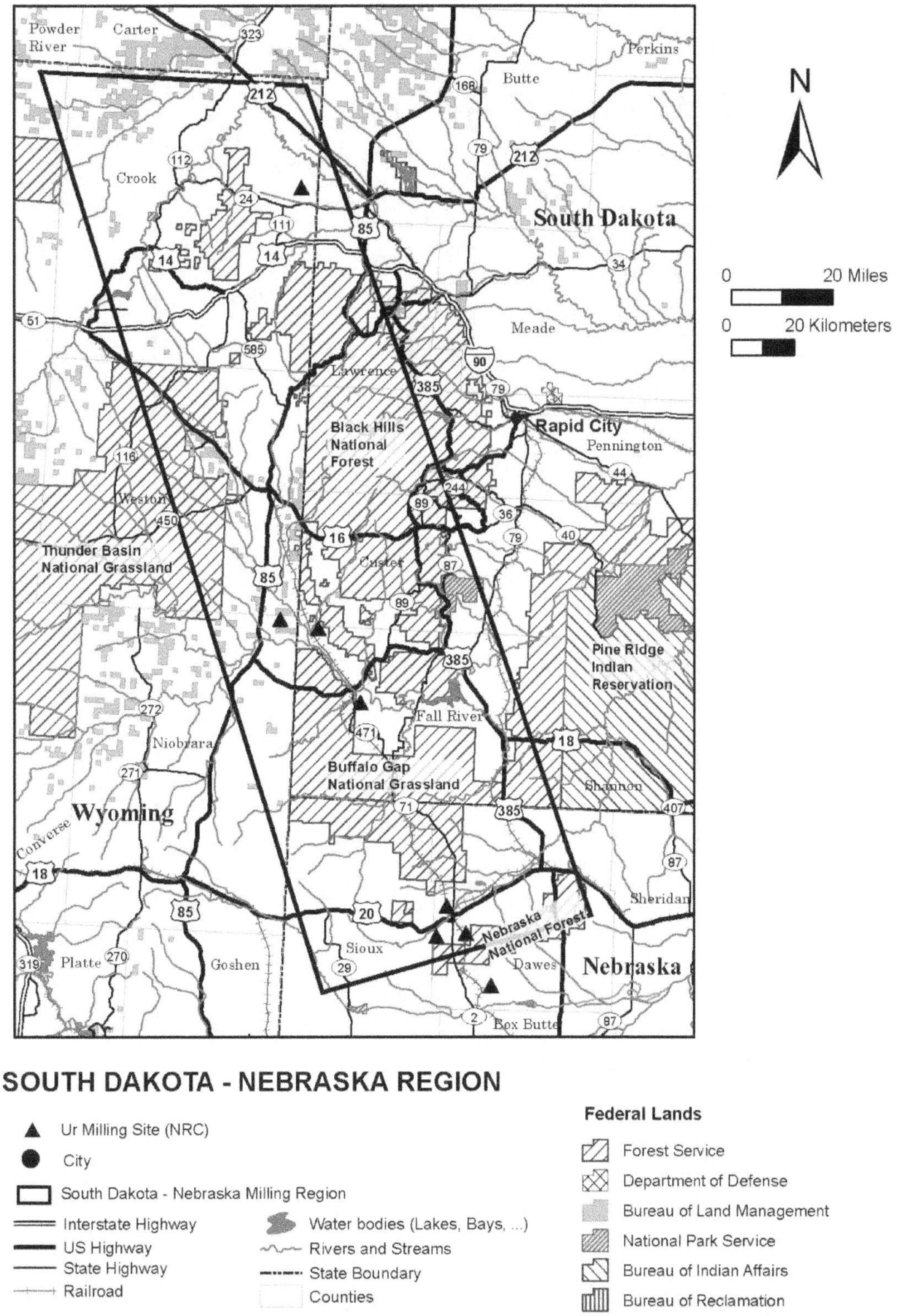
Powder River
Carter
Perkins
Butte
Crook
South Dakota
Meade
Lawrence
Black Hills National Forest
Rapid City
Pennington
Weston
Thunder Basin National Grassland
Custer
Pine Ridge Indian Reservation
Fall River
Niobrara
Buffalo Gap National Grassland
Wyoming
Converse
Shannon
Sheridan
Nebraska National Forest
Sioux
Platte
Goshen
Dawes
Nebraska
Box Butte
N
0 20 Miles
0 20 Kilometers
SOUTH DAKOTA - NEBRASKA REGION
Ur Milling Site (NRC)
City
South Dakota - Nebraska Milling Region
Interstate Highway
US Highway
State Highway
Railroad
Water bodies (Lakes, Bays, ...)
Rivers and Streams
State Boundary
Counties
Federal Lands
Forest Service
Department of Defense
Bureau of Land Management
National Park Service
Bureau of Indian Affairs
Bureau of Reclamation

Figure 5.1-1. Nebraska-South Dakota-Wyoming Uranium Milling Region General Map With Current (Crow Butte, Nebraska) and Potential Future Uranium Milling Site Locations. Source: Modified from NRC (2009a).

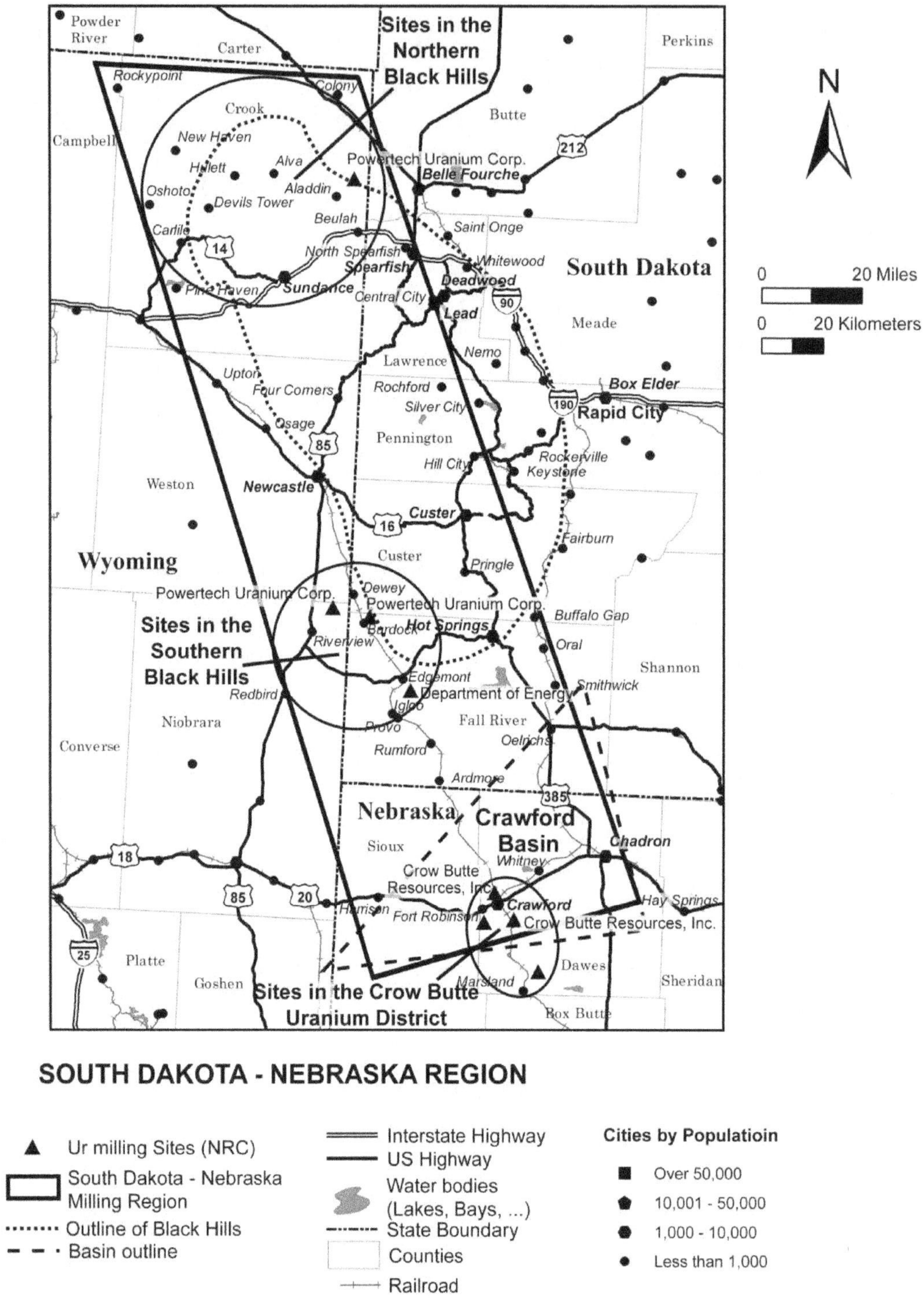

1

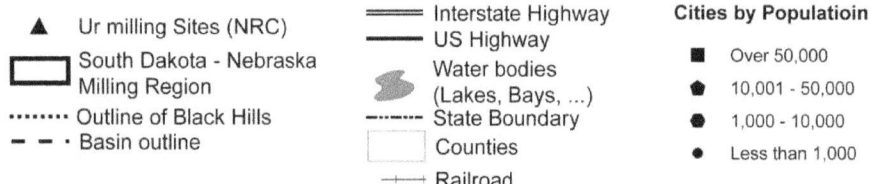

Figure 5.1-2. Map Showing the Nebraska-South Dakota-Wyoming Uranium Milling Region and Uranium Milling Sites in the Black Hills Uranium Districts in South Dakota and Wyoming and in the Crow Butte Uranium District in Nebraska.
Source: Modified from NRC (2009a).

2

1

Table 5.1-1. Past, Existing, and Potential Uranium Recovery Sites in the Nebraska-South Dakota-Wyoming Uranium Milling Region*

Site Name	Company/Owner	Type	County, State	Status[†]	Approximate Distance km (mi)	Direction
North Trend	Cameco (Crow Butte Resources, Inc.)	ISR— Expansion	Dawes County, Nebraska	Potential site— license application received June 2007 (under NRC review)	95 (59)	SSE
Three Crow	Cameco (Crow Butte Resources, Inc.)	ISR— Expansion	Dawes County, Nebraska	Potential site	101 (63)	SSE
Marsland	Cameco (Crow Butte Resources, Inc.)	ISR – Expansion	Dawes County, Nebraska	Potential site	129 (80)	SSE
Crow Butte	Cameco (Crow Butte Resources, Inc.)	ISR— Commercial scale	Dawes County, Nebraska	Operating	105 (65)	SSE
Edgemont	DOE	Conventional uranium mill	Fall River, South Dakota	UMTRCA[†] Title II disposal site	26 (16)	SSE
Dewey-Burdock	Powertech (USA) Inc.	ISR— Commercial scale	Fall River and Custer, South Dakota	Potential site— license application submitted to NRC in August, 2009	0	—
Dewey Terrace	Powertech (USA) Inc.	ISR— Expansion	Niobrara, Wyoming	Potential site	13 (8)	WNW
Aladdin	Powertech (USA) Inc.	ISR— Expansion	Crook, Wyoming	Potential site	137 (85)	NNW

*Sources: NRC (2009a, 2012); Powertech (2009b)
[†]Status: Uranium Mill Tailings Radiation Control Act (UMTRCA) Title II sites are uranium mill processing or tailings sites that have been decommissioned. The U.S. Department of Energy is the long-term custodian of these sites.

2
3   The proposed Dewey-Burdock ISR Project is located within the Edgemont Uranium District on
4   the southwestern flank of the Black Hills uplift. Uranium in the Edgemont Uranium District was
5   first discovered in 1951 and mined until 1972. The district derived its name from the town of
6   Edgemont, South Dakota, which was the closest population center to the district. Uranium was
7   extracted from small conventional underground and surface mines in sandstone deposits within
8   the Inyan Kara Group. The uranium ore was shipped to conventional mills for processing. The
9   only uranium mill built in South Dakota was at Edgemont. The Edgemont uranium mill

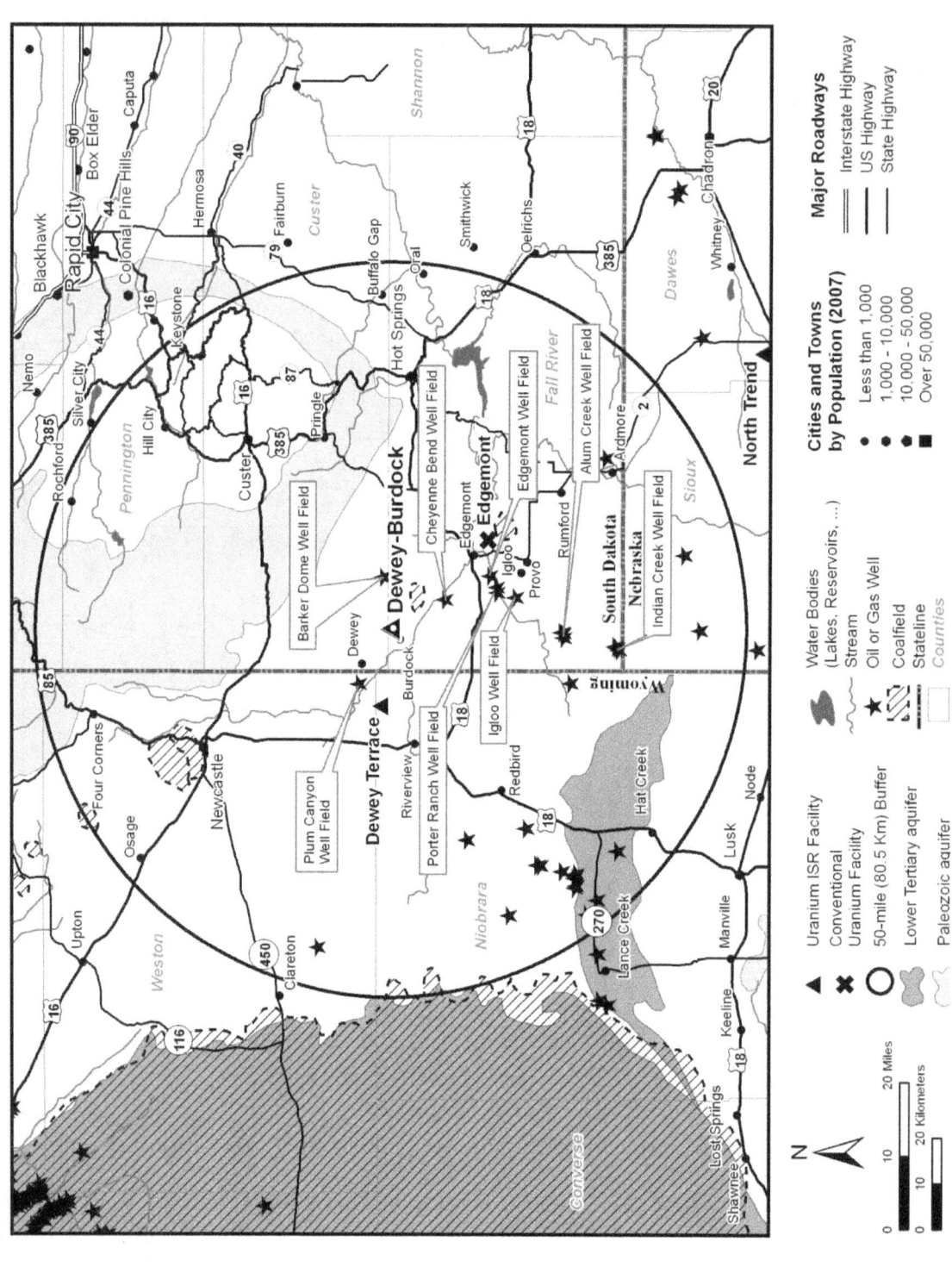

**Figure 5.1-3. Oilfields, Coalfields, and Uranium Occurrences Near the Proposed Dewey-Burdock ISR Project.
Sources: ESRI (2008); National Atlas of the United States (2009); WYOGCC (2012); NOGCC (2012); SDDENR (2012a).**

1

1   processed 1.78 million metric tons [1.98 million short tons] of ore and produced 3.11 million kg
2   [6.86 million lb] of uranium oxide as U_3O_8 before it ceased production in 1974 (SDDENR, 2010).
3   Approximately half the ore {0.9 million metric tons [1.0 million short tons] of ore containing about
4   1.45 million kg [3.2 million lb] of U_3O_8} processed at Edgemont was produced from deposits in
5   South Dakota, and the other half came from out of state.
6
7   Most of the historic uranium mining operations within the Edgemont Uranium District were
8   abandoned prior to the 1970s because they became uneconomical. Abandoned open pits and
9   overburden piles associated with historic surface mining occur in the eastern portion of the
10   proposed Dewey-Burdock ISR Project site (see Figure 3.2-3). Many of the abandoned mine
11   sites in the Edgemont Uranium District are on USFS-managed property. In recent years USFS
12   has reclaimed several abandoned mines in Fall River County, such as the Blue Lagoon,
13   Gladiator, and Dead Horse mines (SDDENR, 2010).
14
15   The Tennessee Valley Authority (TVA) reclaimed the uranium mill at Edgemont from 1986 to
16   1989. The areas excavated during cleanup of the mill site at Edgemont were backfilled with
17   clean soil, graded for proper drainage, and revegetated (SDDENR, 2010). Contaminated
18   uranium mill buildings, tailings sands and slimes, and contaminated soil from the mill site and
19   nearby areas were removed and placed in an engineered disposal site southeast of Edgemont
20   (Figure 5.1-3) (SDDENR, 2010). The Edgemont disposal site is an UMTRCA Title II site owned
21   and administered by U.S. Department of Energy (DOE) under a general NRC license for the
22   custody and long-term care of uranium pursuant to 10 CFR Part 40.28.
23
24   Silver King Mines, Inc. (as Darrow Lease operator and manager for TVA) drilled approximately
25   4,000 exploration holes in the Dewey-Burdock area during the mid-1970s. TVA's uranium
26   exploration activities in the Dewey-Burdock area ended in the early 1980s and did not result in
27   conventional uranium mining or ISR uranium extraction (Powertech, 2009a).
28
29   **5.1.1.2   Coal Mining**
30
31   As discussed in GEIS Section 5.3.3, active or former coal mines have not been identified in the
32   Nebraska-South Dakota-Wyoming Uranium Milling Region (NRC, 2009a). Based on information
33   exchanged with BLM staff during a site visit to the project area in December 2009, past
34   resource development in the region included exploitation of small bituminous coal deposits
35   located east and south of the proposed Dewey-Burdock ISR Project site (NRC, 2009b). This
36   information is consistent with isolated mapped coal fields located approximately 3 km [2 mi]
37   southeast of the proposed project and approximately 6 km [4 mi] southeast of Edgemont
38   (Figure 5.1-3).
39
40   Unlike the sedimentary formations that host commercially extractable coal deposits in the
41   Powder River Basin in Campbell and Converse Counties, Wyoming (i.e., the Wasatch and
42   Fort Union Formations), the sedimentary formations beneath the counties comprising the
43   Nebraska-South Dakota-Wyoming Uranium Milling Region do not contain thick, continuous coal
44   beds (NRC, 2009a). SEIS Section 3.4.1 describes the lithology of sedimentary formations
45   beneath the proposed Dewey-Burdock ISR Project area as unable to support large-scale
46   commercial coal mining.
47
48

1   **5.1.1.3   Oil and Gas Production**
2
3   Regional oil and gas exploration, production, and pipeline construction could potentially
4   generate cumulative impacts. Coal bed methane gas extraction removes natural gas from coal
5   beds. This form of mining is common in the Powder River Basin located 80 km [50 mi] west of
6   the proposed Dewey-Burdock ISR Project (see Figure 5.1-3). Because the Nebraska-South
7   Dakota-Wyoming Uranium Milling Region does not contain commercially viable coal beds, no
8   ongoing or planned coal bed methane production occurs within an 80-km [50-mi] radius of the
9   proposed site (Figure 5.1-3).
10
11   The status of permitted oil and gas wells in Fall River and Custer Counties in South Dakota and
12   Niobrara and Weston Counties in Wyoming is provided in Table 5.1-2. In Fall River County,
13   11 oil wells are actively producing (SDDENR, 2012a). One producing oil well, one underground
14   injection control (UIC) permitted well for salt water disposal, and six plugged and abandoned
15   wells are located in the Cheyenne Bend oilfield 11 km [7 mi] southeast of the proposed site
16   (Figure 5.1-3).   The 10 remaining oil wells in production are located within the Edgemont,
17   Porter Ranch, Igloo, and Alum Creek oilfields (Figure 5.1-3). The Edgemont, Porter Ranch, and
18   Igloo oilfields are located immediately southwest of the city of Edgemont. The Alum Creek
19   oilfield is located approximately 23 km [14 mi] southwest of Edgemont. All Fall River County
20   producing wells are operating within the Minnelusa Formation at depths ranging from 1,081 m
21   [3,547 ft] at the Alum Creek oilfield to 786 m [2,580 ft] at the Cheyenne Bend oilfield
22   (SDDENR, 2012a).
23
24   In Custer County, four oil wells are in active production (SDDENR, 2012a). All four producing
25   wells are located at the Barker Dome oilfield located 6 km [4 mi] east of the proposed site
26   (Figure 5.1-4). The Barker Dome oilfield also contains one UIC permitted well for salt water
27   disposal, one well that has been converted to water supply, and 18 plugged and abandoned
28   wells. Three of the producing oil wells at Barker Dome are located in the Minnelusa Formation
29   at total depths of 423 to 433 m [1,387 to 1,420 ft].   The fourth producing well is located in the
30   Madison Formation at a total depth of 588 m [1,928 ft] (SDDENR, 2012a).
31
32   Weston and Niobrara Counties in Wyoming contain many more completed oil and gas
33   production wells than Fall River and Custer Counties (Table 5.1-2). The closest producing wells
34   to the proposed project are in the Plum Canyon oilfield 5 km [3 mi] to the northwest in Niobrara
35   County (Figure 5.1-4) (WYOGCC, 2012). The Plum Canyon oilfield contains 4 producing wells,
36   which are all located in the Leo Sandstone of the Minnelusa Formation at depths ranging from
37   approximately 785 to 823 m [2,575 to 2,700 ft]. The total depths of completed wells generally
38   increase from east to west across Weston and Niobrara Counties. For example, within the
39   Powder River Basin, which encompasses the southwestern part of Weston County and the
40   northwestern part of Niobrara County, many completed wells reach total depths of more than
41   1,981 m [6,500 ft] (WYOGCC, 2012).
42
43   Demand for drilling permits for oil and gas exploration in the vicinity of the proposed project has
44   been low. Since 2005, South Dakota Department of Environment and Natural Resources
45   (SDDENR) has issued 16 permits for oil and gas exploration drilling in Fall River County and no
46   permits in Custer County (SDDENR, 2012b).
47
48   The potential effects of oil well drilling include the need to build temporary access roads to reach
49   and construct 1.2-ha [3-ac] drill pads for each drill site (BLM, 2009a). The length of time
50

Table 5.1-2. Status of Permitted Oil and Gas Wells in Fall River and Custer Counties, South Dakota, and Niobrara and Weston Counties, Wyoming

County, State	Number of Plugged and Abandoned Wells	Number of Completed Wells	Number of New Permits to Drill	Permits Issued*
Fall River, South Dakota	342	11	2	396
Custer, South Dakota	72	4	0	86
Niobrara, Wyoming	1,661	383	21	2,281
Weston, Wyoming	5,252	1,568	7	7,317

Sources: SDDENR (2012a); WYOGCC (2012)
*The "Permits Issued" category includes wells currently being drilled, wells never drilled, Underground Injection Control (UIC) permitted wells, wells converted to water supply, dormant wells, and wells with expired permits

1
2   required for drilling varies with the depth of each drillhole. Seven tracts of USFS-managed land
3   are available for oil and gas leasing in Custer County in the vicinity of the project area (BLM,
4   2009a). All the tracts are located within Township 6 South, Range 1 East immediately east of
5   Dewey (see Figure 3.2-4). Two of the tracts (SDM79010BO and SDM79010BN) border the
6   perimeter of the proposed project (Figure 5.1-4). If lease applications were filed and approved
7   by USFS and if the leaseholders apply for SDDENR drilling permits, it is expected that
8   exploratory drilling for oil would be conducted.
9
10   **5.1.1.4   Wind Power**
11
12   Because of the proximity of currently operating wind energy projects, the potential exists for the
13   development of wind power facilities in the Nebraska-South Dakota-Wyoming Uranium
14   Milling Region, and these facilities would contribute to meeting forecasted electric power
15   demands. There are wind energy projects currently operating and under construction in
16   South Dakota, Wyoming, and Nebraska (see Table 5.1-3). South Dakota's wind resource is
17   882,412 megawatts (MW), which ranks 5th in the United States (AWEA, 2012b). Wyoming's
18   wind resource is 552,073 MW, which ranks 8th in the United States (AWEA, 2012c). Nebraska's
19   wind resource is 917,999 MW, which ranks 4th in the United States (AWEA, 2012a). The
20   current online capacity of wind energy projects is 784 MW in South Dakota, 1,412 MW in
21   Wyoming, and 337 MW in Nebraska (AWEA, 2012a–c).
22
23   Wind projects in South Dakota, Wyoming, and Nebraska range in capacity from one turbine
24   producing 0.1 MW to 105 turbines producing 210 MW (AWEA, 2012d). The wind power
25   projects closest to the proposed Dewey-Burdock project site are 161 km [100 mi] to the
26   west-southwest near Glenrock in Converse County, Wyoming. Wind power projects in
27   Wyoming are located primarily in the southeastern part of the state (AWEA, 2012c). In
28   South Dakota, wind power projects are located in the central and eastern parts of the state more
29   than 241 km [150 mi] from the proposed Dewey-Burdock site (AWEA, 2012b). Wind power
30   projects in Nebraska are located primarily in the north-central and eastern parts of the state and
31   are also more than 241 km [150 mi] from the proposed Dewey-Burdock site (AWEA, 2012a).
32
33   The Dewey-Burdock Wind Association, LLC is a landowner group formed to explore the
34   possibility of a wind farm (referred to herein as the Dewey-Burdock Wind Project) on privately
35   owned land within and surrounding the proposed Dewey-Burdock ISR Project site (Powertech,
36
37
38

1
2
3
4

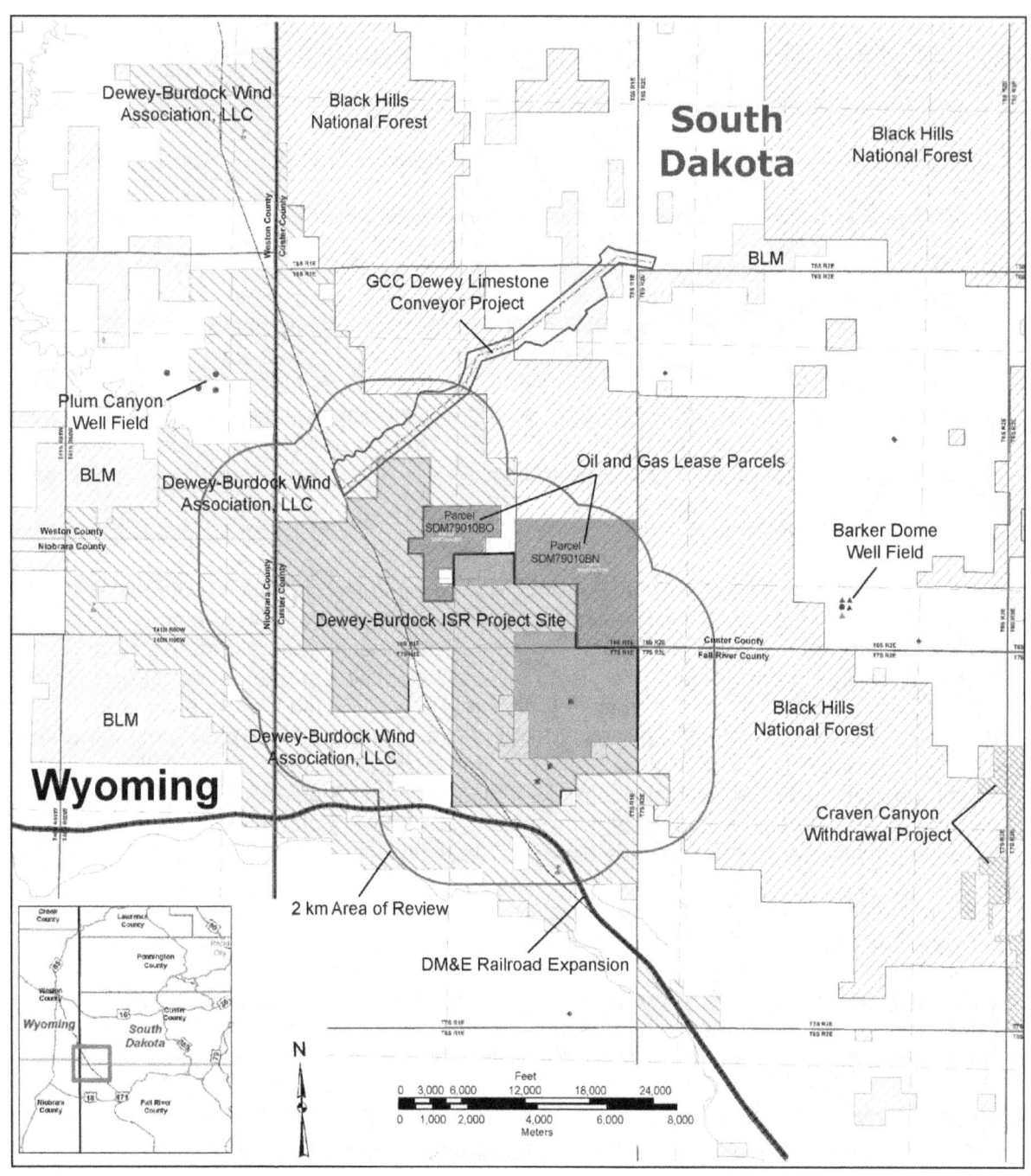

**Figure 5.1-4. Existing, Pending, and Future Projects Within and in the Vicinity of the
Proposed Dewey-Burdock ISR Project.
Source: Modified from Powertech (2010).**

5
6

1

Table 5.1-3. Summary of Wind Energy in South Dakota, Wyoming, and Nebraska

State	Current Online Capacity (MW)	Capacity Added in 2010 (MW)	Wind Resource (MW at 80 m Hub Height)	U.S. Wind Resource Rank
South Dakota	784	396	882,412	5th
Wyoming	1,412	311	552,073	8th
Nebraska	337	60	917,999	4th
Source: AWEA, 2012a–c				

2

3   2010). Land designated as having potential for wind power electrical generation is shown in
4   Figure 5.1-4. The Dewey-Burdock Wind Project is in the conceptual phase.
5
6   The development of wind energy projects in the Nebraska-South Dakota-Wyoming Uranium
7   Milling Region is limited by availability of transmission lines to end users. Existing transmission
8   capacity for wind-generated power is low, and there are no plans to expand existing or construct
9   new transmission corridors in the Nebraska-South Dakota-Wyoming Uranium Milling Region
10 (AWEA, 2012d).
11

12 **5.1.1.5   Transportation Projects**
13

14 *Dewey Conveyor Project*
15

16 In 2007, GCC Dacotah Inc. submitted an Application for Transportation and Utility Systems and
17 Facilities on Federal Lands for the Dewey Conveyor Project. If constructed, the Dewey
18 Conveyor Project will transport limestone mined from the Minnekahta Limestone to a rail load-
19 out facility near Dewey, South Dakota (BLM, 2009a). The conveyor project lies north of the
20 Dewey-Burdock Project area in portions of Township 5 South, Range 1 East, Section 36;
21 Township 6 South, Range 1 East, Sections 1, 2, 9, 10, 11, 12, 15, 16, 17, 18, 19, and 20; and
22 Township 5 South, Range 2 East, Section 31 (Figure 5.1-4). The area proposed for limestone
23 quarrying operations is several kilometers [miles] north, where the Minnekahta Limestone lies at
24 or close to the ground surface (BLM, 2009a). The town of Dewey is located along the existing
25 Burlington Northern Santa Fe (BNSF) Railroad transportation corridor.
26

27 The proposed conveyor route crosses BLM-administered public lands, USFS-administered
28 National Forest System land, and GCC Dacotah Inc.'s privately owned land (Figure 5.1-4). The
29 project anticipates construction of an elevated, enclosed conveyor 10.6-km [6.6-mi] in length, a
30 one-lane service road, and access points (BLM, 2009a). The elevated conveyor would be about
31 5 m [16 ft] high and would provide a minimum vertical clearance of 2 m [6 ft] beneath the
32 structure. Depending on terrain, structural supports would be required at intervals of 7.6 to
33 12 m [25 to 40 ft]. BLM and USFS will evaluate the application and decide whether to approve
34 it, grant GCC Dacotah Inc. a right-of-way (ROW) to allow the conveyor to cross federal lands,
35 and issue a special use permit. BLM and USFS will decide whether stipulations or mitigation
36 measures must be attached to the ROW grant and special use permit.
37

38 *Powder River Basin Expansion Project*
39

40 The Dakota Minnesota and Eastern (DM&E) railroad filed an application to construct the Powder
41 River Basin (PRB) Expansion Project with the federal Surface Transportation Board (STB) in
42 February 1998. The project seeks approval to construct and operate a new rail line and

1   associated facilities in east-central Wyoming and southwest South Dakota (STB, 2001). If
2   approved and completed, the project will add rail coal-hauling capacity and establish a
3   dedicated, direct route to transport coal from the Powder River Basin to Midwest markets.
4   DM&E's proposed rail expansion will extend DM&E's existing northern line near Wall, South
5   Dakota, southwest to Edgemont, then northwest to Burdock, and finally west into Wyoming.
6   The extension will add 418 km [260 mi] of rail line and connect the northern DM&E line to
7   operating coal mines located south of Gillette, Wyoming (see Figure 5.1-5). The proposed rail
8   expansion route is south of the proposed Dewey-Burdock ISR Project site (see Figure 5.1-4).
9
10 At this time, Canadian Pacific—DM&E's parent company—has not yet decided whether to build
11 the extension. The decision to build is contingent on several factors: (i) acquiring the
12 necessary ROW to build the line, (ii) executing agreements with Powder River Basin mining
13 companies for the right of DM&E to operate loading tracks and facilities, (iii) securing
14 contractual commitments from prospective coal shippers to ensure revenues from the proposed
15 line are economical, and (iv) arranging financing for the project.
16
17 **5.1.1.6   Other Mining**
18
19 Gold mining is not extensive in South Dakota; however, gold is the leading mineral commodity
20 by dollar value. Only Wharf Resources Inc. actively mines gold in the state, and it holds four
21 permits for gold operations in the northern Black Hills (Holm, et al., 2008). Wharf Resources is
22 the only company to report silver production, which is a byproduct of its gold recovery process.
23 Sand and gravel are the major nonmetallic mineral commodities produced in South Dakota.
24 Sand and gravel are quarried in every county in South Dakota, mainly for road construction
25 projects. Limestone is quarried in the Black Hills, primarily for the production of cement for use
26 in construction projects.
27
28 **5.1.1.7   Environmental Impact Statements as Indicators of Past, Present, and**
29 **   Reasonably Foreseeable Future Actions**
30
31 Indicators of present and reasonably foreseeable future actions are draft and final EISs federal
32 agencies prepare within a recent time period. Using information in GEIS Section 5.2.2 (NRC,
33 2009a) and other publicly available information, several EISs were identified for the
34 Nebraska-South Dakota Wyoming Uranium Milling Region (see Table 5.1-4). A majority of
35 EISs in Table 5.1-4 are related to resource management actions in the Black Hills National
36 Forest (BHNF) or associated management units. These EISs are for actions that are focused
37 on improving natural resource conditions and reducing adverse impacts from various
38 human-related activities. Three exceptions are the draft EIS that BLM prepared for the Dewey
39 Conveyor Project (BLM, 2009a), the final programmatic EIS that BLM prepared for wind energy
40 development on BLM-administered lands in the western United States (BLM, 2005b), and the
41 final EIS that the STB prepared for the DM&E proposal to build the PRB Rail Expansion Project
42 (STB, 2001).
43
44 **5.1.2   Methodology**
45
46 In calculating and assessing potential cumulative impacts, the NRC staff developed a
47 methodology that follows CEQ guidance (see NRC, 2009a and CEQ, 1997).
48
49 1.   Identify the potential environmental impacts of the federal action, and evaluate the
50   incremental impact of the action when added to other past, present, and reasonably

1
2
3
4
5

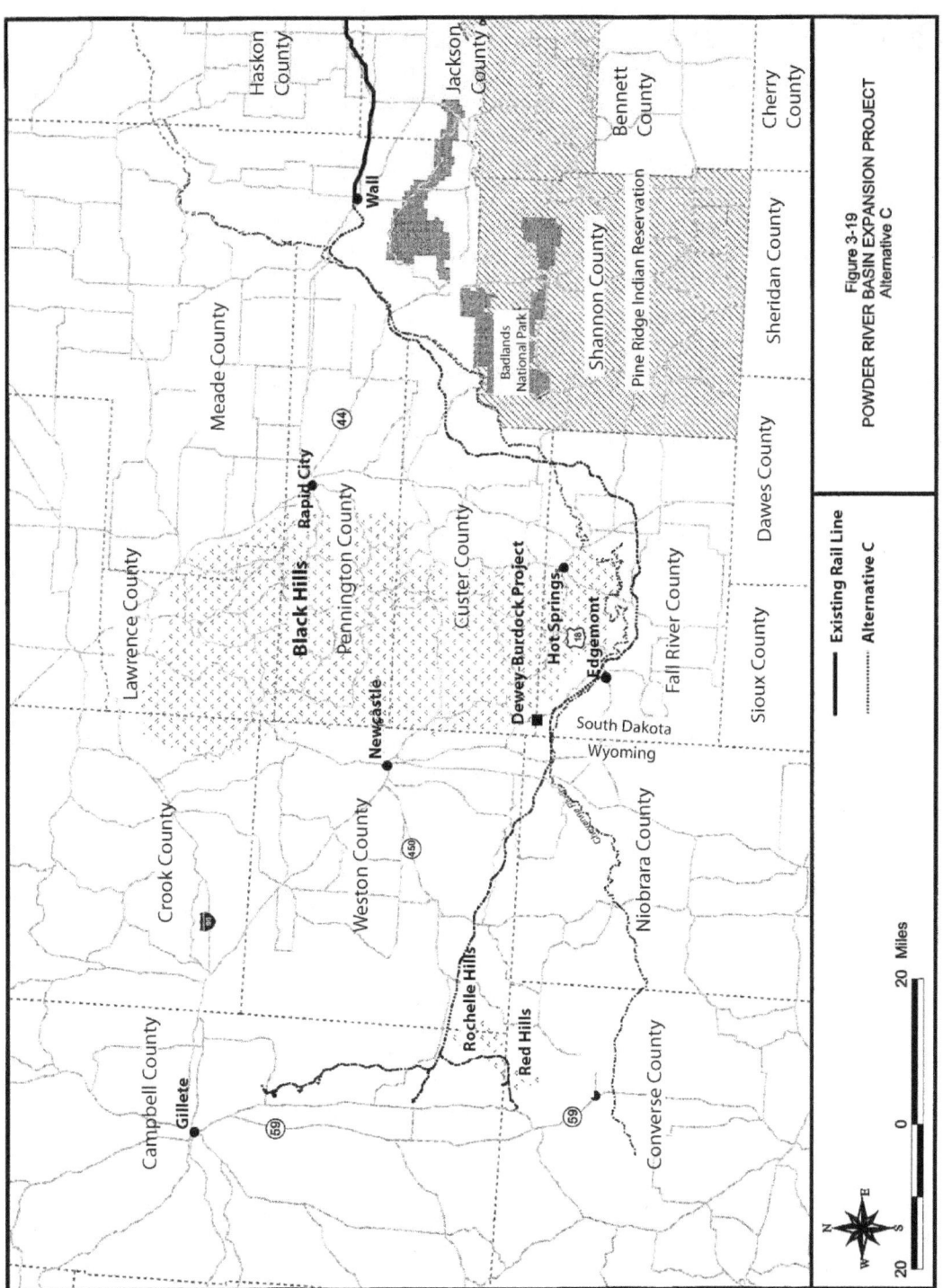

Figure 5.1-5. Map Showing Proposed New Rail Construction (Alternative C) for DM&E's Powder River Basin Expansion Project From Wall, South Dakota, West Into the Powder River Basin of Wyoming. Source: Modified from STB (2001).

6
7

**Table 5.1-4. Draft and Final NEPA Documents Related to the
Nebraska-South Dakota-Wyoming Uranium Milling Region**

November 19, 2001	Surface Transportation Board (STB), Final EIS, Dakota, Minnesota and Eastern Railroad Corporation Powder River Basin Expansion Project
June, 2005	BLM, Final Programmatic EIS on Wind Energy Development on BLM-Administered Lands in the Western United States
June 3, 2005	USFS, Final EIS, Dean Project Area, Proposes To Implement Multiple Resource Management Actions, BHNF, Bearlodge Ranger District, Sundance, Crook County, WY (resource management)
August 12, 2005	USFS, Final EIS, Black-Tailed Prairie Dog Conservation and Management on the Nebraska National Forest and Associated Units, Implementation, Dawes, Sioux, Blaine, Cherry, Thomas Counties, NE, and Custer, Fall River, Jackson, Pennington, Jones, Lyman, Stanley Counties, SD (resource management—prairie dog)
October 28, 2005	National Park Service, Draft EIS, Badlands National Park/North Unit General Management Plan, Implementation, Jackson, Pennington, and Shananon Counties, SD (resource management)
November 20, 2005	USFS, Final EIS, Deerfield Project Area, Proposes To Implement Multiple Resource Management Actions, Mystic Ranger District, BHNF, Pennington County, SD (resource management)
November 25, 2005	USFS, Final EIS, Bugtown Gulch Mountain Pine Beetle and Fuels Projects, To Implement Multiple Resource Management Actions, BHNF, Hell Canyon Ranger District, Custer County, SD (resource management)
January 13, 2006	USFS, Final EIS, Black Hills, National Forest Land and Resource Management Plan Phase II Amendment, Proposal To Amend the 1997 Land and Resource Management Plan, Custer, Fall River, Lawrence, Meade, Pennington Counties, SD; Crook and Weston Counties, WY (resource management)
February 3, 2006	USFS, Final EIS, Black-Tailed Prairie Dog Conservation and Management on the Nebraska National Forest and Associated Units, Implementation, Dawes, Sioux, Blaine, Cherry, Thomas Counties, NE, and Custer, Fall River, Jackson, Pennington, Jones, Lyman, Stanley Counties, SD (resource management—prairie dog)
May 12, 2006	USFS, Final SEIS, Dean Project Area, Proposes To Implement Multiple Resource Management Actions, New Information To Disclose Direct, Indirect, and Cumulative Environmental Impacts, BHNF, Bearlodge Ranger District, Sundance, Crook County, WY (resource management)
June 1, 2007	USFS, Final EIS, Norwood Project, Proposes To Implement Multiple Resources Management Actions, BHNF, Hell Canyon Ranger District, Pennington County, SD, and Weston, Crook Counties, WY (resource management)

1

**Table 5.1-4. Draft and Final NEPA Documents Related to the
Nebraska-South Dakota-Wyoming Uranium Milling Region (continued)**

June 8, 2007	USFS, Draft EIS, Nebraska and South Dakota Black-Tailed Prairie Dog Management, To Manage Prairie Dog Colonies in an Adaptive Fashion, Nebraska National Forest and Associated Units, Including Land and Resource Management Plan Amendment 3, Dawes, Sioux, Blaine Counties, NE, and Custer, Fall River, Jackson, Pennington, Jones, Lyman, Stanley Counties, SD (resource management—prairie dog)
June 29, 2007	USFS, Final EIS, Mitchell Project Area, To Implement Multiple Resource Management Actions, Mystic Ranger District, BHNF, Pennington County, SD (resource management)
September 14, 2007	USFS, Final EIS, Citadel Project Area, Proposes To Implement Multiple Resource Management Actions, Northern Hills Ranger District, BHNF, Lawrence County, SD (resource management)
February 22, 2008	USFS, Draft EIS, Upper Spring Creek Project, Proposes To Implement Multiple Resource Management Actions, Mystic Ranger District, BHNF, Pennington County, SD (resource management)
January 2009	BLM, Draft EIS, Dewey Conveyor Project

1
2   foreseeable future actions for each resource area. Potential environmental impacts are
3   discussed and analyzed in Chapter 4 of this SEIS.
4
5 2.   Identify the geographic scope for the analysis for each resource area. This scope will
6   vary from resource area to resource area, depending on the geographic extent to which
7   the potential impacts of the resource area could be at issue.
8
9 3.   Identify the timeframe for assessing cumulative impacts. The NRC staff use the period
10   from 2009 to 2030 for identifying and assessing cumulative effects. The timeframe
11   begins with NRC acceptance of the application for an NRC source material license to
12   operate the Dewey-Burdock ISR Project in October 2009. The cumulative impact
13   analysis timeframe ends in 2030, the date estimated for license termination after
14   completion of the decommissioning period (see Figure 2.1-1).
15
16   NRC source material licenses for ISR facilities are typically granted for a 10-year period.
17   The proposed Dewey-Burdock ISR Project has an estimated 17-year operational
18   lifespan (see Figure 2.1-1). If NRC grants a source material license, the applicant must
19   apply for license renewal before the initial license period expires to continue operations.
20
21 4.   Identify ongoing and prospective projects and activities that take place or may take place
22   in the area surrounding the project site. These projects and activities are described in
23   Section 5.1.1 of this chapter.
24
25 5.   Assess the cumulative impacts for each resource area from the proposed action and
26   reasonable alternatives, and other past, present, and reasonably foreseeable future
27   actions. This analysis would take into account the environmental impacts of concern
28   identified in Step 1 and the resource-area-specific geographic scope identified in Step 2.
29
30

1   The following terms describe the level of cumulative impact:

3   SMALL: The environmental effects are not detectable or are so minor that they
4   would neither destabilize nor noticeably alter any important attribute of the
5   resource considered.

7   MODERATE: The environmental effects are sufficient to alter noticeably, but not
8   destabilize, important attributes of the resource considered.
9   LARGE: The environmental effects are clearly noticeable and are sufficient to
10   destabilize important attributes of the resource considered.

12 The NRC staff recognize that many aspects of the activities associated with the proposed
13 Dewey-Burdock ISR Project would have SMALL impacts on the affected resources. It is
14 possible, however, that an impact that may be SMALL by itself, but could result in a
15 MODERATE or LARGE cumulative impact when considered in combination with the impacts of
16 other actions on the affected resource. Likewise, if a resource is regionally declining or
17 imperiled, even a SMALL individual impact could be important if it contributes to or accelerates
18 the overall resource decline. The NRC staff determined the appropriate level of analysis that
19 was merited for each resource area potentially affected by the proposed action and
20 alternatives. The level of analysis was determined by considering the impact level to that
21 resource, as described in Chapter 4, as well as the likelihood that the quality, quantity, and
22 stability of the given resource could be affected.

23 Table 5.1-5 summarizes the cumulative impacts of the proposed Dewey-Burdock Project on
24 environmental resources NRC staff identified and analyzed. The cumulative impacts are based
25 on analyses the NRC staff conducted and take into account the other past, present, and
26 reasonably foreseeable activities identified in SEIS Section 5.1.1.

Table 5.1-5. Cumulative Impacts on Environmental Resources

Resource Category	Cumulative Impacts	Comment
Land Use	MODERATE	The proposed project will have a SMALL incremental impact when added to the MODERATE cumulative impacts to land use.
Transportation	MODERATE	The proposed project will have a SMALL to MODERATE incremental impact when added to the MODERATE cumulative impacts to transportation.
Geology and Soils	MODERATE	The proposed project will have a SMALL incremental impact when added to the MODERATE cumulative impacts to geology and soils.

29

Table 5.1-5. Cumulative Impacts on Environmental Resources (continued)

Resource Category	Cumulative Impacts	Comment
Water Resources		
Surface Waters and Wetlands	MODERATE to LARGE	The proposed project will have a SMALL incremental impact when added to the MODERATE to LARGE cumulative impacts to surface waters and wetlands.
Groundwater	MODERATE	The proposed project will have a SMALL incremental impact when added to the MODERATE cumulative impacts on groundwater.
Ecological Resources		
Terrestrial Ecology	MODERATE	The proposed project will have a SMALL incremental impact when added to the MODERATE cumulative impacts to terrestrial ecological resources.
Aquatic Ecology	SMALL	The proposed project will have a SMALL incremental impact when added to the SMALL cumulative impacts to aquatic ecological resources.
Threatened and Endangered Species	MODERATE	The proposed project will have a SMALL incremental impact when added to the MODERATE cumulative impacts to threatened and endangered species.
Air Quality	MODERATE	The proposed project will have a MODERATE incremental impact on air quality when added to the MODERATE cumulative impacts.
Noise	MODERATE	The proposed project will have a SMALL incremental impact on noise when added to the MODERATE cumulative impacts.
Historic and Cultural Resources	MODERATE to LARGE	The proposed project will have a SMALL to LARGE incremental impact on historical and cultural resources when added to the MODERATE to LARGE cumulative impacts.

1

Table 5.1-5. Cumulative Impacts on Environmental Resources (continued)

Resource Category	Cumulative Impacts	Comment
Visual and Scenic Resources	MODERATE to LARGE	The proposed project will have a SMALL incremental impact on visual and scenic resources when added to the MODERATE to LARGE cumulative impacts to the viewshed.
Socioeconomics	SMALL to MODERATE	The proposed project will have a SMALL to MODERATE incremental impact on socioeconomic resources when added to the SMALL to MODERATE cumulative impacts.
Environmental Justice	SMALL	The proposed project will have a SMALL incremental impact on environmental justice when added to the SMALL cumulative impacts.
Public and Occupational Health and Safety	SMALL	The proposed project will have a SMALL incremental impact on public and occupational health when added to the SMALL cumulative impacts.
Waste Management	SMALL to MODERATE	The proposed project will have a SMALL to MODERATE incremental impact on waste management when added to the SMALL to MODERATE cumulative impacts.

1
2 ## 5.2   Land Use
3
4   NRC staff assessed cumulative impacts on land use within a 16-km [10-mi] radius of the
5   proposed Dewey-Burdock ISR Project permit boundary, which includes parts of Custer and
6   Fall River Counties, South Dakota, and Weston and Niobrara Counties, Wyoming. Land use
7   impacts result from interruption to, reduction, or impedance of livestock grazing areas, open
8   wildlife areas, and land access. The assessment of cumulative impacts on land use beyond
9   16 km [10 mi] was not undertaken, because at this distance the impacts on land use from the
10 proposed project will be minimal. The timeframe for the analysis of cumulative impacts is 2009
11 to 2030, as described in SEIS Section 5.1.2.
12
13 The majority of land within the 16-km [10-mi] radius of the proposed project is in private
14 ownership; however, USFS manages tracts of forest, grassland, and recreational land in the
15 vicinity (see Figures 5.1-1 and 5.1-4). The BHNF borders the project to the north and east,
16 and the Buffalo Gap National Grassland is 8 km [5 mi] south of the project. USFS-managed
17 lands provide recreational activities, including camping, hiking, fishing, and hunting.
18 BLM-administered lands are distributed among other federal and private lands to the north,
19 west, and south of the proposed project site. Cattle grazing is the predominant land use on
20 both public and private rangeland.
21

1   Short-term cumulative impacts from the loss of rangeland include a decrease in the area for
2   foraging, temporary loss of animal unit months (AUMs), and temporary loss of water-related
3   range improvements (e.g., improved springs, water pipelines, stock ponds). These impacts
4   would be reduced after an area had been reclaimed. Long-term cumulative impacts result from
5   the permanent loss of forage and forage/cropland productivity in un-reclaimed areas. Other
6   impacts could include dispersal of noxious and invasive weed species both within and beyond
7   areas where the surface had been disturbed, which reduces the area of desirable forage by
8   livestock. The proposed Dewey-Burdock ISR Project will disturb 98 ha [243 ac] if Class V deep
9   injection wells are used to dispose of liquid wastes or 566 ha [1,398 ac] if land application is
10 used to dispose of liquid wastes (see SEIS Section 4.2.1). These amounts of land are small in
11 comparison to the available grazing land within the land use study area {i.e., land within a 16-km
12 [10-mi] radius of the proposed project site}. These amounts of land will also be fenced from
13 grazing at different times over the life of the project.
14
15 Past, ongoing, and future conventional uranium mines and ISR facilities in the vicinity of the
16 proposed Dewey-Burdock ISR Project and within the broader regional area are described in
17 SEIS Section 5.1.1. The Crow Butte ISR facility lies 105 km [65 mi] to the south-southeast in
18 Dawes County, Nebraska, and is the closest operational ISR facility to the Dewey-Burdock site.
19 Three ISR expansion or satellite projects are in the planning or licensing stages in the
20 immediate vicinity of the Crow Butte ISR facility (North Trend, Three Crow, and Marsland) (see
21 SEIS Section 5.1.1.1).
22
23 In the land use study area, the applicant has identified a potential ISR project at Dewey Terrace.
24 The Dewey Terrace project would be located approximately 13 km [8 mi] west of the proposed
25 project area in Weston and Niobrara Counties, Wyoming (Figure 5.1-3). If developed, the
26 potential Dewey Terrace project will have impacts on land use (i.e., surface disturbances and
27 fencing to restrict livestock grazing) within the land use study area. To assess the projected
28 land area that will be affected by the development of the potential Dewey Terrace project, the
29 NRC staff assumed that approximately the same area affected by the proposed action {98 to
30 566 ha [243 to 1,398 ac]} will also apply to other potential ISR projects. Like the proposed
31 Dewey-Burdock ISR Project, this amount of land area is small in comparison to the land use
32 study area.
33
34 Land disturbed by past conventional surface mining is present in the eastern part of the
35 proposed Dewey-Burdock site, where abandoned open mine pits and mine waste overburden
36 piles are found (see SEIS Section 5.1.1.1). Wellfields are planned within these areas (see
37 Figure 3.2-3). If wellfields in the mine waste areas are constructed and operated, additional
38 land disturbance and access restrictions will occur.
39
40 Impacts on land use from oil and gas drilling include building temporary access roads and
41 constructing 1.2-ha [3-ac] drill pads for each drill site (BLM, 2009a). There are no active oil- and
42 gas-producing wells within the proposed Dewey-Burdock permit area. SEIS Section 3.2.3
43 identifies three plugged and abandoned oil and gas wells in the Burdock portion of the site in
44 Fall River County. There are few producing oil wells in the land use study area {i.e., within a
45 16-km [10-mi] radius of the proposed Dewey-Burdock project area}. The Barker Dome oilfield
46 in Custer County and the Plum Canyon oilfield in Weston County each have four producing oil
47 wells (see Figures 5.1-3 and 5.1-4). The Cheyenne Bend oilfield in Fall River County has one
48 producing oil well (see Figure 5.1-3). In addition, demand for oil and gas leasing in the vicinity
49 of the proposed project is low (see SEIS Section 5.1.1.3). The majority of active oil and gas
50 development in the region takes place on USFS-managed land (see Figure 5.1-3). This

1 development occurs west and south of Edgemont and in the Powder River Basin, which is more
2 than 80 km [50 mi] west of the proposed project (see Figure 5.1-3).
3
4 Ongoing and proposed coal bed methane operations and wind energy operations in the region
5 are located in the Powder River Basin west of the cumulative impacts land use study area (see
6 SEIS Sections 5.1.1.2 and 5.1.1.4). Sedimentary formations hosting potential coal bed methane
7 reserves are not present in the land use study area. The nearest existing wind power projects
8 to the land use study area are located approximately 161 km [100 mi] to the west-southwest
9 near Glenrock in Converse County, Wyoming. The potential Dewey-Burdock Wind Project is in
10 the conceptual phase and would be located within and surrounding the proposed Dewey-
11 Burdock site (Figure 5.1-4). If developed, the wind project will be constructed on ridges to
12 exploit the best wind conditions rather than low areas where uranium deposits within and in the
13 vicinity of the proposed project tend to be located (e.g., see Figure 4.5-1). Development of wind
14 energy projects is generally compatible with other land uses, including livestock grazing,
15 recreation, wildlife habitat conservation, and oil and gas production activities (BLM, 2005b).
16
17 Two proposed transportation projects are within the cumulative impacts land use study area: the
18 GCC Dacotah Inc.'s Dewey Conveyor Project and the DM&E PRB Expansion Project (see SEIS
19 Section 5.1.1.5).
20
21 Lands along the route of the Dewey Conveyor Project are owned by GCC Dacotah and private
22 landowners or are public lands managed by BLM or USFS. About 16.2 ha [40 ac] of land
23 disturbance will be created during the 1-year conveyor construction phase, resulting in
24 temporary loss of forage. After construction, about 6.5 ha [16 ac] of land disturbance will
25 remain, resulting in long-term losses in available forage. These long-term losses will be
26 confined to the conveyor and maintenance road footprints. The conveyor will be designed to
27 allow livestock and wildlife to freely cross beneath. Adequate signage will be posted to prevent
28 potential trespass by GCC Dacotah employees, and GCC Dacotah employees will be trained
29 regarding property boundaries. The conveyor project is designed so as not to interfere with the
30 operation and maintenance of existing electric transmission and oil and gas distribution lines. In
31 addition, changes in road easements and other infrastructure are not expected. (BLM, 2009a)
32
33 The proposed DM&E PRB Expansion Project will have a significant impact on use of private
34 agricultural land by farmers and ranchers, grazing allotments leased by ranchers on federal
35 lands, and mineral and mining rights on federal lands in western South Dakota and Wyoming.
36 State-owned land and utility corridors are also expected to have impacts. Construction of the
37 rail extension will involve direct and indirect takings of privately held land and the destruction of
38 wells, windmills, corrals, fencing, outbuildings, irrigation systems, fire prevention and
39 suppression systems, and other capital improvements. Access roads, hauling roads, and
40 borrow pits will be built. DM&E will be required to mitigate adverse environmental impacts to
41 private agricultural and ranch lands, federal lands, state lands, and utility corridors. DM&E will
42 negotiate these mitigation measures with landowners and federal and state agencies. DM&E
43 will be required to restore all federal, state, and privately held agricultural lands disturbed by the
44 project to pre-construction conditions as promptly and fully as possible. (STB, 2001)
45
46 The NRC staff have determined that the cumulative impact on land use within the land use
47 study area (i.e., Fall River, Custer, Weston, and Niobrara Counties) resulting from all past,
48 present, and reasonably foreseeable future actions is MODERATE. This finding is based on the
49

1 assessment of existing and potential impacts on land use within the study area from the
2 following actions:
3
4 • Land disturbance from past conventional surface mining in the eastern portion of the
5   proposed Dewey-Burdock site
6
7 • Surface disturbance and restrictions on livestock grazing and recreational activities
8   (e.g., hunting and off-road vehicle use) from development of potential ISR projects, such
9   as the potential Dewey Terrace project
10
11 • Land disturbance from development of the proposed Dewey Conveyor Project
12
13 • Direct and indirect taking of privately held land tied to construction of the DM&E PRB
14   Expansion Project, with resulting destruction of wells, windmills, corrals, fencing,
15   outbuildings, irrigation systems, fire prevention and suppression systems, and other
16   capital improvements
17
18 Other ongoing and reasonably foreseeable future actions are not expected to have a
19 significant impact on land use within the cumulative impacts study area. There are few
20 producing oil wells within the study area, and demand for oil and gas leasing is low. Coal bed
21 methane reserves are not present within the study area. Potential wind energy projects, such
22 as the Dewey-Burdock Wind Project, are generally compatible with the primary land uses in
23 the study area, including livestock grazing, recreation, and wildlife habitat conservation
24 (BLM, 2005b).
25
26 The NRC staff conclude the proposed Dewey-Burdock ISR Project will have a SMALL
27 incremental effect on land use after evaluating its effects and those of all the other past,
28 present, and reasonably foreseeable future actions in the land use study area. As discussed in
29 SEIS Section 4.2.1, land use impacts related to the proposed Dewey-Burdock ISR Project will
30 be SMALL for all stages of the project lifecycle. The estimated land disturbance of 98 to 566 ha
31 [243 to 1,398 ac] for the proposed action is a small amount of land in comparison to the
32 cumulative impacts study area. About this same amount of land will be fenced over the life of
33 the proposed project to restrict livestock grazing and public access to the ISR facilities and to
34 infrastructure, wellfields, and potential land application areas. Fencing around wellfields will be
35 temporary. As wellfield production ends, fencing will be removed and the land reclaimed in
36 accordance with applicable BLM and SDDENR requirements. At the end of operations, the
37 applicant will decommission the site and restore the land to its previous use (with the possible
38 exception of access roads that land owners may request to remain) in accordance with an
39 NRC-approved decommissioning plan (see SEIS Section 2.1.1.1.5).
40
41 ## 5.3   Transportation
42
43 Cumulative impacts on transportation systems of Custer and Fall River Counties, South Dakota,
44 and Weston and Niobrara Counties, Wyoming, were identified and evaluated. Local highways,
45 existing county roads, and access roads were the focus of this analysis over the 2009–2030
46 timeframe (see SEIS Section 5.1.2 for the estimated operating life of the facility).
47
48 As described in SEIS Section 4.3.1, the impacts to heavily traveled regional and local highways
49 will be SMALL during all phases of the proposed Dewey-Burdock ISR Project. Dewey Road,
50

1   the principal access road to the Dewey-Burdock site, will be used throughout the project
2   lifecycle. As described in SEIS Section 4.3.1, daily traffic on Dewey Road will increase
3   sixteenfold during the construction phase and fivefold during the operations phase of the
4   proposed project. The increase in traffic will accelerate the degradation of the road surface,
5   increase fugitive dust emissions, and increase the potential for traffic accidents and wildlife or
6   livestock kills, resulting in a MODERATE impact. Secondary access roads connecting Dewey
7   Road with the proposed plant facilities and the plant facilities within the wellfields will also
8   experience long-term transportation impacts. However, the transportation impacts to secondary
9   access roads are not considered permanent, because the land will ultimately be returned to its
10   natural condition when production and decommissioning are complete (Powertech, 2009b).
11
12   In the cumulative impacts study area, transportation will be impacted by ongoing and
13   reasonably foreseeable future activities. These include impacts to livestock grazing, uranium
14   exploration and mining, and oil and gas exploration and development. The many unimproved,
15   two-track dirt roads and one lane gravel roads in the cumulative impacts transportation study
16   area were constructed to access livestock grazing lands, to facilitate natural resource
17   exploration and extraction, to provide access to recreational areas, and for off-road vehicle
18   recreational activities. County roads in the transportation study area have intermittently
19   provided access for uranium exploration and mining, as well as oil and gas exploration activities,
20   since the mid-1970s. Reasonably foreseeable future uranium, oil, and gas exploration will result
21   in additional trucks and heavy equipment using existing county roads. For example,
22   the potential Dewey Terrace uranium project would be located 13 km [8 mi] west of the
23   Dewey-Burdock ISR Project area in Weston and Niobrara Counties, Wyoming (see SEIS
24   Section 5.1.1.1). If developed, the Dewey Terrace project may contribute to additional traffic on
25   Dewey Road from commuting workers, construction and operations deliveries, and yellowcake
26   and byproduct transport. These future activities may require or benefit from the construction of
27   new road surfaces or the improvement of existing county roads, including Dewey Road.
28
29   As noted in SEIS Section 5.1.1, other reasonably foreseeable future projects, such as wind
30   energy and transportation projects, contribute to the analysis of cumulative impacts.
31
32   Wind energy projects will impact transportation on local roads; however, these impacts would be
33   temporary. During the 1- to 2-year construction period for a wind energy project, the vehicles of
34   100 to 150 workers and vehicles used to transport construction equipment, blades, turbine
35   components, and other materials to the site will cause a relatively short-term increase in the use
36   of local roadways. Shipments of materials, such as gravel, concrete, and water, are not
37   expected to significantly affect local primary and secondary road networks. Shipments of
38   overweight and/or oversized loads are expected to cause temporary disruptions on primary and
39   secondary roads used to access construction sites. It is possible that local roads might require
40   fortification of bridges and removal of obstructions to accommodate overweight and oversized
41   shipments. Once completed, wind energy projects will require a relatively low number of
42   workers to operate and maintain. For example, the operation and maintenance of a
43   180-megawatt capacity wind energy project with about 150 turbines will require 10 to
44   20 workers. Consequently, transportation activities will be limited to a small number of daily
45   trips by pickup trucks, medium-duty vehicles, or personal vehicles. Shipments of large
46   components required for equipment replacement in the event of major mechanical breakdowns
47   are expected to be infrequent. Transportation activities during site decommissioning will be
48   similar to those during construction. Heavy equipment will be required for dismantling turbines
49   and towers, breaking up tower foundations, and regrading and recontouring the site.
50   (BLM, 2005b)

1
2   The proposed Dewey Conveyor will not impact transportation on heavily traveled regional and
3   local roadways but will temporarily impact transportation on Dewey Road. Dewey Road is the
4   primary transportation corridor along the 10.6 km [6.6 mi] length of the proposed conveyor
5   alignment (Figure 5.3-1). Dewey Road continues both north and south of the proposed
6   conveyor project. The construction workforce for the conveyor project will come primarily from
7   Hot Springs, Custer, and Edgemont and use Dewey Road to access the site from the south.
8   Construction of the conveyor will involve approximately 50 workers and take 1 construction
9   season. During construction, deliveries and commuting workers will increase traffic counts
10 on Dewey Road between Edgemont and Dewey. Following construction, approximately
11 12 workers will oversee quarrying, transport, and load-out operation related to the project. Due
12 to the short duration of construction and relatively low number of workers needed to operate the
13 conveyor operation, the proposed Dewey Conveyor Project is not expected to have a significant
14 impact on transportation in the cumulative impacts study area. (BLM, 2009a)
15
16 The proposed DM&E PRB Expansion Project will have temporary impacts on transportation in
17 western South Dakota and Wyoming. The project will require the construction of temporary
18 roads to access the rail line ROW. In the cumulative impacts study area for transportation, the
19 rail line will parallel the BNSF rail line from Edgemont to Burdock before turning west toward
20 Wyoming (see Figure 5.1-4). Therefore, the project will have an impact on Dewey Road from
21 commuting workers and deliveries of equipment and materials during construction of the rail
22 line. DM&E has proposed mitigation measures as part of the proposed PRB Expansion Project
23 to address potential adverse impacts to transportation. To the extent possible, DM&E will
24 confine all project-related construction traffic to a temporary access road within the ROW or
25 established public roads. Any temporary access roads constructed outside the rail line ROW
26 will be removed and the land reclaimed upon completion of construction. As a result of road
27 closures after construction and during operation of railyards, DM&E will provide or develop
28 alternative access for the safe movement of farm and ranch equipment and livestock to fields
29 and pastures. (STB, 2001)
30
31 The NRC staff have determined that the cumulative impact on transportation within the
32 transportation study area resulting from all past, present, and reasonably foreseeable future
33 actions is MODERATE. Regional and local highways in the transportation study area have
34 sufficient capacity to accommodate the traffic of ongoing actions and increases in traffic from
35 other reasonably foreseeable future actions. However, county roads will be impacted. County
36 roads have been used to access uranium exploration and mining and oil and gas exploration
37 activities in the transportation study area since the mid-1970s. Reasonably foreseeable future
38 uranium, oil, and gas exploration and development in the transportation study area will result in
39 additional trucks and heavy equipment using existing county roads. Construction and operation
40 of potential wind energy and transportation projects will also impact county roads in the
41 transportation study area. For example, the potential Dewey-Burdock Wind Project and the
42 proposed Dewey Conveyor Project and DM&E PRB Expansion Project would utilize
43 Dewey Road. Transportation impacts will be most significant during the construction phase of
44 wind energy and transportation projects because construction activities involve more workers
45 and deliveries of materials and equipment.
46
47 The NRC staff have concluded that the proposed Dewey-Burdock ISR Project will have a
48 SMALL to MODERATE incremental effect on transportation when considered with all the other
49 past, present, and reasonably foreseeable future actions in the transportation study area. As
50 described in SEIS Section 4.3.1, increased vehicular traffic associated with the proposed
51

1
2
3

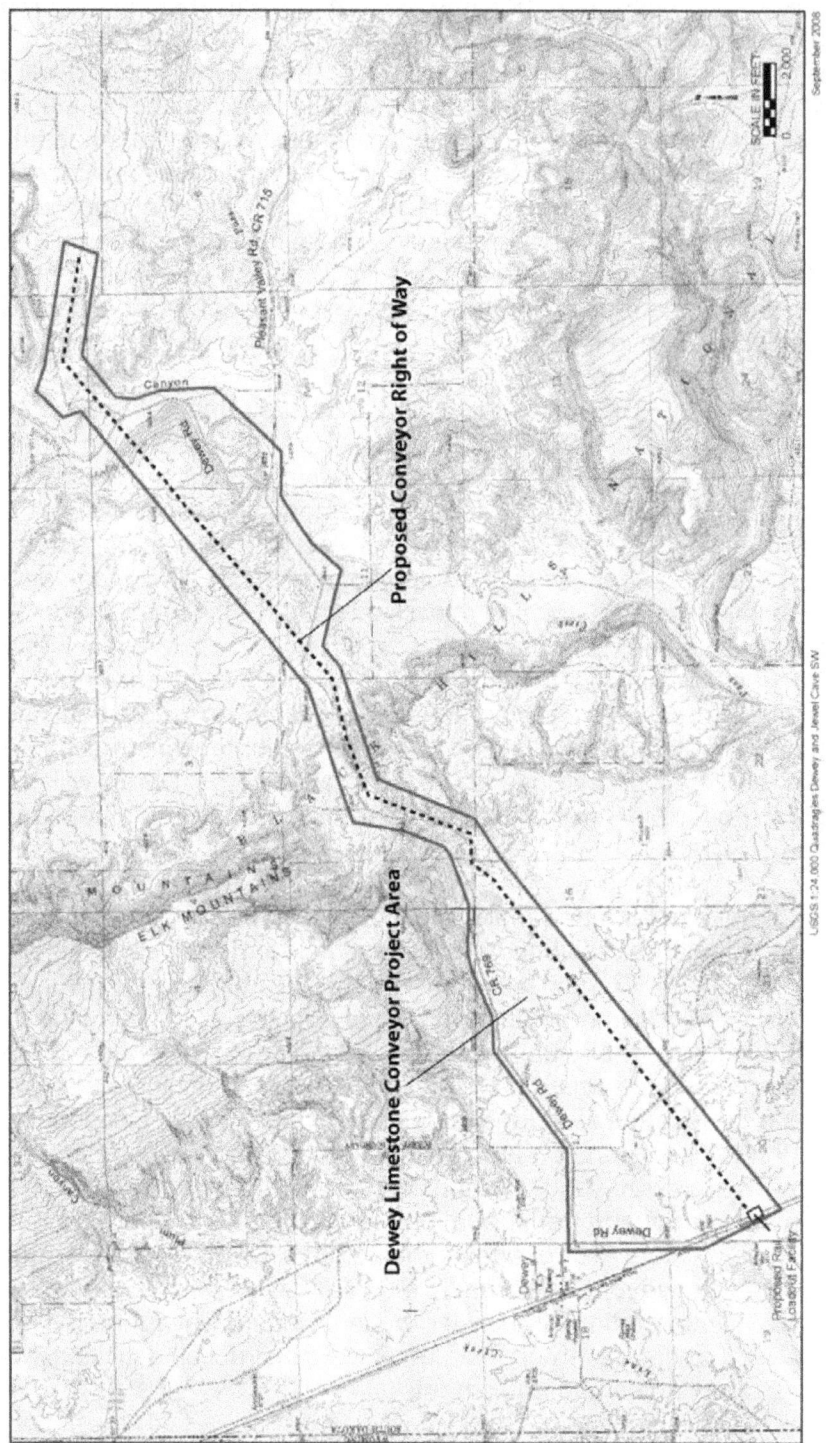

**Figure 5.3-1. Map Showing Location of Dewey Road and Pass Creek in
Relation to the Proposed Dewey Conveyor Project.
Source: Modified from BLM (2009a).**

1   Dewey-Burdock ISR Project will have a SMALL to MODERATE impact. Because regional and
2   local roadways have sufficient capacity to accommodate traffic associated with the proposed
3   project, the proposed Dewey-Burdock project will have a SMALL incremental impact on regional
4   and local roadways within the transportation study area. As described in SEIS Section 4.3.1,
5   Dewey Road would experience a sixteenfold increase in daily traffic during the construction
6   phase and a fivefold increase in daily traffic during the operations phase of the proposed
7   Dewey-Burdock ISR Project. Therefore, the proposed Dewey-Burdock ISR Project will have a
8   MODERATE incremental impact on Dewey Road within the transportation study area.
9
10 ## 5.4   Geology and Soils

11 Cumulative impacts on geology and soils within Custer and Fall River Counties, South Dakota,
12 and Weston and Niobrara Counties, Wyoming, were identified and evaluated focusing on
13 an area within a 16-km [10-mi] radius of the proposed Dewey-Burdock ISR Project site. This
14 area was chosen for the assessment of potential cumulative impacts on geology and soils
15 because the uranium mineralization at other potential uranium deposits within 16 km [10 mi] of
16 the proposed site would be located in the same geologic unit (the Inyan Kara Group). The
17 timeframe for the analysis is 2009 to 2030 (see SEIS Section 5.1.2 for the estimated operating
18 life of the facility).
19
20 As assessed in SEIS Section 4.4.1, all phases of the proposed Dewey-Burdock ISR Project will
21 have a SMALL impact on geology and soils. The primary impacts on geology and soils will
22 result from earthmoving activities. Earthmoving activities that might impact soils include the
23 clearing of ground and topsoil and preparing surfaces for the Burdock central processing plant,
24 Dewey satellite facility, header houses, access roads, drilling sites, and associated structures.
25 Excavating and backfilling trenches for pipelines and cables, and preparing surfaces for
26 potential land application of process-related liquid wastes, will also impact soils. Operations at
27 the proposed site may produce spills of process fluids or chemical materials that may
28 contaminate soils. Best management practices (BMPs) and required monitoring and mitigation,
29 such as spill prevention and cleanup programs, will reduce these potential soil impacts.
30 Subsurface impacts, such as subsidence and activation of nearby faults, will not occur at the
31 proposed project site, because of the relatively small net withdrawal of fluids from production
32 zone aquifers and because of the low pressures during operations relative to those needed to
33 produce small earthquakes. As described in SEIS Section 3.5.3.2, data from aquifer pumping
34 tests indicated a hydraulic connection between the Lakota and Fall River Formations through
35 the intervening Fuson Shale in the Burdock area resulting from unidentified structural features
36 or old, unplugged exploration holes.
37
38 Historical, present, and future natural resource development activities that relate to geology and
39 soils in the geological and soil resources study area include stock grazing, uranium
40 exploration/mining, and oil and gas exploration. Geologic formations hosting potential coal
41 bed methane reserves are not present in the immediate vicinity of the proposed project.
42 Surface-disturbing activities related to uranium, oil, and gas exploration activities, such as
43 construction of new access roads and drill pads, will have direct effects on geological resources.
44 During construction of these roads and drill pads, direct impacts on geology will be limited to
45 excavation and relocation of disturbed bedrock and unconsolidated surficial materials
46 associated with surface disturbances. Impacts from these activities include loss of soil
47 productivity due primarily to wind erosion, changes to soil structure from soil handling, sediment
48 delivery to surface water resources (i.e., runoff), and compaction from equipment and livestock
49 pressure. No geological mineral resources will be lost due to grazing. BMPs and reclamation

1 and restoration of soils disturbed by historic livestock grazing and exploration activities will
2 mitigate loss of soil and soil productivity. However, indirect long-term effects, such as
3 cross-contamination of aquifers, may occur if boreholes associated with uranium, oil, and gas
4 exploration are not properly abandoned.
5
6 Geology and soil resources have been impacted by past conventional uranium mining in the
7 eastern part of the proposed Dewey-Burdock site, where abandoned open mine pits and mine
8 waste overburden piles are found (see SEIS Section 5.1.1.1). Radiological conditions of soils in
9 the areas of past conventional uranium mining are discussed in SEIS Section 3.12.1. There are
10 underground mine workings associated with four former shallow underground uranium mines
11 and two open pit adits (horizontal tunnels). The underground mines consist of declines
12 (downward sloping ramps) ranging from 0 to 24 m [0 to 80 ft] below ground surface. The adits
13 were driven into the sidewalls of the open pits. All of the underground workings were within
14 sandstones of the Fall River Formation. At this time, there are no plans to reclaim or restore the
15 abandoned open mine pits and mine waste overburden piles.
16
17 Development of future ISR projects in the geological and soil resources study area, such as the
18 potential Dewey Terrace project, will have impacts on geology and soils due to increased
19 vehicle traffic, clearing of vegetated areas, soil salvage and redistribution, discharge of ISR-
20 produced groundwater, and construction and maintenance of project facilities and infrastructure
21 (e.g., roads, well pads, pipelines, industrial sites, and associated ancillary facilities). The NRC
22 staff assume that development of future ISR projects within the cumulative impacts study area
23 will be similar to the proposed Dewey-Burdock site, with similar potential for surface impacts to
24 geology and soils. The construction and operation of the infrastructure for these future projects,
25 however, will be subject to the same monitoring, mitigation, and response programs required to
26 limit potential surface impacts (e.g., erosion and contamination from spills) as at the proposed
27 Dewey-Burdock ISR Project. With respect to compaction and surface subsidence, the
28 groundwater will be from the same aquifers and at similar depths as those at Dewey-Burdock,
29 with a small net withdrawal. BMPs and reclamation and restoration of disturbed areas will
30 mitigate loss of soil and soil productivity associated with ISR activities. Salvaged and replaced
31 soil will become viable soon after vegetation is established.
32
33 Other reasonably foreseeable future activities in the vicinity of the proposed Dewey-Burdock
34 ISR Project site that may impact geological resources and soils include wind energy projects
35 (see SEIS Section 5.1.1.4), and proposed transportation projects, such as the Dewey Conveyor
36 Project and the DM&E PRB Expansion Project (see SEIS Section 5.1.1.5).
37
38 Impacts to geological resources and soils from wind energy projects, such as the potential
39 Dewey-Burdock Wind Project, include use of geologic resources (e.g., sand and gravel),
40 activation of geologic hazards (e.g., landslides and rockfalls), and increased soil erosion. Sand
41 and gravel and/or quarry stone will be needed for access roads. Concrete will be needed for
42 buildings, substations, transformer pads, wind tower foundations, and other ancillary structures.
43 These materials will be mined as close to the potential wind energy site as possible. Tower
44 foundations will typically extend to depths of 12 m [40 ft] or less. The diameter of tower bases is
45 generally 5 to 6 m [15 to 20 ft], depending on the turbine size. Construction activities can
46 destabilize slopes if they are not conducted properly. Soil erosion will result from (i) ground
47 surface disturbance to construct and install access roads, wind tower pads, staging areas,
48 substations, underground cables, and other onsite structures; (ii) heavy equipment traffic; and
49 (iii) surface runoff. Any impacts to geology and soils will be largely limited to the project site.
50 Erosion controls that comply with county, state, and federal standards will be applied.

1   Operators will identify unstable slopes and local factors that can induce slope instability.
2   Implementation of BMPs will limit the impacts from earthmoving activities. Foundations and
3   trenches will be backfilled with originally excavated material, and excess excavation material will
4   be stockpiled for use in reclamation activities. (BLM, 2005b)
5
6   The construction of the proposed Dewey Conveyor Project will have direct impacts on
7   geological resources, although these will be limited to surface disturbances associated with
8   excavation and relocation of disturbed bedrock and unconsolidated surficial materials along the
9   various ROWs during construction. The surface disturbances resulting from construction of the
10 conveyor will not result in any loss of known mineral resources. Approximately 16.2 ha [40 ac]
11 of soils along the conveyor route will be directly impacted due to excavation and disturbance.
12 These impacts would include loss of soil to wind and water erosion and decreased soil
13 biological activity. Implementation of BMPs and revegetation of disturbed areas and stockpiled
14 topsoil will minimize soil erosion. (BLM, 2009a)
15
16 The proposed DM&E PRB Expansion Project will have a significant impact on the geology and
17 soils of western South Dakota and Wyoming. Along the route of the proposed rail line, geology
18 and soils will be disturbed by increased traffic, clearing of vegetated areas, and soil salvage and
19 redistribution. To limit the impacts, DM&E has proposed mitigation measures as part of the
20 proposed PRB Expansion Project to address potential adverse impacts on geology and soils.
21 DM&E will limit ground disturbance to only the areas necessary for project-related construction
22 activities and will commence reclamation of disturbed areas as soon as practicable after
23 project-related construction ends. During project-related earthmoving activities, DM&E will
24 stockpile topsoil for application during reclamation to minimize erosion. DM&E will implement
25 appropriate erosion control measures at stockpiles to prevent erosion. DM&E will be required to
26 restore and revegetate soils disturbed by the project to pre-construction conditions as promptly
27 and fully as possible. (STB, 2001)
28
29 The NRC staff determined that the cumulative impact on geology and soils within the study area
30 resulting from all past, present, and reasonably foreseeable future actions is MODERATE.
31 Past conventional underground and open pit surface mining has impacted geology and soils in
32 the eastern part of the proposed Dewey-Burdock site, where abandoned open pits and mine
33 waste overburden piles are not reclaimed or restored. Surface-disturbing activities associated
34 with ongoing and reasonably foreseeable future uranium and oil and gas exploration and
35 development, wind energy, and transportation projects would have direct impacts on geology
36 and soils. Direct impacts will result from increased traffic, clearing of vegetated areas, soil
37 salvage and redistribution, and construction of project facilities and infrastructure. Indirect
38 impacts, such as cross-contamination of aquifers, may also occur if boreholes associated with
39 uranium and oil and gas exploration are not properly abandoned.
40
41 The NRC staff conclude that the proposed Dewey-Burdock ISR Project will have a SMALL
42 incremental effect on geology and soils when considered with all the other past, present, and
43 reasonably foreseeable future actions in the study area. As described in SEIS Section 4.4.1,
44 limited areas of the proposed project site will be disturbed by construction, and implementation
45 of BMPs will limit soil erosion and compaction. Systems and procedures will be in place to
46 monitor and clean up soil contamination resulting from spills and leaks. EPA will evaluate the
47 suitability of deep geologic formations proposed for deep well disposal of liquid wastes prior to
48 granting a Class V UIC deep injection well permit. The EPA UIC Class V permit will impose an
49 upper limit to the allowable injection pressure and will not allow injection at or above the fracture
50 pressure of the injection zone formations. In potential land application areas, the applicant will
51 be required to routinely collect and monitor soils for contamination and comply with discharge

1 limits for treated liquid wastes applied to irrigation areas. When production and aquifer
2 restoration are complete at the proposed project, reclamation and decommissioning will return
3 the site to preproduction conditions through return of topsoil, removal of contaminated soils, and
4 reestablishment of vegetation.
5
6 ## 5.5   Water Resources
7 The impact to surface and groundwater resources was evaluated within an 80-km [50-mi] radius
8 of the proposed Dewey-Burdock ISR Project (Figure 5.1-3). The 80-km [50-mi] radius for the
9 water resources study area encompasses the watersheds, including the Beaver Creek, Upper
10 Cheyenne, and Angostura Reservoir watersheds, that would be potentially impacted by past,
11 present, and reasonably foreseeable future actions (see Figure 3.5-1). The timeframe for the
12 analysis is 2009 to 2030 (see Section 5.1.2 for the estimated operating life of the facility).
13
14 ### 5.5.1   Surface Waters and Wetlands
15
16 The proposed Dewey-Burdock ISR Project is located in the Beaver Creek and Pass Creek
17 watersheds (see SEIS Section 3.5.1). Beaver Creek is a perennial stream, while Pass Creek is
18 dry for most of the year. Both creeks have ephemeral tributaries that flow after snowmelt or
19 heavy rains. Pass Creek joins Beaver Creek southwest of the project area. Beaver Creek flows
20 into the Cheyenne River 4.8 km [3 mi] south of this confluence, which eventually flows into the
21 Missouri River. The U.S. Army Corps of Engineers (USACE) identified four jurisdictional
22 wetlands within the proposed site (see SEIS Section 3.5.2). The jurisdictional sites were
23 Beaver Creek, Pass Creek, and an ephemeral tributary to each. As described in SEIS
24 Section 4.5.1.1, under Section 404 of the Clean Water Act the applicant must obtain a permit
25 from USACE for any activities that may potentially impact jurisdictional wetlands. Prior to
26 operations, the applicant must obtain construction and industrial storm water National Pollutant
27 Discharge Elimination System (NPDES) permits from SDDENR. The NPDES permits will
28 include plans and programs for spill prevention and cleanup, erosion control, and runoff control,
29 which will mitigate the impacts to surface waters and wetlands.
30
31 There are no operating ISR facilities located within 80 km [50 mi] of the proposed site, which is
32 the cumulative impacts surface water study area. Several abandoned open pits and overburden
33 waste piles associated with past surface mining activities are located in the Burdock portion of
34 the site (see SEIS Figure 3.2-3). Radiation surveys reveal that soils near old surface mines
35 have higher than background radiation levels (see SEIS Section 3.12.1). Runoff from snowmelt
36 and heavy rains may leach and transport contaminants from the waste piles associated with
37 these mines to surface waters and wetlands in the Beaver Creek and Pass Creek watersheds
38 (Powertech, 2009c). The Comprehensive Environmental Response, Compensation, and
39 Liability Act (CERCLA), commonly known as Superfund, has been used to clean up
40 uncontrolled or abandoned legacy uranium mines in western Colorado and eastern Utah. EPA
41 is authorized to implement Superfund. Superfund site identification, monitoring, and response
42 activities in South Dakota would be coordinated through SDDENR.
43
44 The potential Dewey Terrace ISR project in Weston and Niobrara Counties, Wyoming, would be
45 located 13 km [8 mi] west of the Dewey-Burdock ISR Project site. This potential future project
46 will necessitate new roads, power lines, facilities construction, underground piping, and well
47 drilling, all of which may have adverse impacts on surface waters and wetlands. As discussed
48 previously for the Dewey-Burdock ISR Project, potential impacts to surface waters and wetlands
49 at the potential Dewey Terrace ISR project site will also be subject to mitigation through BMPs,

1   required NPDES storm water permits, and permits from USACE for any activities that may
2   potentially disturb jurisdictional wetlands identified at the site.
3
4   Surface water quality within the 80-km [50-mi] area of the proposed site may be impacted by
5   conventional oil and gas development, rangeland grazing, wind energy projects, and
6   transportation projects. Cattle grazing is a source of nonpoint pollution to streams and wetlands
7   in the Beaver Creek and Pass Creek drainages. SEIS Section 3.5.1.1 describes Beaver Creek
8   as impaired for all beneficial uses because of high total dissolved and suspended solids, high
9   salinity, presence of fecal coliform, high conductivity levels, and high water temperature. A
10   water quality data report points to livestock as the source of fecal coliform in Beaver Creek
11   (SDDENR, 2008). Poor management of livestock grazing may restrict flow in intermittent
12   streams such as Pass Creek due to erosion and sedimentation resulting from decreased
13   vegetative cover in the drainage area.
14
15   Oil wells within 80 km [50 mi] of the proposed Dewey-Burdock ISR Project site are shown in
16   Figure 5.1-3. As discussed in SEIS Section 5.1.1.3, no producing oil and gas wells are located
17   within the proposed Dewey-Burdock permit boundary and, at present, there is low demand for
18   oil and gas leasing within the project boundary and in its immediate vicinity. Within 80 km [50
19   mi] of the proposed project site, oil wells are clustered west of the site in Weston and Niobrara
20   Counties, southwest of Edgemont in Fall River County, and east of the site at Barker Dome in
21   Custer County. Impacts to surface waters and wetlands from oil and gas exploration activities
22   will be from surface runoff as new access roads and drill pads are constructed. Runoff
23   degrades surface water quality, causes erosion, and leads to siltation of streambeds
24   and wetlands.
25
26   Licensees must obtain construction and industrial NPDES permits from the Wyoming
27   Department of Environmental Quality (WDEQ) in Wyoming and SDDNER in South Dakota prior
28   to conducting oil and gas exploration and production activities. NPDES permits include plans
29   and programs for spill prevention and cleanup, erosion control, and runoff control. These plans
30   and programs significantly mitigate the potential impacts to surface sediment load and turbidity
31   from exploration activities. USACE Section 404 permits are also required for any disturbances
32   in or near jurisdictional wetlands. Section 404 permits include provisions that must be followed
33   to mitigate impacts when conducting activities in and near jurisdictional wetlands.
34
35   Impacts to surface waters and wetlands from potential wind energy projects in the western
36   United States, such as the Dewey-Burdock Wind Project, may include changes in water quality
37   and alteration of natural flow systems. The quality of surface water could be degraded by soil
38   erosion and runoff from construction activities that disturb the ground surface, and by heavy
39   equipment traffic. Surface water flow may be diverted by access road systems or storm water
40   control systems. Operation of a wind energy project uses very small amounts of water and
41   results in virtually no discharges to surface water. Operators of these facilities implement storm
42   water management plans to ensure compliance with applicable regulations and prevent offsite
43   migration of contaminated storm water or increased soil erosion. (BLM, 2005b)
44
45   The proposed Dewey Conveyor Project is located principally within the Pass Creek drainage.
46   Pass Creek and Hell Canyon merge near the southeast portion of the project area and flow
47   southwest to the confluence of Beaver Creek (see Figure 5.3-1). The proposed conveyor
48   project crosses several ephemeral tributaries within the Pass Creek drainage. Some sediment
49   runoff from road and general construction activities associated with the 10.6-km [6.6-mi]-long
50   conveyor is expected, and this could impact surface water bodies. Expected runoff
51   contaminants will predominantly be in the form of suspended or dissolved solids and increases

1 in turbidity. These impacts will be partially mitigated by the fact that many area streambeds in
2 the vicinity of the project area are dry for most of the year. Runoff potential will also be
3 mitigated by the implementation of BMPs for runoff control. (BLM, 2009a)
4
5 The DM&E PRB Expansion Project will have a significant impact on surface water and
6 wetlands, if completed. The new rail line will pass south of the proposed Dewey-Burdock ISR
7 Project site (see Figure 5.1-4), through the Beaver Creek and Pass Creek watersheds. DM&E
8 has proposed mitigation measures to address potential adverse impacts on surface waters and
9 wetlands within the PRB Expansion Project area. Before project-related construction could
10 begin, DM&E must obtain all federal permits, including Clean Water Act Section 404 permits
11 and USACE permits required for project-related alteration or encroachment of wetlands,
12 streams, and rivers. In addition, DM&E must obtain NPDES permits for regulation of storm
13 water discharges to surface waters. DM&E will employ BMPs, such as silt screens and straw
14 bale dikes, to minimize soil erosion, sedimentation, runoff, and surface instability during
15 project-related construction. These mitigation measures will minimize sedimentation into
16 streams and wetlands. (STB, 2001)
17
18 The NRC staff have determined that the cumulative impact on surface water and wetlands
19 within the surface water study area resulting from past, present, and reasonably foreseeable
20 future actions is MODERATE to LARGE. Leaching and transport of contaminants from
21 overburden waste piles associated with past conventional uranium mining in the eastern part of
22 the proposed Dewey-Burdock site may impact surface waters and wetlands in the Beaver Creek
23 and Pass Creek watersheds. Livestock grazing will continue to have the potential to degrade
24 water quality in streams within the study area. Construction activities associated with other
25 ongoing and reasonably foreseeable future actions, including uranium and oil and gas
26 exploration and development, wind energy projects, and transportation projects, will have
27 impacts on surface water and wetland resources. All of these actions will necessitate
28 construction of new roads, power lines, facilities, and infrastructure, which could degrade water
29 quality and alter natural surface water flow systems.
30
31 The NRC staff conclude that the proposed Dewey-Burdock Project will have a SMALL
32 incremental effect on surface water and wetlands when added to all other past, present, and
33 reasonably foreseeable future actions in the surface water study area. As described in SEIS
34 Section 4.5.1, potential impacts to surface waters at the proposed Dewey-Burdock site will be
35 mitigated through proper planning and design of facilities and infrastructure, the use of proper
36 construction methods, and implementation of BMPs. Prior to initiating ISR operations at the
37 proposed project, the applicant must also obtain a construction and industrial storm water
38 National Pollutant Discharge Elimination System (NPDES) permit from SDDENR. The NPDES
39 permit will include plans and programs for spill prevention and cleanup, erosion mitigation, and
40 runoff control. In addition, to comply with Section 404 of the Clean Water Act, the applicant
41 must obtain a permit from USACE for any activities that may potentially disturb the four
42 jurisdictional wetlands identified within the proposed project area.
43
44 ## 5.5.2   Groundwater
45
46 As described in SEIS Section 3.5.3.3, ISR methods will be used to extract uranium from
47 sandstone-hosted uranium ore bodies in the Fall River and Lakota aquifers at the proposed
48 Dewey-Burdock site. The combined Fall River and Lakota aquifers are referred to as the
49 Inyan Kara Group aquifer. Consumptive water use during construction at the Dewey-Burdock
50 site will be generally limited to dust control, cement mixing, pump tests, delineation drilling, and

1   well drilling and completion. The applicant estimated that groundwater consumption during the
2   construction phase in the Dewey and Burdock areas will be 0.8×10^5 m^3 and 1.2×10^5 m^3
3   [21.8×10^6 and 30.6×10^6 gal], respectively (Powertech, 2010). Initially, water for construction
4   activities will be withdrawn from existing wells in the Inyan Kara Group aquifer. The applicant's
5   estimated consumptive groundwater use during the construction phase is of the same
6   magnitude as current withdrawals for domestic and livestock water use from the Inyan Kara
7   Group aquifers within a 2-km [1.2-mi] radius of the proposed project (see Section 4.5.2.1.2.2).
8   The applicant plans to install wells in the deeper Madison aquifer early in the construction
9   phase, and once available, Madison water will become the primary water source for the
10   construction, operation, and aquifer restoration phases (Powertech, 2010).
11
12   Assessments of environmental impacts to groundwater resources at the proposed
13   Dewey-Burdock ISR Project are discussed in SEIS Section 4.5.2. Impacts to groundwater are
14   most likely to occur during the operations and aquifer restoration phases of the ISR facility's
15   lifecycle, but may occur during other phases. Potential groundwater impacts during the
16   operations phase of the proposed project will be mitigated and reduced through implementation
17   of leak detection and cleanup programs, mechanical integrity testing of wells, and adherence to
18   EPA UIC permit requirements. During operations, the applicant commits to monitoring all
19   domestic wells within 2 km [1.2 mi] of the project boundary and providing replacement wells to
20   the well owners in the event of significant drawdown or degradation of water quality in these
21   wells. The applicant's excursion monitoring program will ensure the protection of water quality
22   in aquifers underlying production zone aquifers. After uranium production and aquifer
23   restoration are completed and groundwater withdrawals are terminated at the proposed project,
24   groundwater levels will recover with time. Groundwater restoration will also restore impacted
25   aquifers to acceptable water quality levels. The proposed injection zones for the UIC Class V
26   deep disposal wells are the Deadwood Formation and the Minnelusa Formation. EPA will not
27   authorize injection into the Class V deep disposal wells unless the permittee demonstrates the
28   well is properly sited, such that confinement zones and proper well construction minimize the
29   potential for migration of fluids outside of the approved injection zone.
30
31   Rural population growth, oil and gas exploration development, and ISR uranium extraction are
32   expected to contribute to the cumulative impact on groundwater resources within an 80-km
33   [50-mi] radius of the Dewey-Burdock site. These activities create an increased demand for
34   groundwater and have been the subject of the Black Hills Hydrology Study (USGS, 2010). The
35   U.S. Geological Survey (USGS) conducted this study during 1992–2002 to assess the quantity,
36   quality, and distribution of groundwater in the Black Hills area of South Dakota and to evaluate
37   alternatives for management of water resources in the area. This study is used by federal,
38   state, and local government agencies to set water development policy and protect area
39   groundwater resources.
40
41   Groundwater in the Black Hills area of South Dakota is used for residential, municipal, industrial,
42   and recreational purposes. Forty-five percent of the recent population growth in the Black Hills
43   area of South Dakota has taken place in unincorporated areas without municipal water supply
44   systems (Carter, et al., 2003). Population has grown mainly around Rapid City, but has
45   occurred in rural areas in the southwestern Black Hills. Custer Highlands is a new housing
46   development built approximately 16 km [10 mi] northeast of the proposed Dewey-Burdock
47   site. Recent residential developments 19 to 24 km [12 to 15 mi] east of Dewey-Burdock include
48   the Fundamentalist Church of Jesus Christ of Latter Day Saints facility (NRC, 2009c). The
49   Southern Black Hills Water System proposes constructing a 24-km [15-mi] water transmission
50   pipeline along Argyle Road northwest of Hot Springs, which will serve rural customers in
51   south-central Custer County. The western extension of the pipeline will be 24 km [15 mi] east of

1   the Dewey-Burdock site boundary. The pipeline will transmit water pumped from a Madison
2   aquifer well near Buffalo Gap, South Dakota, 72 km [45 mi] east of the Dewey-Burdock site
3   (Figure 5.1-3).
4
5   The Madison aquifer is the most important regional aquifer supplying Rapid City, Edgemont,
6   and numerous communities in southwestern South Dakota (see Figures 3.5-4 and 3.5-5). As
7   described in SEIS Section 4.5.2, the applicant submitted an application for a water appropriation
8   permit to SDDENR to pump groundwater from the Madison aquifer during ISR construction,
9   operations, and aquifer restoration (Powertech, 2010). Edgemont is the closest community to
10 the project site that obtains municipal water supply from the Madison aquifer. Edgemont lies
11 21 km [13 mi] southeast of the Dewey-Burdock site, and it is expected that any impacts on
12 groundwater levels in the Madison aquifer at a regional level from the proposed project will be
13 SMALL (SEIS Section 4.5.2). The applicant's excursion monitoring program described in SEIS
14 Section 4.5.2.1.1.2 will ensure the protection of water quality in aquifers underlying the
15 production zone. The Madison aquifer is separated from the Deadwood Formation, one of the
16 proposed injection zones for the applicant's UIC Class V deep disposal wells, by the Englewood
17 Formation (see Figure 3.5-5). The Englewood Formation is expected to provide confinement
18 above the proposed Deadwood Formation injection zone (Naus, et al., 2001). The Minnelusa
19 Formation is the other proposed injection zone for the UIC Class V deep disposal wells.
20 Confining units at the base of the Minnelusa Formation are expected to provide hydraulic
21 separation between the Minnelusa Formation and the Madison aquifer. Locally, these confining
22 layers may be absent or provide ineffective confinement, which could allow hydraulic
23 communication between the Minnelusa aquifer and the underlying Madison aquifer (Naus, et al.,
24 2001). Although the Madison aquifer has far greater hydraulic pressure than the Minnelusa
25 aquifer, EPA will not authorize injection into the Class V deep disposal wells unless the
26 permittee demonstrates that there are adequate confining zones above and below the proposed
27 injection zones.
28
29 Aquatic recreational areas, such as Cascade Springs and Keith Springs, are located
30 approximately 40 km [25 mi] east-southeast of the proposed project site. These springs
31 discharge groundwater from the Madison and/or Minnelusa aquifers (Driscoll, et al., 2002).
32 Because Cascade Springs and Keith Springs are located 40 km [25 mi] from the project site, it
33 is expected that estimated withdrawals of water from the Madison aquifer for operations and
34 aquifer restoration at the proposed project will have a SMALL impact on groundwater discharge
35 at Cascade Springs and Keith Springs. The applicant's excursion monitoring program will
36 ensure the protection of water quality in aquifers underlying production zone aquifers.
37
38 Within an 80-km [50-mi] radius of the proposed project, ongoing and planned ISR facilities, oil
39 and gas exploration, wind energy projects, and transportation projects activities may contribute
40 to impacts on groundwater resources.
41
42 The applicant has identified a potential ISR project at Dewey Terrace in Wyoming (Powertech,
43 2009b). The Dewey Terrace project would be located about 13 km [8 mi] west of the
44 Dewey-Burdock ISR Project area in Weston and Niobrara Counties, Wyoming (Figure 5.1-3). If
45 future ISR operations occurred at Dewey Terrace, there will be uranium extraction from the
46 same aquifer (i.e., the Inyan Kara aquifer) as the proposed Dewey-Burdock ISR Project. The
47 combined ISR projects may impact groundwater levels in the ore zone aquifer and impact the
48 water quality of the ore zone aquifer at the two sites. Licensees of ISR facilities are required to
49 implement excursion detection, control, mitigation, and remediation plans under NRC
50 regulations to reduce the potential impact on groundwater quality and quantity.

1   Impacts on groundwater resulting from the interaction of ISR activities and oil and gas
2   exploration and production are not likely because these activities are conducted in
3   stratigraphically separated aquifers. ISR activities at the Dewey-Burdock ISR Project will take
4   place in sandstone aquifers of the Fall River and Lakota aquifers at depths of 61 to 244 m
5   [200 to 800 ft] (see SEIS Section 3.4.1.2). Oil and gas producing wells in Fall River and
6   Custer Counties are located in the Minnelusa and Madison Formations at depths ranging from
7   423 to 1,081 m [1,387 to 3,547 ft] (see SEIS Section 5.1.1.3). In Wyoming, the producing wells
8   closest to the project are in Niobrara County and are located in the Leo Sandstone of the
9   Minnelusa Formation at depths ranging from approximately 785 to 823 m [2,575 to 2,700 ft] (see
10 SEIS Section 5.1.1.3). The NRC-required excursion monitoring programs at ISR facilities will
11 ensure that water quality in aquifers underlying production zone aquifers, including the Madison,
12 Minnelusa, and Deadwood aquifers, would be protected.
13
14 Deep well injection of process-related water is a disposal method ISR and oil production
15 facilities use. For deep well disposal in South Dakota, the applicant must obtain UIC permits for
16 the targeted deep aquifer from the EPA. The applicant has proposed injecting process-related
17 effluents from the Dewey-Burdock Project into the Deadwood and Minnelusa Formations, below
18 the Morrison Formation (see Figure 3.5-5), using Class V (nonhazardous) wells (Powertech,
19 2010). EPA will evaluate the suitability of the proposed deep injection wells and would only
20 grant a permit if the deep disposal practice is safe for public health and safety and will not
21 impact potential underground sources of drinking water. To ensure water quality, the liquid
22 waste injected via Class V wells into deep aquifers must not be classified as hazardous under
23 the Resource Conservation and Recovery Act and must be treated to meet NRC release
24 standards in 10 CFR Part 20, Subparts D and K and Appendix B.
25
26 Impacts to groundwater from potential wind energy projects in the western United States, such
27 as the Dewey-Burdock Wind Project, will not be significant. During construction, water is
28 required for mixing of concrete and dust control along access roads and other areas of
29 disturbance around the turbines, but these uses will be temporary. Development and
30 construction of wind energy projects will include BMPs to mitigate impacts to both groundwater
31 and surface water. Once a wind energy project is operating, minimal quantities of water are
32 needed. (BLM, 2005b)
33
34 Groundwater for the Dewey Limestone Conveyor project will likely be used to suppress dust
35 during road building and use activities, and for the construction of concrete foundation supports
36 for the conveyor along its 10.6-km [6.6-mi] course. In addition, groundwater will be used for
37 dust control/mitigation once the proposed quarry and conveyor are operational. This water
38 demand will be supplied by one or more production wells (one at the quarry site and one at the
39 rail load-out facility). The source for the supply well at the rail load-out facility will likely be
40 developed in the Inyan Kara Group aquifer. This supply well will likely be used solely for dust
41 suppression at the rail load-out area, and therefore the groundwater demand will be quite low,
42 around 94.6 L/min [25 gpm] or less. (BLM, 2009a)
43
44 The proposed DM&E PRB Expansion Project (see SEIS Section 5.1.1.5) will have an impact on
45 groundwater. Groundwater will be used to suppress dust during rail and bridge construction
46 activities. Once operational, the PRB Expansion Project will use negligible amounts of
47 groundwater. Water demand during construction activities will be supplied by existing municipal
48 and private wells. DM&E will ensure that any wells that may be affected by project-related
49 construction or reconstruction activities are appropriately protected or capped to prevent well
50 and groundwater contamination. If wells are located on private land, DM&E will secure
51 permission from the landowner before undertaking any actions. (STB, 2001)

1 The NRC staff have determined that the cumulative impact on groundwater resources within the
2 water resources study area resulting from past, present, and reasonably foreseeable future
3 actions is MODERATE. This finding is based on ongoing and reasonably foreseeable future
4 actions that will (i) increase demand on the regional Madison aquifer, which is used for
5 residential, municipal, and recreational purposes in the study area; (ii) impact groundwater
6 quantity and quality in the Inyan Kara Group aquifer, which hosts uranium deposits surrounding
7 the proposed Dewey-Burdock site; and (iii) potentially impact water quality in deep geologic
8 formations that are used for deep disposal of liquid wastes. In addition, ongoing and reasonably
9 foreseeable future actions will use groundwater for construction of concrete foundations and
10 supports and for dust suppression during construction and operations activities, which will
11 potentially impact water quantity in regional and local aquifers in the study area.
12
13 The NRC staff conclude that the proposed Dewey-Burdock ISR Project will have a SMALL
14 incremental effect on groundwater resources when added to all other past, present, and
15 reasonably foreseeable future actions in the groundwater study area. Based on the foregoing
16 analysis, the potential impact of the proposed project on the existing and future use and quality
17 of water for local and surrounding residential, municipal, and recreational purposes will be
18 minimal. Impacts on groundwater resulting from interaction between ISR activities at the
19 proposed Dewey-Burdock site and oil and gas production are unlikely because the ISR
20 production zone aquifers are separated from underlying oil and gas bearing formations by
21 hundreds to thousands of meters [hundreds to thousands of feet]. EPA permitting requirements
22 will protect groundwater in aquifers used for deep well injection of process-related liquid
23 effluents from the proposed action. The liquid waste injected via Class V wells into deep
24 aquifers will have to be treated to meet NRC release standards in 10 CFR Part 20, Subparts D
25 and K, and Appendix B. After uranium production and aquifer restoration are completed and
26 groundwater withdrawals are terminated at the proposed Dewey-Burdock ISR Project,
27 groundwater levels will recover with time. Groundwater restoration will restore impacted
28 aquifers at the proposed project to acceptable water quality levels. Therefore, the NRC staff
29 conclude that the potential impact on groundwater resources from operating the proposed
30 Dewey-Burdock ISR Project will be SMALL (SEIS Section 4.5.2)..
31

32 ## 5.6   Ecological Resources

33 The cumulative impact to ecological resources was evaluated for the area within an 80-km
34 [50-mi] radius surrounding the proposed Dewey-Burdock ISR Project. The proposed project is
35 located within the Great Plains physiographic province on the edge of the Black Hills uplift. The
36 area under consideration includes the Sagebrush Steppe, Black Hills Foothills, Black Hills
37 Plateau, and Black Hills core highland ecoregions. The timeframe for the analysis of cumulative
38 impacts is 2009 to 2030 (see SEIS Section 5.1.2 for the estimated operating life of the proposed
39 Dewey-Burdock project). Older data are considered where applicable to demonstrate
40 historical trends.
41

42 ### 5.6.1   Terrestrial Ecology
43
44 Activities occurring in the area of the proposed Dewey-Burdock ISR Project boundary include
45 grazing and herd management, hunting, and uranium, oil, and gas exploration. There may be
46 cumulative impacts to ecological resources, including both flora and fauna. These impacts
47 include a reduction in wildlife habitat and forage productivity; modification of existing vegetative
48 communities; and the potential spread of invasive species and noxious weed populations.
49 Concerning wildlife, impacts may involve loss, alteration, or incremental fragmentation

1   of habitat; displacement of and stresses on wildlife; modification of prey and predator
2   communities; and direct or indirect mortalities. Land disturbance resulting from reasonably
3   foreseeable future actions (e.g., potential wind farm and transportation projects discussed in
4   Sections 5.1.1.4 and 5.1.1.5) in the ecological resources cumulative impacts study area will
5   have small ecological impacts, individually, if mitigative measures are employed (BLM, 2005b,
6   2009a; STB, 2001). However, assuming that adjacent habitats for each disturbed parcel of land
7   will be at, or near, carrying capacity, and considering there will be an unavoidable reduction or
8   alteration of the habitats, development activities in the Black Hills Foothills and Sagebrush
9   Steppe ecoregions could cumulatively reduce wildlife and plant populations and alter population
10   structure. For some species that may require specific conditions for their habitats, future use
11   will be strongly influenced by the quality and composition of the remaining habitats.
12   Additionally, grasses and noxious weeds tend to replace sagebrush after disturbances.
13
14   Loss and degradation of native sagebrush shrubland habitats has imperiled much of this
15   ecosystem type as well as sagebrush-obligate species, including the Greater sage-grouse
16   (*Centrocercus urophasianus*). Sage-grouse are found in the sagebrush shrubland habitats, and
17   sagebrush is essential during all seasons and for every phase of their lifecycle (USGS, 2009).
18   Most of the sagebrush lands in the region have been changed by land use, such as livestock
19   grazing, agriculture, or resource extraction. These uses can influence habitats either directly or
20   indirectly, and they can alter the disturbance regime by changing the frequency of fire (USGS,
21   2009). The long-term viability of the sage-grouse rangewide continues to be at risk because of
22   population declines related to habitat loss and degradation. Sage-grouse populations have
23   declined overall from 1965 to 2007 with the greatest decline occurring before the mid-1980s.
24   The total rangewide population decline is estimated at 45 to 80 percent from historic levels
25   (Becker, et al., 2009). Populations have been declining at 2.0 percent per year from 1956 to
26   2003 (Connelly, et al., 2011). Because of its spatial extent, oil and gas resource development is
27   regarded as playing a major role in the decline of the sage-grouse species in the eastern portion
28   of the species' range (Becker, et al., 2009). Future oil and gas development is projected to
29   cause a 7 to 19 percent decline in sage-grouse lek population counts throughout much of the
30   current and historic range of the sage-grouse (Connelly, et al., 2011). As of this writing, the
31   U.S. Fish and Wildlife Service (FWS) has designated the Greater sage-grouse a "candidate
32   species" under the Endangered Species Act (ESA). FWS will consider the bird on an annual
33   basis for listing as a threatened or endangered species. The State of Wyoming is critical for
34   sage-grouse as it currently contains 64 percent of all known sage-grouse habitat and more
35   active leks than any other state (Doherty, et al., 2011).
36
37   According to the South Dakota Department of Game, Fish, and Parks, there are no crucial big
38   game habitats or migration corridors in the ecological resources study. However, the area does
39   contribute habitat for a variety of big game, including deer, antelope, turkeys, elk, and bighorn
40   sheep. Destruction or alteration of portions of this habitat in conjunction with human
41   disturbance associated with ongoing and reasonably foreseeable future actions could result in
42   SMALL incremental impacts to herd animals.
43
44   As discussed in SEIS Section 4.6.1, the proposed Dewey-Burdock Project has the potential to
45   impact vegetation, small- to medium-sized mammals, reptiles, and a number of avian species.
46   These species include raptors, waterfowl, shorebirds, upland game birds, and nongame birds
47   known to occur as seasonal, migratory, or year-round residents. Impacts may occur to species
48   during all phases of the proposed project and are expected to be SMALL to MODERATE.
49   Potential SMALL to MODERATE impacts to avian species (e.g., habitat loss, fragmentation,
50   noise disturbance) will also be likely to occur at other present and reasonably foreseeable future
51   actions (e.g., oil and gas facilities, wind energy projects, and transportation projects) throughout

1 the cumulative impacts study area and potentially impact other localized populations. Wind
2 energy projects, such as the potential Dewey-Burdock Wind Project, have the potential to
3 increase avian mortality resulting from bird collisions. BLM reported that the number of bird
4 collisions at wind energy projects is relatively small, when compared with collisions from other
5 human-made structures (BLM, 2005b).
6
7 The NRC staff have determined that the cumulative impact on terrestrial ecology within the
8 ecological resources study area resulting from all past, present, and reasonably foreseeable
9 future actions is MODERATE. This finding is based on habitat disturbance resulting from
10 actions including (i) uranium and oil and gas exploration and development, (ii) potential ISR
11 projects such as the Dewey Terrace ISR Project in Niobrara and Weston Counties in Wyoming,
12 (iii) potential wind energy projects such as the Dewey-Burdock Wind Project, and (iv) potential
13 transportation projects such as the Dewey Conveyor Project and the DM&E PRB Expansion
14 Project. Habitat disturbance associated with these actions will impact vegetation by promoting
15 the spread of noxious weeds and fragmenting vegetative communities. Impacts to wildlife could
16 include loss, alteration, or incremental fragmentation of habitat; displacement of and stresses on
17 wildlife; and direct and indirect mortalities.
18
19 The NRC staff conclude that the proposed Dewey-Burdock Project will have a SMALL
20 incremental effect on terrestrial ecology when considered with all other past, present, and
21 reasonably foreseeable actions in the ecological resources study area. The proposed action will
22 disturb a maximum of 566 ha [1,398 ac] of habitat with most of the habitat disturbance
23 consisting of scattered, confined drill sites for wells and potential land irrigation areas. These
24 disturbances will not dramatically transform large expanses of habitat from their original
25 character; therefore, no substantial long-term impact will generally be expected. Furthermore,
26 the applicant will control and monitor potential land application areas to reduce impacts to soils
27 and vegetation that could adversely affect flora and fauna. For vegetative species with
28 specialized habitat requirements, future population viability will be strongly influenced by the
29 quality and composition of the remaining habitat. Because the area of disturbed land will be a
30 small percentage of the ecological resources study area, and because of stated mitigative
31 measures the applicant has committed to as described in SEIS Section 4.6.1, impacts on
32 vegetation from the proposed Dewey-Burdock project will have only a SMALL incremental
33 impact when considered with all past, present, and reasonably foreseeable future actions.
34 Although sage-grouse have been present in Fall River County in the past, and although a
35 potential habitat for sage-grouse exists, Greater sage-grouse are not reported within 6.4 km
36 [4 mi] of the proposed project boundary (SEIS Sections 3.6.3 and 4.6.1.1.1.2). Because NRC
37 staff expect that similar habitat is present in the project area that FWS evaluated for the nearby
38 Buffalo Gap National Grassland (see SEIS Sections 3.6.3 and 4.6.1.1.1.2) (Hodorff, 2005), it is
39 unlikely that optimum canopy coverage of sagebrush habitat is present to support breeding and
40 wintering populations within the proposed project area.
41
42 **5.6.2   Aquatic Ecology**
43
44 Potential impacts to aquatic species at the proposed Dewey-Burdock project site will occur
45 primarily along Beaver Creek, Pass Creek, scattered stock ponds, and drainages. As described
46 in SEIS Section 4.6.1.1.2, because of the limited and ephemeral nature of surface water at the
47 proposed Dewey-Burdock Project, the occurrence of aquatic species is also limited. Beaver
48 Creek is a perennial stream that supports aquatic habitat but does not support sensitive aquatic
49 species due to annual low flow conditions. Further, U.S. Environmental Protection Agency
50 (EPA) lists Beaver Creek as an impaired water body partially due to high dissolved and

1   suspended solids, high salinity, and fecal coliform (SDDENR, 2008). Therefore, ISR activities at
2   the proposed Dewey-Burdock Project site are unlikely to further degrade the water quality of
3   perennial streams in the areas. Pass Creek is an ephemeral stream that supports some
4   intermittent habitat. However, Pass Creek does not provide a year-round source of surface
5   water sufficient to maintain a population of aquatic species. No loss of aquatic habitat will result
6   from planned construction activities or land application sites at the proposed Dewey-Burdock
7   Project (Powertech, 2009a). In addition, no surface water will be diverted, no process water will
8   be discharged into an aquatic habitat, and storm water runoff will be managed through the
9   NPDES permit (as discussed in SEIS Section 4.5.1.1.1.2). Therefore, during all phases of the
10   proposed Dewey-Burdock Project lifecycle, the potential impacts to aquatic species and habitats
11   will be SMALL.
12
13   The NRC staff determined that the cumulative impact on aquatic ecology resulting from all past,
14   present, and reasonably foreseeable future actions is SMALL. Cumulative impacts from oil and
15   gas exploration and development, other ISR activities, wind energy projects, and transportation
16   projects will not affect the aquatic ecosystem across the ecological resources study area. This
17   conclusion is based on the limited and ephemeral nature of surface water in and surrounding
18   the study area. The Beaver Creek and Pass Creek systems are the main surface water
19   drainages in the study area. As discussed previously, Beaver Creek does not support sensitive
20   aquatic species and is impaired due to high dissolved and suspended solids, high salinity, and
21   fecal coliform (SDDENR, 2008). Pass Creek, on the other hand, does not provide a year-round
22   source of water sufficient to maintain a population of aquatic species. In addition, all proposed
23   activities in the study area will employ BMPs and comply with federal and state water quality
24   regulations, which will reduce impacts on aquatic ecology.
25
26   The NRC staff have concluded that the proposed Dewey-Burdock Project will have a SMALL
27   incremental effect on aquatic ecology when considered with all other past, present, and
28   reasonably foreseeable actions in the study area. This conclusion is based on the limited and
29   ephemeral nature of Beaver Creek and Pass Creek and other surface water features on the
30   proposed Dewey-Burdock ISR Project site, and on the existing impaired status of Beaver Creek.

31   **5.6.3   Protected Species**

32   As discussed in SEIS Section 4.6.1.1.4, no federally listed species are present within the
33   proposed Dewey-Burdock Project license area. Potentially suitable habitat for migrating
34   whooping cranes exists where standing water is present, which will occur primarily along
35   Beaver Creek and Pass Creek and their drainages, and old mine pits. Direct impacts are
36   unlikely because whooping cranes are not known to breed in South Dakota; however, the
37   proposed project could distress migrating cranes.
38
39   Potential suitable habitat for the black-footed ferret (*Mustela nigripes*) exists in the form of a
40   black-tailed prairie dog (*Cynomys ludovicianus*) complex. However, no evidence of the
41   presence of black-footed ferrets has been observed at the proposed site. Furthermore, it is
42   unlikely that the species will recolonize the immediate area in the foreseeable future without
43   FWS reintroducing it to the area. The prairie dog colony located on the proposed site will
44   experience some unavoidable, direct disturbance. Displacement of the prairie dog colony could
45   impact several federal- or state-listed species, including the mountain plover (*Charadrius
46   montanus*), ferruginous hawk (*Buteo regalis*), swift fox (*Vulpes velox)*, and burrowing owl
47   (*Athene cunicularia*), which utilize burrows as habitat and/or prairie dogs as prey.
48

1   As described in SEIS Section 3.6.1, as of this writing, FWS has designated the Greater
2   sage-grouse as a "candidate species" under the ESA and will consider the bird on an annual
3   basis for listing as a threatened or endangered species. The State of Wyoming is critical for
4   sage-grouse as it currently contains 64 percent of all known sage-grouse habitat and more
5   active leks than any other state (Doherty, et al., 2011). No sage-grouse or leks were observed
6   within a 6.4-km [4-mi] perimeter of the Dewey-Burdock site during wildlife surveys (Powertech,
7   2009a). As discussed in SEIS Section 3.6.3, although sage-grouse have appeared in Fall River
8   County in the past, and although potential sage-grouse habitat is present, the small stands of
9   sagebrush surrounded by grasslands and pine breaks in the project counties do not provide
10 optimum canopy coverage to support breeding and wintering populations.
11
12 Rangewide, the long-term viability of the sage-grouse continues to be at risk because of
13 population declines related to habitat loss and degradation. Because of its spatial extent, oil
14 and gas resource development is regarded as playing a major role in the decline of the
15 sage-grouse species in the eastern portion of species' range (Becker, et al., 2009). Future oil
16 and gas development is projected to cause a 7 to 19 percent decline in sage-grouse lek
17 population counts throughout much of the current and historic range of the sage-grouse
18 (Connelly, et al., 2011).
19
20 The NRC staff determined that the cumulative impact on protected species within the ecological
21 resources study area resulting from all past, present, and reasonably foreseeable future actions
22 is MODERATE. This finding is based on habitat disturbance to potential protected species
23 resulting from actions including (i) uranium and oil and gas exploration and development,
24 (ii) potential ISR projects such as the Dewey Terrace ISR expansion project in Niobrara and
25 Weston Counties in Wyoming, (iii) potential wind energy projects such as the Dewey-Burdock
26 Wind Project, and (iv) potential transportation projects such as the Dewey Conveyor Project and
27 the DM&E PRB Expansion Project. Impacts to protected and threatened species from these
28 actions could include loss, alteration, or incremental fragmentation of habitat; displacement of
29 and stresses on species; and direct and indirect mortalities.
30
31 The NRC staff have concluded that the proposed Dewey-Burdock Project will have a SMALL
32 incremental effect on protected species when considered with all other past, present, and
33 reasonably foreseeable actions in the study area. No federally listed protected species are
34 present within the proposed Dewey-Burdock Project license area, and the proposed license
35 area does not contain critical habitat for any protected species. Furthermore, habitat
36 disturbance at the proposed project site will consist primarily of scattered, confined drill sites for
37 wells and potential land irrigation areas that will not result in large expanses of habitat being
38 dramatically transformed, lost, or degraded.
39

40 ## 5.7   Air Quality

41 Cumulative impacts to air quality were assessed primarily for the portions of the
42 Black Hills-Rapid City Intrastate Air Quality Control Region located within an 80-km [50-mi]
43 radius of the proposed Dewey-Burdock ISR Project. This area, hereafter called the air quality
44 region of influence, covers the majority of Custer and Fall River Counties, the eastern portion
45 of Pennington County (excluding Rapid City), and a very small portion of southwestern
46 Lawrence County (see Figure 5.1.3).
47
48

1   ## 5.7.1   Non-Greenhouse Gas Emissions
2
3   As described in Section 5.1.1, past, present, and foreseeable activities that may contribute to
4   pollutant emissions include uranium exploration and extraction, oil and gas exploration and
5   production, coal mining and coal bed methane operations, wind energy projects, the proposed
6   Dewey Conveyor Project, and the proposed DM&E PRB Expansion Project. Air pollutants
7   emitted by these sources potentially have a cumulative impact within the region and include, but
8   are not limited to, carbon monoxide (CO) and nitrogen oxides (NO_x) from internal combustion
9   engines used at natural gas pipeline compressor stations; CO, NO_x, particulates, SO_2, and
10   volatile organic compounds (VOCs) from gasoline and diesel vehicle tailpipe emissions; dust
11   generated by vehicle traffic on unpaved roads and agricultural activities; NO_2 and particulate
12   emissions from railroad locomotives; and air pollutants transported from emission sources
13   located outside the region. The contribution of past and present activities will be addressed
14   first. Then the analyses will examine the foreseeable activities.
15
16   The past and present contributions of projects in the region that emit air pollutants are
17   represented in the ambient air quality monitoring results described in SEIS Section 3.7.2.
18   These monitoring results indicate the air quality is in attainment for all NAAQS. Table 3.7-3
19   contains data primarily from Wind Cave National Park, the nearest ambient air quality
20   monitoring station, and a Prevention of Significant Deterioration (PSD) Class I site. This
21   monitoring station was established in 2005 to determine air pollution background levels and
22   whether the site was impacted by the long-range transport of air pollutants, such as pollution
23   from the increase in oil and gas development in Colorado, Wyoming, and Montana
24   (SDDENR, 2009). According to the South Dakota Ambient Air Monitoring Annual Network
25   Plan (SDDENR, 2009), the annual PM_{10} concentrations at the Wind Cave site are the lowest in
26   the state and the annual $PM_{2.5}$ concentrations are some of the lowest in the state. The nitrogen
27   dioxide (NO_2) and sulfur dioxide (SO_2) annual concentrations are very low and are at the
28   monitoring equipment's detection limit (i.e., the ability of the equipment to detect the presence of
29   a compound). The 8-hour average ozone levels at the Wind Cave station are similar to those at
30   the state's other monitoring sites and are below NAAQS. Over the last couple of years, trends
31   at the Wind Cave site, as well as some of the other monitoring sites, show decreasing ozone
32   concentration levels. Ongoing ambient air monitoring, such as that conducted at Wind Cave
33   Nation Park, provides an avenue to continually assess air quality from the cumulative emissions
34   observed at a particular location. The air permitting process provides a mechanism for
35   regulatory authorities such as SDDENR to protect air quality through permit conditions and
36   restrictions. The permitting process, including the Prevention of Significant Deterioration, is
37   described in SEIS Sections 2.1.1.1.6.1.1 and 3.7.2.
38
39   Regional air modeling and other studies in the region of influence often focus on Wind Cave
40   National Park, the Class I area located in Custer County about 46.7 km [29 mi] from the
41   proposed site. As a Class I area, these analyses examine impacts to visibility. Visibility
42   impairment occurs when the pollution in the air either scatters or absorbs the light. Both natural
43   and man-made sources contribute air pollution, which impairs visibility. Natural sources include
44   windblown dust and smoke from fires. Man-made sources include electric utilities (i.e., power
45   plants), industrial fuel burning, and motor vehicles.
46
47   The South Dakota Department of Environment and Natural Resource Regional Haze State
48   Implementation Plan (SDDENR, 2011) provided pollution emission inventories and modeling
49   results and also indentified the sources of the pollutants that affect the visibility. The plan
50   provided information based on 2002 actual emissions and 2018 projections. This plan identified

1   sulfate, organic carbon, and nitrate as the major contributors to visibility impairment at Wind
2   Cave National Park. The modeling indicates that only about 3 percent of the sulfur dioxide
3   pollution affecting visibility at Wind Cave National Park comes from sources within South Dakota
4   and at most, about 10 percent of the nitrogen dioxide pollution comes from sources within
5   South Dakota. The state that contributes the most sulfur dioxide and nitrogen dioxide pollution
6   that affects visibility at this Class I area is Wyoming. The state that contributes the most organic
7   carbon is South Dakota, with the predominant source coming from natural fires. The state that
8   contributes the coarsest particulate matter is South Dakota, accounting for up to 45 percent of
9   the total. However, between 60 and 71 percent of this coarse particulate matter is attributed to
10 natural sources.
11
12 BLM also evaluated potential long-range air impacts to the Wind Cave National Park from
13 activities in Wyoming, specifically the Powder River Basin west of the proposed Dewey-Burdock
14 ISR Project. Emission sources for these activities included coal-related facilities (i.e., mines,
15 power plants, railroads, conversion facilities), permitted sources in Wyoming and Montana, coal
16 bed methane production sources, and miscellaneous (i.e., roads, urban areas, conventional oil
17 and gas, noncoal power plants). Emissions were developed for base year 2004 (NO_2, SO_2,
18 $PM_{2.5}$, and PM_{10}) and were projected for year 2020. For the Wind Cave site, year 2020
19 projected impacts were well below NAAQS standards. All modeled NO_x and SO_2 levels were
20 near or less than 1 percent of the NAAQS, and the highest PM level was about 12 percent of
21 the NAAQS (BLM, 2009b). Visibility impacts were identified for the Wind Cave site. When
22 comparing the year 2004 baseline case to the projected year 2020 impacts, the number of days
23 with greater than a 10 percent change in visibility increases by 31 days per year. (BLM, 2009b)
24
25 The analyses will now consider the various reasonably foreseeable future actions starting with
26 the proposed DM&E PRB Expansion project. This project would impact air quality in western
27 Wyoming and southern South Dakota. Mitigation measures have been recommended as part of
28 the proposed DM&E PRB Expansion Project to address potential adverse impacts to air quality.
29 DM&E would be required to meet EPA emission standards for diesel-electric locomotives
30 (40 CFR Part 92). To the extent practicable, DM&E would adopt fuel-saving practices, such as
31 throttle modulation, dynamic braking, increased use of coasting trains, and shutting down
32 locomotives when not in use for more than an hour, to reduce overall emissions during
33 project-related operations. To minimize fugitive dust emissions during project-related
34 construction activities, DM&E would implement fugitive dust suppression controls, such as
35 spraying water, tarp covers for haul vehicles, and installation of wind barriers. (STB, 2001)
36
37 The only ISR site listed in Table 5.1-1 that occurs within the entire Black Hills-Rapid City
38 Intrastate Air Quality Control Region is the proposed Dewey-Burdock ISR Project. The
39 Edgemont site associated with conventional uranium milling is within the air quality region of
40 influence and currently serves as a UMTRCA Title II disposal site under DOE ownership. As
41 described in SEIS Sections 5.1.1.2 and 5.1.1.3, coal mining and oil and gas well development
42 activities within the air quality region of influence are minimal.
43
44 None of the wind energy projects listed in Table 5.1-3 are within the air quality region of
45 influence. The nearest existing wind power project is located about 161 km [100 mi]
46 west-southwest in Converse County, Wyoming. As described in SEIS Section 5.1.1.4, a
47 landowner group has organized to explore the possibility of a wind farm on privately owned land
48 within and surrounding the proposed Dewey-Burdock ISR Project (see Figure 5.1-4). For wind
49 energy projects, such as the potential Dewey-Burdock Wind Project, the construction phase
50 would generate more air emissions than the operation phase (BLM, 2005b). Multiple concurrent

1   construction projects could contribute to regional pollutant emissions loads from construction
2   and worker vehicle exhaust emissions. Localized incidences of fugitive dust along unpaved
3   roads could occur if multiple construction projects occurred simultaneously. However,
4   programmatic BMPs would include mitigation measures to reduce airborne dust at project sites.
5   The dust emission contribution to cumulative impacts to regional air quality would be minimal,
6   because they would be localized and temporary. Air emissions from vehicles involved in
7   operational activities at wind energy projects would be minimal because of the small number of
8   employees needed onsite at any one time (see SEIS Section 5.3). The small number of
9   employees and associated trips during project operations would not have a noticeable effect on
10   cumulative regional air quality (BLM, 2005b).
11
12   The proposed Dewey Limestone Conveyor project has the potential to cumulatively impact air
13   quality in the vicinity of the proposed project. The aboveground conveyor system would be fully
14   enclosed, preventing material and very little dust from escaping into the atmosphere. Fugitive
15   dust would be monitored during construction and during the initial stages of operation using
16   particulate dust collectors (PM_{10} and PM_{25} samplers). The State of South Dakota's Air Quality
17   permit requires this monitoring for various facilities associated with the conveyor project. The
18   rail load-out facility located approximately 1.6 km [1 mi] from the northwestern boundary of the
19   proposed project site would require an air quality permit from SDDENR, which would include
20   requirements for minimizing dust generation by using air pollution control equipment and other
21   applicable operational BMPs (BLM, 2009a).
22
23   The NRC staff determined that the cumulative impact on air quality within the study area
24   resulting from past, present, and reasonably foreseeable future actions is MODERATE. The
25   current ambient air pollution concentrations relate to the air quality impacts from past and
26   present actions. As described in SEIS Section 3.7.2, the area is classified as in attainment for
27   each of the NAAQS pollutants. However, the Regional Haze State Implementation Plan and
28   BLM regional analyses discussed in this SEIS section indicate that Wind Cave National Park
29   does experience visibility impacts.
30
31   The NRC staff conclude that the proposed Dewey-Burdock ISR Project will have a MODERATE
32   incremental effect on climate and air quality when added to all other past, present, and
33   reasonably foreseeable future actions in the study area. On a local scale, fugitive emissions
34   impact air quality from localized dust emissions that are short term and intermittent in nature.
35   As discussed earlier in this section, the regional-scale air modeling and studies often focus on
36   Wind Cave National Park and visibility impacts have been identified at this location. Fugitive
37   dust contributes to visibility impacts. The pollutant with the largest emission levels from the
38   proposed Dewey-Burdock ISR Project is fugitive dust, and cumulative visibility impacts
39   (i.e., increasing regional haze) are possible. The fugitive dust emissions are not included in the
40   modeling performed on the initial emission inventory. The applicant committed to perform air
41   dispersion modeling using the revised emission inventory before the final SEIS is prepared
42   (Powertech, 2012). The final SEIS analyses would be based on this updated modeling. SEIS
43   Section 4.7.1 describes the scope of this update which would include Air Quality Related Values
44   modeling for Wind Cave National Park. As described in Section C.4.2, the modeling results
45   from a similar project are used to estimate the potential impacts from the proposed project.
46   Similarities between the two projects include distance to the nearest Class I area and fugitive
47   dust emission levels. The Dewey-Burdock peak year fugitive dust emission levels are about
48   68 percent for PM_{10} and 31 percent for $PM_{2.5}$ of the levels from the other project. Potential
49   changes to regional haze are calculated in terms of a perceptible "just noticeable change in
50   visibility" when compared to background conditions. The potential visibility impacts from the
51   other project to the Class I areas are predicted to be below the "just noticeable visibility change"

1 threshold (BLM, 2005a). This supports the notion that the proposed Dewey-Burdock ISR
2 Project would contribute to visibility impacts, but the magnitude of the impact would be small.
3
4 **5.7.2   Global Climate Change and Greenhouse Gas Emissions**
5
6 NRC staff determined that a meaningful approach to address the cumulative impacts of
7 greenhouse gas emissions, including carbon dioxide, is to recognize that (i) such emissions
8 contribute to climate change, (ii) climate change is best characterized as the result of numerous
9 and varied sources, each of which might seem to make a relatively small addition to global
10 atmospheric greenhouse gas (GHG) concentrations, (iii) carbon footprint is a relevant factor in
11 evaluating potential impacts of an alternative, and (iv) analysis may include both the proposed
12 action's contribution to atmospheric GHG levels and the potential effects of climate change to
13 the proposed action. These concepts are reflected in Sutley (2010).
14
15 GHG emissions are described in SEIS Sections 2.1.1.1.6.1.1, 3.7.2, and 4.7. As described in
16 SEIS Section 4.7.1.1.1, the operation phase emissions bound the other phases in terms of GHG
17 levels generated. The operation phase GHG annual emission estimate of 55,764 metric tons
18 [61,469 short tons] is roughly evenly split between electrical consumption and mobile sources
19 with a small amount attributed to stationary sources (Table 4.7-1). These mobile sources
20 include equipment associated with the drilling activity with the primary contributor being the drill
21 rig (Table C–12). As described throughout SEIS Section 4.7.1.2, NRC staff do not expect to
22 see any appreciable difference in the overall greenhouse gas emission levels between the land
23 disposal option and the deep well disposal option.
24
25 As described in SEIS Section 3.7.2, South Dakota accounted for approximately 36.5 million
26 metric tons [40.2 short tons] of gross carbon dioxide equivalent (CO_2e) emissions in 2005 and
27 forecast levels of 39.1 and 46.6 million metric tons [43.1 and 51.4 short tons] in 2010 and 2020,
28 respectively (Center for Climate Strategies, 2007). The 2005 total is reduced to 34.9 million
29 metric tons [38.5 short tons] as a result of annual sequestration (removal) due to forestry and
30 other land uses (Center for Climate Strategies, 2007). The proposed Dewey-Burdock ISR
31 Project emission estimate at 55,764 metric tons [61,469 short tons] equates to less than
32 1 percent (0.15 percent) to the overall GHG emissions for South Dakota in 2005. The low level
33 of GHG emissions from the proposed Dewey-Burdock Project relative to the state estimates
34 provides the basis for the NRC staff conclusion that the proposed Dewey-Burdock ISR Project
35 would have a SMALL incremental impact on air quality in terms of GHG emissions when added
36 to the MODERATE cumulative impacts anticipated from other GHG emissions from past,
37 present, and reasonably foreseeable future actions.
38
39 NRC also examined the potential effect of climate change on the proposed Dewey-Burdock ISR
40 Project. While there is general agreement in the scientific community that some climate change
41 is occurring, considerable uncertainty remains in the magnitude and direction of some of the
42 changes, especially predicting trends in a specific geographic location. As described in SEIS
43 Section 3.7.2, the recent report from GCRP served as a source for climate change
44 information (GCRP, 2009). The average temperature in the Great Plains increased by
45 approximately 0.83 °C [1.5 °F] from the 1961 to 1979 baseline. South Dakota and the
46 proposed Dewey-Burdock site are considered to be part of the Great Plains in this study. The
47 projected change in temperature over the period from 2000 to 2020, which encompasses the
48 period the proposed Dewey-Burdock ISR Project would be licensed, ranges from a decrease of
49 approximately 0.28 °C [0.5 °F] to an increase of approximately 1.1 °C [2 °F]. Although GCRP
50 did not incrementally forecast a change in precipitation by decade, it did project a change in

1   spring precipitation from the baseline period (1961 to 1979) to the next century (2080 to 2099).
2   For the region of South Dakota where the proposed Dewey-Burdock ISR Project would be
3   located, GCRP forecasted a 10 to 15 percent increase in spring precipitation (GCRP, 2009).
4
5   Based on the previous analyses, the overall effect of projected climate change on the proposed
6   Dewey-Burdock ISR Project is SMALL. The predicted increases in temperature and
7   precipitation over the next decade are small. Much of the activity associated with ISR milling
8   occurs below ground, whereas the listed climate change parameters are associated with the
9   surficial and atmospheric environments. The predicted increase in precipitation and subsequent
10 infiltration into the groundwater could result in an increase in recharge to the aquifer in future.
11 This could affect the proposed project by increasing the volume of groundwater in the ore body
12 and improving the effectiveness of the aquifer restoration process. Similarly, potential changes
13 to the site environment and resources, such as ecology during the period when the proposed
14 activities would be conducted, would not be sufficient to alter the environmental conditions at
15 the proposed site in a manner that would change the magnitude of the environmental impacts
16 from what has already been evaluated in this SEIS.
17
18 ## 5.8   Noise
19
20 Cumulative impacts from noise were assessed within an 8-km [5-mi] radius of the proposed
21 Dewey-Burdock ISR Project. This area served as the cumulative assessment geographic
22 boundary and was chosen because noise dissipates quickly from the source. GEIS
23 Section 4.4.7 stated that sound levels as high as 132 dBA will taper to the lower limit of human
24 hearing (20 dBA) at a distance of 6 km [3.7 mi] in this region, so a larger 8-km [5-mi] study area
25 will be appropriate to evaluate potential cumulative impacts on noise (NRC, 2009a). The
26 timeframe for the analysis is 2009 to 2030 (see SEIS Section 5.1.2 for the estimated operating
27 life of the facility).
28
29 Noise associated with the proposed Dewey-Burdock ISR Project includes the operation of
30 equipment such as trucks, bulldozers, and compressors; traffic due to commuting workers or
31 material/waste shipments; and wellfield, central processing plant, and satellite facility activities
32 and equipment. Other noises would include traffic noise from nearby roads and railroads. As
33 detailed in SEIS Section 4.8.1, noise impacts to onsite and offsite residential and wildlife
34 receptors and onsite workers from ISR activities at the proposed project would be SMALL for all
35 stages of the project lifecycle.
36
37 Present and reasonably foreseeable future noise-generating activities in the vicinity of the
38 proposed Dewey-Burdock ISR Project would primarily be from operating heavy equipment and
39 traffic noise associated with (i) uranium and oil and gas exploration and development, (ii) wind
40 energy projects, and (iii) transportation projects.
41
42 Oil and gas operations generate noise during construction, well drilling, and operation of
43 compressor stations. However, noise levels from these activities are reduced to ambient levels
44 at distances of approximately 488 m [1,600 ft] (BLM, 2003). Noise-related impacts are generally
45 limited to the 610 m [2,000 ft] immediately surrounding each discrete source (e.g., drill rig,
46 compressor station). Within the cumulative impacts from noise study area, there are four
47 producing oil wells at the Barker Dome oilfield 6 km [4 mi] east of the proposed Dewey-Burdock
48 site and another four producing oil wells at the Plum Canyon oilfield 5 km [3 mi] northwest of the
49 proposed Dewey-Burdock site (see Figure 5.1-4). As described in SEIS Section 5.1.1.1,
50 demand for oil and gas leasing in the vicinity surrounding the proposed Dewey-Burdock ISR

1   project area is low and the level of oil and gas exploration and development is not anticipated to
2   increase significantly in the foreseeable future.
3
4   At this time, no future ISR projects have been identified within the cumulative noise impacts
5   study area {i.e., within a 8-km [5-mi] radius of the proposed Dewey-Burdock site}. The applicant
6   has identified a potential ISR project at Dewey Terrace located 13 km [8 mi] east of the
7   Dewey-Burdock site (see SEIS Section 5.1.1.1). If developed, Dewey Road may be used to
8   access the potential Dewey Terrace project from Edgemont, which is the nearest community to
9   the south. Therefore, the potential Dewey Terrace project may contribute to noise within the
10   study area from additional traffic on Dewey Road from commuting workers, construction and
11   operations deliveries, and yellowcake and byproduct transport.
12
13   Construction of a wind energy project, such as the potential Dewey-Burdock Wind Project, will
14   produce noise from activities including access road construction, grading, drilling and blasting
15   (for tower foundations), construction of ancillary structures, cleanup, and revegetation. In
16   general, construction activities will last for a short period (1 to 2 years at most) and will occur
17   during the day; accordingly, their potential impacts will be temporary and intermittent in nature.
18   Noise generated by turbines, substations, transmission lines, and maintenance activities during
19   the operational phase of a wind energy project will approach typical background levels for rural
20   areas at distances of 610 m [2,000 ft] or less. Like construction activities, decommissioning
21   activities will occur during the day and would last for a short period compared with wind turbine
22   operation, and therefore the potential impacts will be temporary and intermittent in nature.
23   (BLM, 2005b)
24
25   Noise sources associated with the proposed Dewey Conveyor Project include the conveyor,
26   conveyor drive motors, locomotives, and diesel-powered loaders. Noise levels from the
27   proposed Dewey Conveyor Project are predicted to be below the EPA guideline of 55 dBA
28   within 21 m [70 ft] from the conveyor drive motors and below the estimated existing 40 dBA
29   within 111 m [365 ft] from the conveyor drive motors. Noise levels due to the rail load-out are
30   predicted to meet the EPA guidelines of 55 dBA within 320 m [1,050 ft] from equipment and
31   meet the existing ambient 40 dBA within 1,288 m [4,225 ft] from equipment. Mitigation
32   measures the conveyor operator, GCC Dacotah, proposes to reduce noise impacts include
33   installing high-grade mufflers on diesel-powered equipment, combining noisy operations to
34   occur for short durations, and limiting rail loading to daytime hours. (BLM, 2009a)
35
36   The proposed DM&E PBR Expansion Project will have a significant impact on noise in western
37   South Dakota and Wyoming. Noise will be produced by heavy equipment use and vehicular
38   traffic during construction and by locomotive engine and wheel/rail noise during rail line
39   operations. DM&E has proposed mitigation measures as part of the proposed expansion
40   project to address potential adverse impacts on noise. DM&E will maintain project-related
41   construction and maintenance vehicles in good working condition with properly functioning
42   mufflers to control noise. DM&E will comply with Federal Railroad Administration regulations
43   (49 CFR Part 210) for decibel limits for train operations. DM&E will mitigate train wayside noise
44   (locomotive engine and wheel/rail noise) for noise-sensitive receptors along project-related new
45   rail line construction to within 70 dBA. To minimize noise, DM&E will properly maintain rails and
46   regularly service locomotives, keeping mufflers in good working order to control noise.
47   (STB, 2001)
48
49   The NRC staff have determined that the cumulative impact on noise within the noise study area
50   resulting from all past, present, and reasonably foreseeable future actions is MODERATE.

1   Operation of reasonably foreseeable future actions, such as the Dewey Conveyor Project and
2   DM&E PBR Expansion Project, would have significant noise impacts within the cumulative
3   impacts study area. Noise associated with operation of the conveyor project will include the
4   conveyor, conveyor drive motors, locomotives, and diesel-powered loaders. Locomotive engine
5   and wheel/rail noise will have long-term noise impacts during operation of the DM&E rail line
6   project. In addition, the potential Dewey Terrace ISR project may contribute to noise along
7   Dewey Road from commuting workers, equipment and materials deliveries, and yellowcake and
8   byproduct transport. Other ongoing and reasonably foreseeable future actions are not expected
9   to have a significant impact on noise within the cumulative impacts study area. There are only
10 eight producing oil wells within the study area, and demand for oil and gas leasing is low. Coal
11 bed methane reserves are not present within the study area. Potential wind energy projects,
12 such as the Dewey-Burdock Wind Project, are generally compatible with the primary land uses
13 in the study area, including livestock grazing, recreation, and wildlife habitat conservation (BLM,
14 2005b). During operation of a wind energy project, noise generated by turbines, substations,
15 transmission lines, and maintenance activities will approach typical background levels for rural
16 areas at distances of 610 m [2,000 ft] or less (BLM, 2005b).
17
18 The NRC staff have concluded that the proposed Dewey-Burdock Project would have a
19 SMALL incremental effect on noise when considered with all other past, present, and
20 reasonably foreseeable actions in the noise study area. There are few sensitive noise receptors
21 (e.g., residences, communities) in the cumulative impacts noise study area. As described in
22 SEIS Section 4.8.1, noise generated by construction and operational activities at the proposed
23 Dewey-Burdock ISR Project will dissipate or be reduced by mitigation measures before reaching
24 onsite and offsite residential and sensitive wildlife receptors. Additionally, noise levels will be
25 mitigated by administrative and engineering controls to maintain noise levels in work areas
26 below Occupational Safety and Health Administration (OSHA) regulatory limits.
27
28 **5.9   Historic and Cultural Resources**
29
30 Cumulative impacts on historic and cultural resources were assessed within a 16-km [10-mi]
31 radius of the proposed Dewey-Burdock ISR Project. This area delineates the geographic
32 boundary utilized for the cumulative analysis of historic and cultural resources and will be
33 collectively referred to as the "historic and cultural resources study area." The assessment of
34 cumulative impacts on historic and cultural resources beyond 16 km [10 mi] was not undertaken
35 because at this distance the impacts on historic and cultural resources from the proposed
36 Dewey-Burdock ISR Project on other past, present, and reasonably foreseeable future actions
37 will be minimal. The timeframe for the analysis is 2009 to 2030 (see SEIS Section 5.1.2 for the
38 estimated operating life of the facility). In 2009, the applicant submitted a license application to
39 NRC; year 2030 represents the license termination at the end of the decommissioning period.
40
41 Potential impacts to cultural and historic resources could result from energy development,
42 erosion, and grazing activities. These impacts would result primarily from the loss or damage to
43 historical, cultural, and archaeological resources, but also from temporary restrictions on access
44 to these resources. Applicants for ISR facilities would conduct appropriate historic and cultural
45 resource surveys as part of prelicense application activities. Impacts to cultural resources are
46 often minimized for projects located on federal or tribal lands or that are part of a federal action,
47 because such projects are subject to the National Historic Preservation Act (NHPA), the
48 Section 106 consultation process, and other applicable statutes.
49

1   Cultural resources may be affected indirectly by the consequences of nearby projects, such as
2   erosion, destabilization of land surfaces, increased area access, and increased vibration from
3   locomotive and heavy truck traffic. As discussed in SEIS Section 4.9, the impact of the
4   proposed ISR project on historic and cultural resources in the Dewey-Burdock project area has
5   been categorized as SMALL to LARGE, depending on the phase of the facility lifecycle.
6
7   The analysis of cumulative impacts on historic and cultural resources at the proposed project
8   focused on identification and the assessment and implementation of mitigative measures to
9   protect resources within the area of potential effect (APE). As described in SEIS Section 3.9,
10   the APE is defined as the area that may be directly or indirectly impacted by construction,
11   operations, aquifer restoration, and decommissioning activities associated with the proposed
12   action. As described in SEIS Section 4.9.1, 18 historic sites listed or recommended as eligible
13   for listing on the National Register of Historic Places (NRHP), including two sites with burial or
14   cairn features, are located within the proposed project area (see Tables 4.9-1 and 4.9-2). In
15   addition, Tables 4.9-2 and 4.9-3 list sites with burial and cairn features and sites within 76 m
16   [250 ft] of project activity areas that are unevaluated. Mitigative measures that will be
17   implemented to protect the NRHP-eligible and unevaluated sites are described in SEIS Section
18   4.9.1. Efforts to identify properties of religious and cultural significance to Native American
19   tribes at the proposed Dewey-Burdock site through Section 106 consultation involving NRC,
20   SD SHPO, BLM, tribal representatives and the applicant are ongoing, but have not been
21   completed (see SEIS Section 1.7.3.5 and Appendix A). The NRC cannot determine effects to
22   these properties at this time.
23
24   The applicant stated that site avoidance is the goal during development and production of the
25   proposed project (Powertech, 2009a, Section 3.8.1). Sites in areas of activity where ground
26   disturbance is planned will be fenced to avoid accidental disturbance. Furthermore, personnel
27   will be made aware of the presence of sites prior to the start of ground-disturbing activities
28   (Powertech, 2009a). If it is determined that NRHP-eligible or unevaluated sites listed in Tables
29   4.9-1, 4.9-2, and 4.9-3 cannot be avoided, then treatment plans will require that the applicant
30   complete mitigation prior to construction. As described in SEIS Section 4.9.1, treatment plans
31   will be established following the development of an agreement between the applicant, NRC,
32   SD SHPO, interested federal and state agencies (e.g., BLM and EPA), and interested Native
33   American tribes. Prior to construction, the applicant will also develop an Unexpected Discovery
34   Plan that would outline the steps required in the event that unexpected historical and cultural
35   resources are encountered.
36
37   The rock art sites in Craven Canyon are the most significant cultural resource that has been
38   identified in the vicinity of the proposed Dewey-Burdock ISR Project. Craven Canyon is located
39   approximately 10 km [6 mi] east of the proposed Dewey-Burdock ISR Project boundary (see
40   Figure 5.1-3). The rock art in Craven Canyon consists of both petroglyphs, the oldest form of
41   rock art, and pictographs. Recently, there have been increased prohibitions on the extraction of
42   uranium and other minerals in the Craven Canyon area, which is designed to protect cultural
43   resources such as rock art.
44
45   Past, present, and reasonably foreseeable future actions that have the potential for cumulative
46   effects on historic and cultural resources identified in the cumulative impacts study area include
47   uranium exploration and extraction, oil and gas exploration, wind energy projects (e.g., the
48   Dewey-Burdock Wind Project), and transportation projects (e.g., the proposed Dewey
49   Conveyor Project and the proposed DM&E PRB Expansion Project) (see SEIS Sections 5.1.1.1
50   through 5.1.1.5).

1   Uranium extraction, and oil and gas exploration and drilling have occurred in the cumulative
2   impacts study area, and additional drilling is likely to occur in the future. In the case of oil and
3   gas exploration, areas have been proposed for lease sales, but neither applications nor permits
4   to drill have been filed to date (see SEIS Section 5.1.1.3). Activities associated with exploration
5   drilling will include access road and drill pad construction. All access roads and drill sites
6   proposed for any type of exploration drilling will need to be surveyed for historic and cultural
7   resources. Surveys by professional archaeologists and cultural specialists to identify and
8   evaluate NRHP eligibility prior to project construction activities will need to be conducted. In
9   addition, identification of properties of importance to Native American tribes will also need to be
10 undertaken as part of consultation. If NRHP-eligible sites are found, appropriate levels of
11 evaluation and mitigation will be required prior to construction.
12
13 One project that may have a cumulative impact on historic and cultural resources in the vicinity
14 of the proposed Dewey-Burdock ISR Project is the potential Dewey Terrace ISR project. As
15 with the current proposed project, the potential Dewey Terrace ISR project will be surveyed for
16 historic and cultural resources prior to licensing and, if NRHP-eligible sites are indentified,
17 appropriate levels of evaluation and mitigation will be required.
18
19 Surface-disturbing activities from wind energy developments, such as the potential
20 Dewey-Burdock Wind Project, could uncover and destroy cultural resources. However, the
21 development and implementation of programmatic agreements and BMPs will limit the potential
22 impacts at a wind energy project site. For example, a cultural resources management plan will
23 be developed to determine the mitigation activities needed for cultural resources found at a site.
24 Avoidance of the historic and cultural resources will be the preferred mitigation option. Other
25 mitigation options will include archaeological surveys and excavation (as warranted),
26 monitoring, and inadvertent discovery procedures. The programmatic agreements and BMPs
27 will also require consultation under NHPA Section 106, including consultation with SD SHPO
28 and Native American tribes. The implementation of agreements and BMPs would greatly limit
29 impacts from wind energy projects on cultural resources, which are expected to be mainly
30 archaeological sites. However, impacts to cultural resources with a visual component
31 (i.e., sacred landscapes) may occur. (BLM, 2005b)
32
33 As described in SEIS Section 5.1.1.5, the proposed GCC Dacotah Inc. Dewey Conveyor Project
34 would use an elevated, enclosed conveyor to transport limestone quarried from the Minnekahta
35 Limestone to a rail load out facility near Dewey, South Dakota (see Figure 5.3-1). GCC
36 Dacotah Inc. controls minerals rights to areas of potential limestone exploitation north of the
37 proposed conveyor, where the Minnekahta Limestone lies at or near the ground surface (BLM,
38 2009a). These mineral rights are controlled either by ownership or leasing of private lands, or
39 have been acquired by the staking of claims on lands underlain by federally held mineral rights.
40 To date, the location of quarrying operations has not been finalized. However, federal mineral
41 lands acquired by GCC Dacotah Inc. for potential limestone mining have been previously
42 surveyed for cultural resources and over 60 sites were identified (Buechler, 1999; Sundstrom,
43 1999; Winham, et al., 2001). It is expected that many sites would be impacted during quarrying
44 activities. Therefore, appropriate measures would be required to ensure that identified cultural
45 resource sites are avoided and protected during quarrying operations (BLM, 2009a).
46
47 NRHP-eligible historic or cultural resource sites have not been identified along the proposed
48 Dewey Conveyor Project route or within a 30-m [100-ft]-wide buffer zone on either side of the
49 proposed construction zone (see Figure 5.3-1). However, the implementation of alternatives for
50 the proposed Dewey Conveyor Project will result in direct impacts to NRHP-eligible properties.
51 To address these impacts, the following mitigation measures have been proposed: (i) GCC

1   Dacotah Inc. will make a reasonable effort to design the project in a manner to avoid
2   NRHP-eligible properties; (ii) unless authorized by BLM, USFS, and SD SHPO, no surface
3   disturbance will occur within 30 m [100 ft] of the boundary of identified NRHP-eligible properties;
4   and (iii) unless authorized by BLM, USFS, and SD SHPO, no surface disturbance will occur
5   within 30 m [100 ft] of the boundary of 14 unevaluated sites and until their NRHP eligibility has
6   been determined. GCC Dacotah Inc. has also indicated that measures will be taken to ensure
7   that even those sites that are not NRHP-eligible will be avoided and protected, wherever
8   possible. (BLM, 2009a)
9
10   The proposed DM&E PRB Expansion Project will have a significant impact on cultural and
11   historical resources. The project area has a long history of human occupation. Known sites of
12   archaeological and historical significance occur throughout the area. The Department of
13   Transportation Section of Environmental Analysis (SEA) identified 408 cultural resources sites
14   within 0.6 km [1.0 mi] of Alternative C for the proposed DM&E project (see Figure 5.1-5). Of
15   these, 96 sites were in South Dakota and 312 were in Wyoming. Within 0.6 km [1.0 mi] of an
16   alternate route (Alternative B) for the proposed project, SEA identified 298 cultural resources
17   sites, 70 in South Dakota and 228 in Wyoming. SEA determined that the project will have
18   significant impacts to these resources because of the likelihood that construction of the
19   proposed project will encounter significant cultural resources. To address potential adverse
20   impacts on cultural resources, DM&E has proposed mitigation measures, including (i) informing
21   workers of applicable federal, state, and local requirements for the protection of archaeological
22   resources, graves, and other cultural resources and training them on how to recognize and treat
23   resources; (ii) complying with a programmatic agreement and identification plan developed
24   through the NHPA Section 106 consultation process; and (iii) implementing mitigation measures
25   documented in an memorandum of agreement (MOA) developed to ensure that the concerns of
26   Native Americans are considered and addressed. (STB, 2001)
27
28   Because the cumulative impacts study area has a long history of human occupation, it is
29   expected that historic properties of religious and cultural importance to Native American tribes
30   occur throughout the area and that many will be affected by the ongoing and reasonably
31   foreseeable future actions discussed previously. Certain historic properties may be eligible for
32   inclusion in the NRHP because of their association with cultural practices or beliefs of a living
33   community that are rooted in its history and are important in maintaining its continuing cultural
34   identity (National Register Bulletin 38). Historic properties that might be present within the
35   cumulative impacts study area include camp and burial sites, plant collection areas, and sacred
36   and worship sites.
37
38   The NRC staff have determined that the cumulative impact on cultural and historic resources
39   within the cultural and historic resources study area resulting from all past, present, and
40   reasonably foreseeable future actions is MODERATE to LARGE. Archaeological and historic
41   sites and artifacts are present in the area of the proposed site, and any present and future
42   projects could potentially cause adverse impacts to these sites and artifacts.
43
44   The NRC staff conclude that the proposed Dewey-Burdock ISR Project will have a SMALL to
45   LARGE incremental impact on historic and cultural resources when added to the MODERATE
46   to LARGE cumulative impact to these resources expected from other past, present, and
47   reasonably foreseeable future actions. As discussed previously, 18 historic sites listed or
48   recommended as eligible for listing on NRHP are within the proposed Dewey-Burdock project
49   area. ISR activities, especially ground-disturbing activities during the construction phase at the
50   proposed project, may result in a cumulative loss of historic and cultural resources. The

1   mitigation of adverse impacts at the proposed project will be addressed in an agreement
2   between the applicant, NRC, SD SHPO, interested federal and state agencies (e.g., BLM,
3   SDDENR), and interested Native American tribes.
4
5   ## 5.10   Visual and Scenic Resources
6
7   Cumulative impacts to visual and scenic resources were assessed within a 3.2-km [2-mi] radius
8   of the proposed Dewey-Burdock ISR Project. Beyond this distance, any changes to the
9   landscape would be in the background distance zone for the purposes of visual resource
10 management (VRM) defined by BLM, and would be either unobtrusive or imperceptible to
11 viewers (BLM, 1984, 1986). The timeframe for the analysis is 2009 to 2030 (see SEIS
12 Section 5.1.2 for the estimated operating life of the facility).
13
14 As described in SEIS Section 2.1.1.1, the proposed Dewey-Burdock site encompasses 4,282 ha
15 [10,580 ac] of mostly private land in northern Fall River and southern Custer Counties, South
16 Dakota. BLM has not assigned a VRM class to the region that encompasses the proposed
17 project area. However, similar areas adjacent to the proposed project in Wyoming are identified
18 as VRM Classes III and IV (BLM, 2000). At present, human-made features within and in the
19 immediate vicinity of the proposed site include roads, power lines, ranch residences, fence
20 lines, and abandoned open pits and overburden piles associated with past conventional
21 uranium mining. The primary visual feature superimposed on the proposed project landscape is
22 the transportation and utility corridor consisting of Dewey Road, the BNSF railroad, and
23 overhead power lines. The abandoned open pits and overburden piles from historical mining
24 that are located within the eastern and northeastern parts of the proposed project site contribute
25 adversely to the scenic and visual quality of the area. However, the abandoned open pits and
26 overburden piles are not visible from surrounding county roads and highways.
27
28 As described in SEIS Section 4.10.1, potential impacts on visual and scenic resources from the
29 proposed Dewey-Burdock ISR Project will be the contrast of surface facilities and infrastructure
30 (e.g., drilling rigs, powerlines, process buildings, header houses, wellheads, irrigation center
31 pivots) with the existing visual inventory. These types of visual impacts are consistent with the
32 management objectives of the VRM Class III and IV areas that include similar areas adjacent to
33 the proposed project in Wyoming (BLM, 2000). As described in detail in SEIS Section 4.10.1,
34 the impacts to visual and scenic resources from the surface structures and equipment will be
35 SMALL for all phases of the proposed Dewey-Burdock ISR Project. NRC staff base this
36 conclusion on the remote location of the project site and mitigation measures that will be used to
37 reduce potential visual and scenic impacts (e.g., selecting building materials and paint that
38 blend with the natural environment, dust suppression).
39
40 Past, present, and reasonably foreseeable future activities that could have cumulative impacts
41 on the visual and scenic resources in the vicinity of the proposed Dewey-Burdock Project
42 include uranium exploration/extraction, potential oil and gas exploration and development, wind
43 energy projects, and potential transportation projects (i.e., the proposed Dewey Conveyor
44 Project and the proposed DM&E PRB Expansion Project).
45
46 Surface disturbances and fugitive dust emissions associated with access roads and drill pad
47 construction developed for uranium and oil and gas exploration should have only a minor
48 cumulative impact on the visual and scenic resources in the area. Access road segments will
49 be considerably shorter than Dewey Road. Truck and equipment traffic for both construction
50 and drilling activities will be relatively minor, consisting of one or two pieces of equipment per

1   day for construction and two to four pick-up truck trips per day to support drilling activities. All
2   surface disturbances and equipment associated with exploration drilling will be temporary, and
3   the affected ground surface will be fully reclaimed after use. Demand for oil and gas leases is
4   low, and there are no producing oil wells within the 3.2-km [2-mi] radius that could potentially
5   contribute to cumulative impacts related to visual and scenic resources (see SEIS
6   Section 5.1.1.3). Furthermore, there are no reasonably foreseeable future ISR operations in the
7   3.2-km [2-mi] radius that could potentially impact visual and scenic resources (see SEIS
8   Section 5.1.1.1).
9
10   Wind energy projects, such as the potential Dewey-Burdock Wind Project (see Figure 5.1-4),
11   will have an impact on visual and scenic resources within the cumulative impacts study area.
12   The heights, type, and color of turbines, together with their placement with respect to local
13   topography (i.e., on a ridge or mesa), are factors that will contribute to visual intrusion on the
14   landscape. Also, the need for additional transmission lines to connect wind energy projects to
15   the regional power grid could contribute to cumulative impacts. On U.S. government-owned
16   lands, flexibility in locating turbines and transmission line towers to avoid visual impacts to
17   important view sheds will be considered through consultation with the wind energy developer
18   and the managing federal agency (e.g., BLM, USFS) on a project-specific basis. (BLM, 2005b)
19
20   The proposed 10.6-km [6.6-mi]-long Dewey Limestone Conveyor project will have an impact on
21   visual and scenic resources within the cumulative impacts study area (see Figure 5.1-4). The
22   proposed conveyor will consist of elevated 1.5 m by 2.4 m by 12.2 m [5 ft by 8 ft by 40 ft]
23   conveyor segments attached to supporting concrete piers or foundations spaced 7.6 to 12.2 m
24   [25 to 40 ft] apart. The average conveyor height will be 4.9 m [16 ft] with approximately 2.7 m
25   [9 ft] of clearance beneath the conveyor segments. The conveyor alignment is proposed to
26   begin at Dewey Road approximately 1.8 km [1.1 mi] south of the town of Dewey and
27   approximately 1.6 km [1 mi] north-northwest of the proposed Dewey-Burdock Project boundary.
28   The alignment will head east-northeast, progressively away from the proposed Dewey-Burdock
29   Project area. (BLM, 2009a)
30
31   The DM&E PRB Expansion Project will impact visual and scenic resources in the cumulative
32   impacts study area by the visual intrusion of the railroad on the landscape (see Figure 5.1-4).
33   Construction and operation will affect the current scenic character of the cumulative impacts
34   study area as well as the remoteness and feeling of vastness this undeveloped area provides.
35   Some visual mitigation will be accomplished by the use of nonreflective rails and color matching
36   of facilities where possible. For example, DM&E will comply with USFS color coordination
37   requirements for facilities associated with the railroad. Any facility more than 41 cm [16 in] tall
38   will be required to be olive drab, flat tan, or desert brown except where they are required by law
39   to be a specific color. (STB, 2001)
40
41   The NRC staff have determined that the cumulative impact on visual and scenic resources in
42   the study area resulting from all past, present, and reasonably foreseeable future actions is
43   MODERATE to LARGE. This finding is based on the structures and infrastructure from potential
44   future actions that could significantly alter the viewshed within 3.2 km [2 mi] of the proposed
45   Dewey-Burdock ISR Project including (i) turbines and transmission lines associated with future
46   wind energy projects (e.g., the Dewey-Burdock Wind Project), (ii) the elevated conveyor and
47   supporting concrete piers associated with the Dewey Conveyor Project, and (iii) rails and
48   facilities associated with the DM&E PRB Expansion Project.
49

1   The NRC staff have concluded that the proposed Dewey-Burdock ISR Project will have a
2   SMALL incremental impact on visual and scenic resources when considered with all the other
3   past, present, and reasonably foreseeable future actions in the study area. As described in
4   SEIS Section 4.10.1, visual and scenic impacts from the equipment used to construct buildings
5   and drill wells will be temporary and visual impacts from structures and fugitive dust will be
6   mitigated by the rolling topography and BMPs (e.g., color consideration for structures and
7   dust suppression).
8
5.11   Socioeconomics
10
11 As described in SEIS Section 5.1.2, the timeframe for this cumulative impacts analysis for
12 socioeconomics resources begins in 2009 and ends in 2030. The following socioeconomic
13 indicators were evaluated as part of this analysis.
14
15 •   Population
16 •   Employment
17 •   Housing
18 •   School enrollment
19 •   Public services
20 •   Fiscal revenue

21 The geographic boundary varies for the socioeconomic resource indicators listed and is
22 described as part of the analyses for each subcategory. The potential socioeconomic impacts
23 for the proposed Dewey-Burdock ISR Project will be SMALL. These impacts are described in
24 SEIS Section 4.11.
25
5.11.1   Population

27 The geographic boundary for the cumulative population analysis includes Custer and Fall River
28 Counties in South Dakota and Niobrara and Weston Counties in Wyoming. Population change
29 over time is generally an excellent indicator of cumulative social and economic change in a
30 given area. South Dakota's population has grown from 696,004 in 1990 to 814,180 in 2010 and
31 is estimated to decline modestly to 801,939 in 2020 (Brooks, 2008; USCB, 2012). Population in
32 Custer County grew from 6,179 in 1990 to 8,216 in 2010 and is projected to decline slightly to
33 8,186 in 2020 (Brooks, 2008; USCB, 2012). In Fall River County, population decreased slightly
34 from 7,353 in 1990 to 7,094 in 2010 and is projected to increase to 7,423 in 2020 (Brooks, 2008;
35 USCB, 2012). Wyoming population has grown from 453,588 in 1990 to 563,626 in 2010 and is
36 projected to increase to 622,360 in 2020 and 668,830 in 2030 (WDAI, 2011, 2012). Niobrara
37 County population has declined slightly from 2,499 in 1990 to 2,484 in 2010 and is projected to
38 increase to 2,660 in 2020 and 2,710 in 2030 (WDAI, 2011, 2012). Weston County population
39 has grown from 6,518 in 1990 to 7,208 in 2010 and is estimated to increase to 7,900 in 2020
40 and 8,120 in 2030 (WDAI, 2011, 2012).
41
42 The relatively flat county population projections do not take into account the current economic
43 conditions, climate change legislation (including cap and trade components), and future
44 technological changes (e.g., wind energy and clean coal innovations). If the reasonably
45 foreseeable future actions described in SEIS Section 5.1.1 go forward and become functional
46 within the boundary of the cumulative population analysis study area, workers will be required to
47 build and operate these facilities. These future actions include potential wind energy projects,
48 such as the Dewey-Burdock Wind Project, and proposed transportation projects, which include

1   the Dewey Conveyor Project and the DM&E PRB Expansion Project. Additional workers will
2   also be required to staff any expansion in uranium extraction projects, such as the development
3   of the potential Dewey-Terrace project in Weston and Niobrara Counties. It is likely that any
4   additional workers will desire to live closer to their place of employment and become active in
5   their community. The towns of Custer (population 2,067), Hot Springs (population 3,711),
6   Edgemont (population 774), and Newcastle (population 3.532) may see population increases
7   associated with future actions in the population analysis study area. Assuming that energy
8   development and transportation projects are developed and constructed, the addition of new
9   workers in these towns will have a MODERATE cumulative impact on population. The relatively
10 small pool of workers associated with the proposed Dewey-Burdock ISR Project (86 short-term
11 positions during construction, 84 positions during operations, 9 positions during aquifer
12 restoration, and 9 positions during decommissioning) will have only a SMALL incremental
13 impact on population. If a disproportionate number of workers associated with the proposed
14 Dewey-Burdock project elect to reside in small towns like Edgemont, the incremental impact on
15 population could be MODERATE.
16
17 ## 5.11.2   Employment

18 The geographic boundary for the cumulative employment analysis includes Custer and Fall
19 River Counties in South Dakota and Niobrara and Weston Counties in Wyoming. While no
20 individual county employment projections are available, the State of South Dakota is expected
21 to experience modest growth through 2020, with an average annual growth rate of 0.9 percent
22 (SDDLR, 2012). Employment in mining is expected to increase annually by 4 jobs or
23 0.5 percent through 2020, while employment in heavy construction is expected to increase
24 annually by 50 jobs or 1.5 percent through 2020. The State of Wyoming is expected to
25 experience modest growth through 2021, with an average annual growth rate of 1.5 percent
26 (WDWS, 2012). Employment in mining (including oil and gas extraction) is expected to increase
27 annually by 846 jobs or 3.2 percent through 2021.
28
29 The cumulative employment analysis study area may experience an increased rate of
30 employment from ongoing and reasonably foreseeable future actions that may occur (see SEIS
31 Section 5.1.1). If the potential Dewey-Burdock Wind Project and the proposed Dewey Conveyor
32 Project and DM&E PRB Expansion Project are financed and developed, workers will be
33 required to build and operate these projects. Wind energy projects are expected to employ
34 100 to 150 workers during a 1 to 2 year construction period and 10 to 20 workers to operate and
35 maintain the project (BLM, 2005b). The proposed Dewey Conveyor project is expected to
36 employ 50 workers during the 1 year construction period and about 12 workers afterwards to
37 operate the project (BLM, 2009a). The proposed DM&E project will employ more than 900
38 workers over the 2 to 3 year construction phase (STB, 2001). However, only a small portion of
39 the overall construction workforce will be located in a single location at any one time. Once a
40 particular phase of DM&E project is complete, workers will relocate to other job locations (STB,
41 2001). Workers will also be required to staff potential ISR facilities in the study area, such as
42 the potential Dewey-Terrace project. It is assumed that potential ISR facilities in the study
43 area will employ the same number of workers as the proposed Dewey-Burdock ISR Project
44 (86 during construction, 84 during operations, 9 during aquifer restoration, and 9 during
45 decommissioning). This projected growth related to future actions will result in SMALL to
46 MODERATE cumulative impacts to employment in the form of additional job opportunities.
47 Based on the number workers expected at the proposed action, the proposed Dewey-Burdock
48 ISR Project will have a SMALL incremental impact on employment.
49

1 ### 5.11.3   Housing

2 The geographic boundary for the cumulative housing analysis includes Custer and Fall River
3 Counties in South Dakota and Niobrara and Weston Counties in Wyoming. With the projected
4 growth from ongoing and reasonably foreseeable future actions, new employees moving into
5 the study area will require housing. Smaller communities, such as Edgemont, are likely to
6 experience MODERATE cumulative impacts due to limited housing availability. Assuming,
7 however, that new employees relocate to one of the larger communities, such as Custer, Hot
8 Springs, or Newcastle, there should be adequate housing opportunities to absorb the influx of
9 facility workers. Therefore, the cumulative impact will be SMALL. Given the number of
10 Dewey-Burdock ISR facility employees (86 during construction, 84 during operations, 9 during
11 aquifer restoration, and 9 during decommissioning), there will be SMALL incremental impacts to
12 housing markets, prices, and real estate development in larger communities such as Custer,
13 Hot Springs, and Newcastle. However, housing impacts may be MODERATE if a
14 disproportionate number of employees at the proposed Dewey-Burdock ISR project elect to
15 reside in smaller communities, such as Edgemont.
16
17 ### 5.11.4   Education

18 The Custer School District, Hot Springs School District, Edgemont School District, Weston
19 County School District No. 1, and Weston County School District No. 7 represent the
20 geographic boundary for the school enrollment resource analysis. These school districts were
21 selected because most permanent Dewey-Burdock ISR facility employees will be likely to live in
22 one of these districts. Most of the construction workforce, however, is not expected to relocate
23 entire families during the relatively brief construction phase (1 to 2 years). Student enrollment in
24 these school districts totaled 2,915 in 2010 and ranged from 150 students in the Edgemont
25 School District to 882 students in the Custer School District (see Table 3.11-5).
26
27 Most of the construction workforce for the ongoing and reasonably foreseeable future actions
28 described in SEIS Section 5.1.1 is not expected to relocate entire families into the school
29 enrollment study area. The construction phases of future actions, such as wind projects, ISR
30 facilities, and transportation projects, are relatively brief, ranging from 1 to 3 years. During
31 operations of ongoing and reasonably foreseeable future actions, new employees will be more
32 likely to move their families and send their children to schools in the study area. The potential
33 increase in school-aged children will likely be split between the school districts in the school
34 enrollment study area. Based on the number of permanent employees needed to operate
35 reasonably foreseeable future actions (e.g., 84 for ISR facilities, 10 to 20 for wind projects, and
36 about 12 for transportation projects), cumulative impacts to school enrollment are expected to
37 be SMALL. Based on the number of workers (84) needed for the proposed Dewey-Burdock ISR
38 Project, the proposed action will have a SMALL incremental impact on school resources in the
39 larger school districts within the school enrollment study area, such as the Custer and Hot
40 Springs school districts. However, school enrollment impacts may be MODERATE if a
41 disproportionate number of employees at the proposed Dewey-Burdock ISR project elect to
42 reside in smaller communities, such as Edgemont.
43
44 ### 5.11.5   Public Services

45 The geographic boundary for the public services socioeconomic resource cumulative impact
46 analysis includes Custer and Fall River Counties in South Dakota and Niobrara and Weston
47 Counties in Wyoming. There may be incremental impacts to local government facilities and
48 public services as population increases in affected counties and communities, which generally

1   result in across-the-board increases in the demand on services. Even small changes in
2   population size may result in additional demand for health and human services, such as
3   doctors, hospitals, police, and fire response. Additionally, the various reasonably foreseeable
4   future actions described in SEIS Section 5.1.1 may result in increased demand for specific
5   services (e.g., road maintenance). Operational impacts to public services and public
6   infrastructure, as a result of the workers relocating with their families, will be area-specific, and
7   may be long term. As described in SEIS Section 3.11.7, there are a number of existing medical
8   and emergency facilities that will be capable of handling issues related to increased population.
9   Additionally, the State of South Dakota Social Services has offices located throughout the state,
10   including in Custer and Hot Springs. The State of Wyoming has numerous social services
11   offices located throughout the state as well. There is an office for Niobrara and Weston
12   Counties, as well as other local offices located in Newcastle. It is not anticipated that additional
13   population from ongoing and reasonably foreseeable future actions will stress the current social
14   services capabilities in the public services resource study area. Therefore, cumulative impacts
15   to public services are expected to be SMALL. Given the number of workers required for the
16   proposed Dewey-Burdock ISR Project (86 during construction, 84 during operations, 9 during
17   aquifer restoration, and 9 during decommissioning), incremental impacts from the proposed
18   action will have a SMALL impact on public services.
19
20   **5.11.6   Local Finance**

21   The geographic boundary for the local finance socioeconomic resource is Fall River and Custer
22   Counties. Tax revenue will accrue mainly in Fall River and Custer Counties and to the State of
23   South Dakota, and because of the structure of the taxing system, taxes may not accrue or be
24   distributed to the localities proportionate to the population/public service impacts experienced by
25   those entities. The tax system in place helps capture tax revenue during construction,
26   operation, and decommissioning of industrial facilities. Additionally, a county ad valorem tax
27   from current and future mineral extraction operations will contribute to local government
28   revenue. Indirectly, counties and municipalities will benefit from increased sales tax revenue
29   from increases in population and resultant demand for goods and services. If reasonably
30   foreseeable future actions are constructed and operated, there will be a MODERATE
31   cumulative impact on local finance. Given that the proposed Dewey-Burdock ISR Project is only
32   one of numerous potential future projects, contributions from the Dewey-Burdock ISR Project
33   are expected to have a SMALL incremental impact on local finance.
34
35   The NRC staff determined that the cumulative impact on socioeconomic resources resulting
36   from past, present, and reasonably foreseeable future actions ranges from SMALL to
37   MODERATE. Impacts to population and local finance will be MODERATE; impacts to
38   employment will be SMALL to MODERATE, and impacts to housing, education, and public
39   services will be SMALL.
40
41   The NRC staff conclude that the proposed Dewey-Burdock ISR Project will have a SMALL to
42   MODERATE incremental effect on socioeconomic resources when considered with other past,
43   present, and reasonably foreseeable actions. Impacts to population, housing, and education
44   will be SMALL to MODERATE, while impacts to employment, public services, and local finance
45   will be SMALL.
46
47

1   ## 5.12   Environmental Justice
2
3   Impacts relating to environmental justice for the proposed Dewey-Burdock ISR Project are
4   described in detail in SEIS Section 4.12. The geographic boundary for this resource includes
5   Custer and Fall River Counties in South Dakota, Weston County in Wyoming, and the Pine
6   Ridge Indian Reservation in Shannon County, South Dakota. The timeframe for the analysis is
7   2009 to 2030 (see SEIS Section 5.1.2 for the estimated operating life of the proposed project).
8
9   As described in SEIS Section 4.12.1, NRC staff determined that the percentage of minority
10   populations living in affected block groups in the vicinity of the proposed Dewey-Burdock ISR
11   Project site in Custer, Fall River, and Weston Counties does not significantly exceed the
12   percentage of minority populations recorded at the state and county levels and is well below the
13   national level. Furthermore, NRC staff determined the percentage of low-income populations
14   living in affected census tracts in the vicinity of the proposed project site in Custer, Fall River,
15   and Weston Counties does not significantly exceed the percentage of low-income populations
16   recorded at the state or county level. Based on an analysis of potential impacts to minority and
17   low-income populations described in SEIS Section 4.12.2, NRC concluded that there will be no
18   disproportionally high or adverse impacts to minority or low-income populations residing near
19   the proposed project area.
20
21   In GEIS Section 6.4, NRC staff identified the Native American Oglala Sioux Tribe as a minority
22   population in the Nebraska-South Dakota-Wyoming Milling Region and the Pine Ridge Indian
23   Reservation as a low-income population (NRC, 2009a). The Pine Ridge Indian Reservation is
24   located in Shannon County, South Dakota, approximately 80 km [50 mi] from the proposed
25   Dewey-Burdock ISR Project. Environmental justice impacts related to the protection of cultural
26   and religious resources of significance to the Oglala Sioux Tribe and other potentially affected
27   Native American tribes are being addressed through the NHPA Section 106 consultation
28   process as described in SEIS Sections 1.7.3.5 and 4.9.1. As described in SEIS Section 4.12.1,
29   environmental justice impacts to Native American tribes will primarily be no different than those
30   experienced by other populations within the vicinity of the project area. Although the proposed
31   action may potentially affect certain sites of religious or cultural significance to the tribes, the
32   impacts to such sites would be reduced through mitigation strategies developed during Section
33   106 consultations.
34
35   Because the economic base of the study area is largely ranching and resource extraction, low
36   income areas are not only widely dispersed but small in size. Furthermore, it is unlikely that
37   race and poverty characteristics in regions surrounding the proposed Dewey-Burdock ISR
38   Project area will change significantly as a result of past, present, and reasonably foreseeable
39   future projects discussed in Section 5.1.1. For reasonably foreseeable future actions, the extent
40   to which there will be potential environmental impacts (e.g., visual impacts of wind turbines and
41   transmission infrastructure associated with wind energy projects) and health and safety risks
42   that create an environmental justice concern will depend on the precise location of low-income
43   and minority populations in relation to specific projects. Full analysis of the potential impacts
44   of specific projects on low-income and minority populations will be undertaken as part of
45   site-specific environmental justice reviews of each proposed development site.
46
47   Based on available minority and low income population information and the analysis of human
48   health and environmental impacts presented in Chapters 4 and 5, NRC staff conclude that the
49   potential for adverse incremental impacts within the study area will be SMALL. The NRC staff
50   also conclude that the proposed project will have a SMALL incremental impact on

1 environmental justice populations when added to the SMALL cumulative impacts from other
2 past, present, and reasonably foreseeable future actions.
3
4 ## 5.13   Public and Occupational Health and Safety
5
6 Cumulative impacts on public and occupational health and safety were evaluated within a
7 105-km [65-mi] radius of the proposed Dewey-Burdock site. This distance was chosen because
8 the nearest operating ISR facility to the proposed Dewey-Burdock site is located approximately
9 105 km [65 mi] south at Crow Butte in Dawes County, Nebraska. The timeframe for the
10 analysis is 2009 to 2030 (see SEIS Section 5.1.2 for the estimated operating life of the facility).
11
12 The public and occupational health and safety impacts from the proposed Dewey-Burdock ISR
13 Project will be SMALL and are discussed in detail in SEIS Section 4.13.1. During normal
14 activities associated with all phases of the project lifecycle, radiological and nonradiological
15 worker and public health and safety impacts will be SMALL. Annual radiological doses to the
16 population within 105 km [65 mi] of the proposed project will be far below applicable NRC
17 regulations. For accidents, radiological and nonradiological impacts to workers may be
18 MODERATE if the appropriate mitigation measures and other procedures intended to ensure
19 worker safety are not followed.   Typical protection measures, such as radiation and
20 occupational monitoring, respiratory protection, standard operating procedures for spill response
21 and cleanup, and worker training in radiological health and emergency response, will be
22 required as a part of the applicant's NRC-approved Radiation Protection Program (Powertech,
23 2011). These procedures and plans will reduce the overall radiological and nonradiological
24 impacts to workers from accidents to SMALL.
25
26 Past, existing, and anticipated future uranium recovery facilities in the vicinity of the
27 proposed Dewey-Burdock ISR Project and within the broader regional area are described in
28 Section 5.1.1.1. Abandoned open pits and overburden waste piles associated with past surface
29 mining activities occur in the Burdock portion of the proposed site (see Figure 3.2-3). Radiation
30 surveys have revealed that soils in and near the old surface mining works have elevated
31 radiation levels (see SEIS Section 3.12.1), which could potentially increase radiological doses to
32 onsite workers. Within a 105-km [65-mi] radius of the proposed project, there is one operating
33 ISR facility at Crow Butte in Dawes County, Nebraska. In addition, three satellite facilities or
34 ISR expansions for the Crow Butte site are in the planning or prelicensing stages: North Trend,
35 Three Crow, and Marsland. The applicant has also identified a potential ISR project at Dewey
36 Terrace in Niobrara and Weston Counties, Wyoming (Powertech, 2009b). If constructed and
37 operated, all of these facilities will have similar radiological and nonradiological impacts on
38 public and occupational health and safety to those at the proposed Dewey-Burdock site.
39 Potential cumulative impacts from these facilities will result from incremental increases in annual
40 radiological doses to the population when combined with the impacts of the proposed
41 Dewey-Burdock ISR Project.
42
43 As stated in Section 4.13.1, for normal operations, Rn-222 will be the only significant
44 radionuclide anticipated to be released at the proposed Dewey-Burdock ISR Project; the
45 primary sources will be from wellfield venting and releases from within the central plant for
46 process operations (predominantly via vent stacks on the ion-exchange columns and various
47 tanks). As further described in SEIS Section 4.13.1, the maximum expected exposure to a
48 member of the public is located southeast of the Dewey satellite facility within the proposed
49 Dewey-Burdock project permit boundary (see Figure 4.13-1). This maximum exposure is
50 estimated to be 0.06 mSv/yr [6.0 mrem/yr] and is consistent with estimates of expected

1 exposure levels at other operating ISR facilities in the United States (NRC, 2009a). This
2 exposure, combined with exposures from other operating and potential ISR facilities in the study
3 area, will remain far below the 10 CFR Part 20 public dose limit of 1.0 mSv/yr [100 mrem/yr] and
4 have a negligible contribution to the 6.2 mSv [620 mrem] average yearly dose received by a
5 member of the public from all sources.
6
7 As described in SEIS Section 4.13.1, both worker and public radiological exposures are
8 addressed in NRC regulations at 10 CFR Part 20. Licensees are required to implement an
9 NRC-approved radiation protection program to protect occupational workers and ensure that
10 radiological doses are "as low as reasonably achievable" (ALARA). The applicant's radiation
11 protection program includes commitments for implementing management controls, engineering
12 controls, radiation safety training, radon monitoring and sampling, and audit programs
13 (Powertech, 2011). Measured and calculated doses for workers and the public are commonly
14 only a fraction of regulated limits. Analysis of three separate accident scenarios (thickener
15 failure and spill, pregnant lixiviant and loaded resin spills, and yellowcake dryer accident
16 release) will also result in hypothetical exposures that are less than NRC regulatory limits and
17 produce SMALL potential impacts (SEIS Section 4.13.1.1.2.2).
18
19 The types and quantities of chemicals (hazardous and nonhazardous) for proposed use at the
20 Dewey-Burdock ISR Project do not differ from those evaluated in the GEIS. The use of
21 hazardous chemicals at ISR facilities is controlled under several regulations (see SEIS
22 Section 4.13.1.1.2.3 for a list of these regulations) that are designed to provide adequate
23 protection to workers and the public. The handling and storage of chemicals at the facility will
24 follow standard industrial safety standards and practices. Industrial safety aspects associated
25 with the use of hazardous chemicals are regulated by the South Dakota Occupational Safety
26 and Health Administration. Nonradiological worker safety will be addressed through
27 occupational health and safety regulations and practices.
28
29 Other past, present, and reasonably foreseeable future actions in the vicinity of the
30 Dewey-Burdock Project that could contribute to nonradiological public and occupational health
31 and safety include oil and gas exploration, wind energy projects, the proposed Dewey Conveyor
32 Project, and the proposed DM&E PRB Expansion Project (see SEIS Sections 5.1.1.3, 5.1.1.4,
33 and 5.1.1.5). Increased risk to human health and safety will occur during development and
34 operation of these projects from the inherent hazards associated with construction and
35 maintenance activities. However, these risks will be minimized by implementation of BMPs,
36 development and implementation of health and safety programs, safety setbacks to nearest
37 residences, mitigation measures, and compliance with applicable federal and state occupational
38 and public safety regulations (BLM, 2005b, 2009a; STB, 2001). Hazardous materials that are
39 likely to be used during these ongoing and reasonably foreseeable future projects include diesel
40 fuel, gasoline, hydraulic fluids, motor oil/grease, and compressed gasses used for welding
41 (e.g., acetylene or propane). A large-scale release of diesel fuel or several of the other
42 substances used at the projects may have implications for public health and safety. The
43 location of the release will be the primary factor in determining its importance. However, the
44 probability of a release anywhere along a proposed transportation route is extremely low, the
45 probability of a release within a populated area will be even lower, and the probability of a
46 release involving an injury or fatality will be still lower (BLM, 2009a). Therefore, it is not
47 anticipated that a release involving a severe effect on human health and safety will occur during
48 these ongoing and potential future actions. In addition, ongoing and potential future actions will
49 have federal- and/or state-mandated spill prevention and control plans to prevent spills of oil
50 and other petroleum products and other hazardous materials during construction and operation
51 activities (BLM, 2009a; STB, 2001).

1   The NRC staff have determined that the cumulative impact on public and occupational health
2   and safety in the study area resulting from all past, present, and reasonably foreseeable future
3   actions is SMALL. This finding is based on estimates of combined radiological exposures from
4   currently operating and proposed future ISR facilities in the study area, which are estimated to
5   remain far below the regulatory public limit of 1.0 mSv/yr [100 mrem/yr] and have a negligible
6   contribution to the 6.2 mSv [620 mrem] average yearly dose for a member of the public from all
7   sources. Nonradiological exposures to workers and the general public from hazardous
8   chemicals and materials resulting from past, present, and reasonably foreseeable future actions
9   will be minimized by implementation of BMPs, mitigation measures, and compliance with
10   applicable federal and state occupational and public safety regulations.
11
12   The NRC staff conclude that the proposed Dewey-Burdock ISR Project will have a SMALL
13   incremental impact on public and occupational health when considered with all the other past,
14   present, and reasonably foreseeable future actions in the study area. The maximum expected
15   exposure to a member of the public at the proposed Dewey-Burdock Project is estimated to be
16   0.06 mSv/yr [6.0 mrem/yr] and is consistent with estimates of expected exposure levels at other
17   operating ISR facilities in the United States (NRC, 2009a). Because the facility is located in a
18   remote, sparsely populated area, the exposure to members of the public will be limited.
19   Occupational health hazards will be limited because licensees are required to implement an
20   NRC-approved radiation protection program to protect workers. As described in SEIS
21   Section 4.13.1.1.2.3, the handling, storage, and disposal of chemicals at the proposed project
22   would follow standard industrial safety standards and practices and the applicant must comply
23   with EPA, SDDENR, and OSHA regulations regarding the industrial and environmental safety
24   aspects associated with the use of chemicals.
25
26   ## 5.14   Waste Management
27
28   Waste management impacts from the proposed Dewey-Burdock ISR Project would be SMALL
29   to MODERATE and are detailed in SEIS Section 4.14.1. Cumulative impacts on waste
30   management were considered within a 105-km [65-mi] radius of the proposed Dewey-Burdock
31   Project site, and the timeframe for the analysis is 2009 to 2030 (see Section 5.1.2 for the
32   estimated operating life of the facility). This distance was chosen because the nearest
33   operating ISR facility that could generate waste volumes consistent with those projected for the
34   proposed Dewey-Burdock site is located approximately 105 km [65 mi] south at the Cameco
35   Crow Butte operation in Crawford, Nebraska.
36
37   The proposed Dewey-Burdock ISR Project will generate radiological and nonradiological liquid
38   and solid wastes that must be handled and disposed of properly. Waste streams and the
39   types and volumes of wastes to be disposed are described in SEIS Section 2.1.1.1.6. The
40   primary radiological wastes are process-related liquid wastes, waste treatment solids, and
41   process-contaminated structures and soils, all of which are classified as byproduct material
42   waste. As discussed in SEIS Section 4.14.1, liquid byproduct material generated during
43   operations is composed of production bleed, waste brine streams from elution, laundry water,
44   plant washdown water, laboratory chemicals, and aquifer restoration water. Liquid byproduct
45   material will be treated onsite using a combination of ion exchange, reverse osmosis, and
46   radium settling followed by deep disposal in Class V injection wells, land application, or
47   combined deep well disposal in Class V injection wells and land application. State- and
48   federal-permitting actions, NRC license conditions, and NRC and state inspections ensure that
49   proper waste disposal practices will be used to comply with safety and environmental
50   requirements to protect workers, the public, and the environment.

1   As described in SEIS Section 4.14.1, the overall impacts from the disposal of process-related
2   liquid wastes at the proposed Dewey-Burdock ISR Project will be SMALL. In addition, impacts
3   associated with disposal of solid radioactive wastes will be SMALL based on the required
4   preoperational disposal agreements made between the licensee and the licensed byproduct
5   material waste disposal facility. Hazardous waste disposal impacts at the proposed Dewey-
6   Burdock Project will be SMALL based on the low volumes of waste generated. Impacts from
7   disposal of nonradioactive, nonhazardous solid wastes will be SMALL during the construction,
8   operations, and aquifer restoration phases of the proposed project based on estimated
9   volumes and the available capacity of local municipal solid waste landfills. However, impacts
10   from disposal of nonhazardous solid wastes will be SMALL to MODERATE during the
11   decommissioning phase depending on the long-term status of existing local landfill resources. If
12   local landfill capacity is not expanded prior to the proposed decommissioning phase, impacts
13   will be MODERATE because the projected capacity of the local landfill (i.e., the Custer-Fall
14   River landfill) will be insufficient to accommodate all the decommissioning nonhazardous solid
15   waste. If local landfill capacity is expanded prior to the decommissioning phase, impacts from
16   disposal of nonhazardous solid wastes will be SMALL.
17
18   Past, existing, and anticipated future uranium recovery facilities in the vicinity of the proposed
19   Dewey-Burdock ISR Project and within the broader regional area are described in
20   Section 5.1.1.1. Abandoned open pits and overburden waste piles associated with past
21   surface mining activities occur in the Burdock portion of the Dewey-Burdock site (see SEIS
22   Figures 3.2-3). Radiation surveys reveal that soils near the old surface mining works have
23   higher than background radiation levels (Powertech, 2009a). At present, there are no plans to
24   clean up and reclaim the old surface mines. However, potential future state- or federal-funded
25   cleanup and reclamation of the abandoned open pits and overburden waste piles will have an
26   impact on waste management if the radioactive soils require disposal in a licensed byproduct
27   disposal facility. As noted previously, within a 105-km [65-mi] radius of the proposed
28   Dewey-Burdock ISR Project, there is one operating ISR facility at Crow Butte in Dawes County,
29   Nebraska, which will generate waste volumes consistent with those projected for the proposed
30   Dewey-Burdock ISR project. In addition, three satellite facilities or ISR expansions are in the
31   planning and licensing stages at the Crow Butte site: North Trend, Three Crow, and Marsland
32   (see SEIS Section 5.1.1.1). Powertech has also identified a potential ISR project at Dewey
33   Terrace in Niobrara and Weston Counties, Wyoming (Powertech, 2009b). All of these potential
34   ISR facilities will generate solid and liquid waste volumes consistent with those projected for the
35   proposed Dewey-Burdock ISR Project, which could contribute to waste management impacts
36   within the cumulative impacts study area. Generation of nonhazardous solids wastes at the
37   planned and potential ISR facilities could impact landfill resources in the cumulative impacts
38   study area. Impacts to landfill resources will be MODERATE if current landfill capacities are not
39   adequate to accept nonhazardous solid wastes generated by the planned and potential ISR
40   facilities and an expansion is necessary to accommodate added volume. Before ISR operations
41   begin, NRC requires ISR facilities to have an agreement in place with a licensed disposal facility
42   to accept byproduct material. Because radioactive wastes are so closely monitored throughout
43   the United States, the impact on waste management from these potential facilities is anticipated
44   to be SMALL.
45
46   Regarding the potential cumulative impacts of liquid waste disposal, the applicant is seeking
47   permits from EPA for four to eight Class V deep disposal wells for liquid byproduct materials
48   (Powertech, 2011, Appendix 2.7–L). Additional deep disposal well use in the region is
49   anticipated as additional ISR facilities are licensed. The EPA-permitting process for these wells
50   evaluates the suitability of proposals to ensure groundwater resources are protected and
51   potential environmental impacts are limited to acceptable levels. Based on the assumption that

1   EPA will not permit deep injection wells that will have a significant potential to impact
2   groundwater resources, the NRC staff conclude the cumulative impacts of using deep disposal
3   wells for the proposed action along with the potential impacts from present and reasonably
4   foreseeable future actions will be SMALL.
5
6   Other ongoing and reasonably foreseeable future activities in the vicinity of the proposed
7   Dewey-Burdock ISR Project site that may generate nonradiological hazardous wastes include
8   oil and gas exploration, wind energy projects, and proposed transportation projects, such as the
9   Dewey Conveyor Project and the DM&E PRB Expansion Project (see SEIS Sections 5.1.1.3,
10 5.1.1.4, and 5.1.1.5). Each of these projects will require shipment, storage, use, and disposal of
11 hazardous materials and generation of solid and hazardous wastes; however, BMPs addressing
12 these activities will effectively mitigate potential impacts. Each project will also be
13 responsible for complying with applicable federal and state regulations and site-specific
14 license agreements that manage generated wastes. For example, applicants will be required to
15 comply with Department of Transportation Hazardous Materials regulations (49 CFR Parts 171
16 and 179) when handling, storing, and disposing hazardous materials. The types of hazardous
17 substances that will likely be present during activities associated with these projects include
18 diesel fuel, gasoline, hydraulic fluids, motor oil/grease, and compressed gases used for welding
19 (e.g., acetylene, propane). Potential impacts will result from accidental releases of these
20 substances during transportation, or during use and storage. The environmental effects of a
21 release will depend on the substance, quantity, timing, and location of the release. The event
22 could range from a minor oil spill on the project site where cleanup equipment will be readily
23 available, to a severe spill during transport involving a large release of fuel or other hazardous
24 substance. Some of the chemicals could have immediate adverse impacts on water quality and
25 aquatic resources if a spill entered a flowing stream. With rapid cleanup actions, contamination
26 will not result in a long-term impact to soils, surface water, or groundwater.
27
28 The NRC staff have determined that the cumulative impact on waste management in the study
29 area resulting from all past, present, and reasonably foreseeable future actions is SMALL to
30 MODERATE. All present and reasonably foreseeable future actions will implement BMPs to
31 address shipment, storage, use, and disposal of radiological and nonradiological hazardous
32 materials (both liquid and solid) and will be required to comply with applicable federal and state
33 regulations and site-specific license agreements that manage generated wastes. Impacts to
34 landfill resources will be MODERATE if current landfill capacities are not adequate to accept
35 nonhazardous solid wastes generated by the planned and potential ISR facilities and an
36 expansion is necessary to accommodate added volume.
37
38 The NRC staff conclude that the proposed Dewey-Burdock ISR Project will have a SMALL to
39 MODERATE incremental impact on waste management when considered with all the other
40 past, present, and reasonably foreseeable future actions in the study area. The applicant will be
41 required to obtain the necessary permits and contractual agreements for disposing of its solid
42 byproduct material, hazardous waste, and nonradiological, nonhazardous solid and liquid
43 wastes. In addition, the applicant will be required to comply with applicable federal and state
44 regulations and site-specific license agreements for the management and disposal of
45 process-related liquid wastes. Impacts from disposal of nonradioactive, nonhazardous solid
46 wastes will be SMALL during the construction, operations, and aquifer restoration phases of the
47 proposed project based on estimated volumes and the available capacity of local municipal solid
48 waste landfills. However, impacts from disposal of nonhazardous solid wastes will be SMALL to
49 MODERATE during the decommissioning phase depending on the long-term status of existing
50 local landfill resources. If local landfill capacity is not expanded prior to the proposed

1   decommissioning phase, impacts will be MODERATE because the projected capacity of the
2   local landfill (i.e., the Custer-Fall River landfill) will be insufficient to accommodate all the
3   decommissioning nonhazardous solid waste. If local landfill capacity is expanded prior to the
4   decommissioning phase, impacts from disposal of nonhazardous solid wastes will be SMALL.
5

6   ## 5.15   References

7

8   10 CFR Part 20. *Code of Federal Regulations*, Title 10, *Energy*, Part 20. "Standards for
9   Protection Against Radiation." Washington, DC: U.S. Government Printing Office.
10

11 10 CFR Part 40. Code of Federal Regulations, Title 10, *Energy*, Part 40. "*Domestic Licensing
12 of Source Material.*" Washington, DC: U.S. Government Printing Office.
13

14 40 CFR Part 92. *Code of Federal Regulations*, Title 40, *Protection of the Environment*, Part 92,
15 "Control of Air Pollution from Locomotives and Locomotive Engines." Washington, DC: U.S.
16 Government Printing Office.
17

18 40 CFR Part 1500 to 40 CFR Part 1508. *Code of Federal Regulations*, Title 40, *Protection of
19 the Environment*, Parts 1500–1508. "Council on Environmental Quality." Washington, DC:
20 U.S. Government Printing Office.
21

22 49 CFR Part 171. *Code of Federal Regulations,* Title 49, Transportation, Part 171. "General
23 Information, Regulations, and Definitions." Washington, DC: U.S. Government Printing Office.
24

25 49 CFR Part 179. *Code of Federal Regulations*, Title 49, Transportation, Part 179,
26 "Specifications for Tank Cars." Washington, DC: U.S. Government Printing Office.
27

28 49 CFR Part 210. *Code of Federal Regulations*, Title 49, Transportation, Part 210. "Railroad
29 Noise Emission Compliance Regulations." Washington, DC: U.S. Government Printing Office.
30

31 AWEA (American Wind Energy Association). "Wind Energy Facts: Nebraska." ML12243A234.
32 Washington, D.C.: American Wind Energy Association. 2012a. <http://www.awea.org/
33 learnabout/ publications/upload/4Q-11-Nebraska.pdf> (20 August 2012).
34

35 AWEA. "Wind Energy Facts: South Dakota." ML12243A243. Washington, D.C.: American
36 Wind Energy Association. 2012b. <http://www.awea.org/learnabout/publications/upload/4Q-11-
37 South-Dakota.pdf> (20 August 2012).
38

39 AWEA. "Wind Energy Facts: Wyoming." ML12234A254. Washington, D.C.: American Wind
40 Energy Association. 2012c. <http://www.awea.org/learnabout/publications/upload/4Q-11-
41 Wyoming.pdf> (20 August 2012).
42

43 AWEA. "U.S. Wind Energy Market Reports." Washington, D.C.: American Wind Energy
44 Association. 2012d. <http://www.awea.org/learnabout/publications/reports/AWEA-US-Wind-
45 Industry-Market-Reports.cfm> (20 August 2012).
46

1   Becker, J.M., C.A. Duberstein, J.D. Tagestad, and J.L. Downs. "Sage-Grouse and Wind
2   Energy: Biology, Habits, and Potential Effects of Development." Prepared for the
3   U.S. Department of Energy Office of Energy Efficiency and Renewable Energy Wind &
4   Hydropower Technologies Program under Contract DE-AC05-76RL01830. ML12243A257.
5   2009. <http://www.pnl.gov/main/publications/external/technical_reports/pnnl-18567.pdf>
6   (28 April 2010).
7
8   BLM (U.S. Bureau of Land Management). "Draft Environmental Impact Statement, Dewey
9   Conveyor Project." DOI–BLM–MT–040–2009–002–EIS. ML12209A089. Belle Fourche, South
10   Dakota: BLM Field Office, U.S. Department of Interior. January 2009a. <http://www.blm.gov/
11   pgdata/etc/medialib/blm/mt/field_offices/south_dakota/dakotah.Par.28318.File.dat/DeweyDEIS.
12   pdf> (25 July 2012).
13
14   BLM. "Update of Task 3A Report for the Powder River Basin Coal Review Cumulative Air
15   Quality Effects for 2020." ML12243A338. December 2009b. <http://www.blm.gov/pgdata/etc/
16   medialib/blm/wy/programs/energy/coal/prb/coalreview/task_3a_-_2020.Par.91798.File.dat/Task-
17   3A_2020.pdf> (23 April 2010).
18
19   BLM. "Draft Environmental Impact Statement for the Atlantic Rim Natural Gas Field
20   Development Project Carbon County, Wyoming." ML12243A274. Rawlins, Wyoming: BLM
21   Field Office. December 2005a. <http://www.blm.gov/pgdata/etc/medialib/blm/wy/
22   information/NEPA/rfodocs/atlantic_rim.Par.61741.File.dat/00deis.pdf> (20 August 2012).
23
24   BLM. "Chapter 5: Potential Impacts of Wind Energy Development and Analysis of Mitigation
25   Measures." *Final Programmatic Environmental Impact Statement on Wind Energy*
26   *Development on BLM-Administered Lands in the Western United States*. FES 05-11.
27   ML12243A271. Washington, DC: BLM, U.S. Department of the Interior. June 2005b.
28   <http://windeis.anl.gov/documents/fpeis/maintext/Vol1/Vol1Ch5.pdf> (20 August 2012).
29
30   BLM. "Mineral Occurrence and Development Potential Report, Rawlins Resource Management
31   Plan Planning Area." Rawlins, Wyoming: BLM, Rawlins Field Office. 2003. ML12243A327.
32   <http://www.blm.gov/pgdata/etc/medialib/blm/wy/programs/planning/rmps/rawlins/modp.Par.494
33   82.File.dat/Rawlins_Report.pdf> (20 August 2012).
34
35   BLM. "Newcastle Resource Management Plan." ML12209A101. Newcastle, Wyoming: BLM,
36   Newcastle Field Office. 2000. <http://www.blm.gov/rmpweb/Newcastle/rmp.pdf>
37   (16 October 2009).
38
39   BLM. "Visual Resource Inventory." Manual H–8410–1. ML12237A196. Washington, DC:
40   BLM. 1986. <http://www.blm.gov/pgdata/etc/medialib/blm/
41   wo/Information_Resources_Management/policy/blm_handbook.Par.31679.File.dat/H-8410.pdf>
42   (16 October 2009).
43
44   BLM. "Visual Resource Management." Manual 8400. ML12237A194. Washington, DC: BLM.
45   1984. <http://www.blm.gov/pgdata/etc/medialib/blm/wo/Information_Resources_
46   Management/policy/blm_manual.Par.34032.File.dat/8400.pdf> (16 October 2009).
47
48

1   Brooks, T., M. McCurry, and D. Hess. "South Dakota State and County Demographic Profiles."
2   ML12237A222. Brookings, South Dakota: South Dakota Rural Life and Census Data Center.
3   May 2008. <http://www.sdstate.edu/soc/rlcdc/i-o/reports/upload/South-Dakota-State-and-
4   County-Demographic-Profiles-B755.pdf> (22 August 2012).
5
6   Buechler, J.V. "An Intensive (Class III) Cultural Resources Inventory Survey of the
7   Dacotah Cement Land Exchange Proposal in Southwestern Custer County, South Dakota."
8   Project No. 99-9. Rapid City, South Dakota: Dakota Research Services. (Submitted to
9   Dacotah Cement, Rapid City, South Dakota). 1999.
10
11 Carter, J.M., D.G. Driscoll, and J.F. Sawyer. "Ground-Water Resources in the Black Hills Area,
12 South Dakota." U.S. Geological Survey Water Resources Investigations Report 03-4049.
13 ML12243A344. 2003. <http://pubs.usgs.gov/wri/wri034049/wri034049 files/
14 wri034049p0 10.pdf> (20 August 2012).
15
16 Center for Climate Strategies. "South Dakota Greenhouse Gas Inventory and Reference Case
17 Projections 1990–2020." 2007. <http://www.climatestrategies.us/ewebeditpro/items/
18 O25F18227.pdf> (21 December 2009).
19
20 CEQ (Council on Environmental Quality). "Considering Cumulative Effects Under the National
21 Environmental Policy Act." ML13343A349. Washington, DC: Executive Office of the President,
22 CEQ. 1997. <http://energy.gov/sites/prod/files/nepapub/nepa documents/RedDont/G-CEQ-
23 ConsidCumulEffects.pdf> (20 August 2012).
24
25 Connelly, J.W., C.A. Hagen, and M.A. Schroeder. "Characteristics and Dynamics of Greater
26 Sage-Grouse populations." In *Greater Sage-Grouse: Ecology and Conservation of a
27 Landscape Species and Its Habitats.* S. T. Knick and J. W. Connelly, eds. *Studies in Avian
28 Biology.* Vol. 38. pp. 53–67. ML12250A648. Berkeley, California: University of California
29 Press. 2011.
30 <http://wdfw.wa.gov/publications/01310/wdfw01310.pdf> (22 August 2012).
31
32 Doherty, K.E., D.E. Naugle, H. Copeland, A. Pocewicz, and J. Kiesecke. "Energy Development
33 and Conservation Tradeoffs: Systematic Planning for Sage-Grouse in Their Eastern Range." In
34 *Greater Sage-Grouse: Ecology and Conservation of a Landscape Species and Its Habitats.*
35 S. T. Knick and J. W. Connelly, eds. *Studies in Avian Biology.* Vol. 38, pp. 505-516.
36 ML12250A651. Berkeley, California: University of California Press. 2011.
37 <http://www.blm.gov/pgdata/etc/ medialib/blm/wy/programs/energy/og/leasing/protests/
38 2012/may/Appeals.Par.95895.File.dat/AUDExb8.pdf> (22 August 2012).
39
40 Driscoll, D.G., J.M. Carter, J.E. Williamson, and L.D. Putnam. "Hydrology of the Black Hills
41 Area, South Dakota." U.S. Geological Survey Water Resources Investigation Report 02-4094.
42 ML12240A218. 2002. <http://pubs.usgs.gov/wri/wri024094/pdf/mainbodyofreport-1.pdf>
43 (22 December 2009).
44
45 ESRI (Environmental Systems Research Institute). "ArcGIS 9 Media Kit, ESRI Data and
46 Maps 9.3." Redlands, California: ESRI. 2008.
47
48 GCRP (U.S. Global Change Research Program). *Global Climate Change Impacts in the United
49 States.* Washington, DC: Cambridge University Press. 2009.
50

1   Hodorff. "Habitat Assessment and Conservation Strategy for Sage Grouse and Other Selected
2   Species on Buffalo Gap National Grassland." ML120240626. Hot Springs, South Dakota:
3   U.S. Department of Agriculture, Forest Service. September 2005.
4
5   Holm, E.H., T. Cline, Jr., and M. Lees. "South Dakota—2008 Mineral Summary Production,
6   Exploration, and Environmental Issues." ML12243A352. Pierre, South Dakota: South Dakota
7   Department of Environment and Natural Resources, Minerals and Mining Program. 2008.
8   <http://denr.sd.gov/ documents/Goldrpt08a.pdf> (20 August 2012).
9
10 National Atlas of the United States. "Map of the United States." September 17, 2009.
11 <http://nationalatlas.gov> (29 October 2010).
12
13 Naus, C.A., D.G. Driscoll, and J.M. Carter. "Geochemistry of the Madison and Minnelusa
14 Aquifers in the Black Hills Area, South Dakota." ML12240A265. U.S. Geological Survey Water
15 Resources Investigation Report 01-4129. 2001. <http://pubs.usgs.gov/wri/wri014129/
16 pdf/wri014129.pdf> (17 August 2012).
17
18 NOGCC (Nebraska Oil and Gas Conservation Commission). "Oil and Gas Conservation."
19 2012. <http://nogcc.ne.gov> (22 August 2012).
20
21 NRC (U.S. Nuclear Regulatory Commission. "Expected New Uranium Recovery Facility
22 Applications/Restarts/Expansions: Updated August 13, 2012." ML12243A367. 2012.
23 <http://www.nrc.gov/ materials/uranium-recovery/license-apps/ur-projects-list-public.pdf>
24 (20 August 2012).
25
26 NRC. NUREG–1910, "Generic Environmental Impact Statement for *In-Situ* Leach Uranium
27 Milling Facilities." ML091480244, ML091480188. Washington, DC: NRC. May 2009a.
28
29 NRC. "Site Visit to the Proposed Dewey-Burdock Uranium Project, Fall River and Custer
30 Counties, South Dakota, and Meetings with Federal, State, and County Agencies, and Local
31 Organizations, November 30–December 4, 2009." ML093631627. Washington, DC: NRC.
32 2009b.
33
34 Powertech [Powertech (USA) Inc.]. "Dewey-Burdock Project Emissions Inventory Revisions."
35 E-mail (July 31) to B. Werling, Southwest Research Institute® from R. Blubaugh.
36 ML12216A220. Greenwood Village, Colorado: Powertech. 2012.
37
38 Powertech. "Dewey-Burdock Project, Application for NRC Uranium Recovery License Fall River
39 and Custer Counties, South Dakota." Technical Report RAI Responses. ML112071064.
40 Greenwood Village, Colorado: Powertech. June 2011.
41
42 Powertech. "Dewey-Burdock Project, Application for NRC Uranium Recovery License Fall River
43 and Custer Counties, South Dakota ER_RAI Response August 11, 2010." ML102380516.
44 Greenwood Village, Colorado: Powertech. August 2010.
45
46 Powertech. "Dewey-Burdock Project, Application for NRC Uranium Recovery License Fall River
47 and Custer Counties, South Dakota—Environmental Report." Docket No. 040-09075.
48 ML092870160. Greenwood Village, Colorado: Powertech. August 2009a.
49

1   Powertech. "Dewey-Burdock Project, Application for NRC Uranium Recovery License Fall River
2   and Custer Counties, South Dakota—Technical Report." Docket No. 040-09075.
3   ML092870160. Greenwood Village, Colorado: Powertech. August 2009b.
4
5   Powertech. "Dewey-Burdock Project, Supplement to Application for NRC Uranium Recovery
6   License Dated February 2009." Docket No. 040-09075. ML092870160. Greenwood Village,
7   Colorado: Powertech. August 2009c.
8
9   SDDENR (South Dakota Department of Environment and Natural Resources). "Online
10 Oil/Gas/Injection Well Data." Rapid City, South Dakota: Minerals and Mining Program, Oil and
11 Gas. 2012a. <http://www.sddenr.net/oil_gas/> (07 August 2012).
12
13 SDDENR. "Oil and Gas Drilling Permits Issued From 2005-2011." South Dakota Department of
14 Natural Resources. 2012b. <http://denr.sd.gov/des/og/newpermit.aspx> (21 August 2012).
15
16 SDDENR. "South Dakota's Regional Haze State Implementation Program." ML12243A371.
17 2011. <http://denr.sd.gov/des/aq/aqnews/RHSIP20110818.pdf> (31 May 2012).
18
19 SDDENR. "Uranium Question and Answer Fact Sheet." ML12243A369. 2010.
20 <http://denr.sd.gov/powertech/ UraniumQuestion&AnswerSheet92508.pdf>
21 (30 November 2010).
22
23 SDDENR. "South Dakota Ambient Air Monitoring Annual Network Plan 2009." 2009.
24 <http://denr.sd.gov/des/aq/aqnews/South%20Dakota%20AP2009.pdf> (14 December 2010).
25
26 SDDENR. "The 2008 South Dakota Integrated Report for Surface Water Quality Assessment."
27 ML12240A378. Pierre, South Dakota: 2008. <http://denr.sd.gov/documents/08IRFinal.pdf>
28 (05 November 2009).
29
30 SDDLR (South Dakota Department of Labor and Regulation). "South Dakota Industry
31 Employment Projections." Pierre, South Dakota: SDDLR Labor Market Information Center.
32 2012. <http://dlr.sd.gov/lmic/industry_projections.aspx> (25 June 2012).
33
34 STB (Surface Transportation Board). "Final Environmental Impact Statement, Finance Docket
35 No. 33407—Dakota, Minnesota & Eastern Railroad Corporation, Construction into the Powder
36 River Basin, Powder River Basin Expansion Project." ML12243A381. Washington, DC:
37 Surface Transportation Board, Section of Environmental Analysis. 2001.
38 <http://stb.dot.gov/stb/docs/dme/feis/Ch03-1.pdf> (27 May 2010).
39
40 Sundstrom, L. "Living on the Edge: Archaeological and Geomorphological Investigations in the
41 Vicinity of Tepee and Hell Canyons, Western Custer County, South Dakota." Day Star
42 Research, Shorewood, Wisconsin. 1999.
43
44 Sutley, N. "Draft NEPA Guidance on Consideration of Effects of Climate Change and
45 Greenhouse Gas Emissions." Memorandum (February 18) to Heads of Federal Departments
46 and Agencies. Washington, DC: Council on Environmental Quality. 2010.
47
48 USCB (U.S. Census Bureau). "American FactFinder, Census 2000 and 2010, 2006–2010
49 American Community Survey 5-Year Estimate, State and County QuickFacts." ML12248A240.
50 2012. <http://factfinder.census.gov, http://quickfacts.census.gov> (17 April 2012).
51

1   USGS (U.S. Geological Survey). "Black Hills Hydrology Study." Pierre, South Dakota: USGS,
2   South Dakota Water Science Center. 2010. <http://sd.water.usgs.gov/projects/bhhs/Intro.html>
3   (09 November 2010).
4
5   USGS. "Scientific Information for Greater Sage-Grouse and Sagebrush Habitats."
6   U.S. Department of Interior, United States Geological Survey, Briefing Paper. ML12250A713.
7   September 29, 2009. <http://sagemap.wr.usgs.gov/Docs/SAGRBriefingPaper1.pdf>
8   (22 August 2012).
9
10   WDAI (Wyoming Department of Administration and Information). "Population of Wyoming,
11   Counties and Municipalities: 1980 to 1990." ML12250A719. Cheyenne, Wyoming: WDAI,
12   Economic Analysis Division. 2012. <http://eadiv.state.wy.us/pop/C&SC8090.pdf>
13   (25 June 2012).
14
15   WDAI. "Population of Wyoming, Counties, Cities, and Towns: 2010 to 2030 ." ML12250A716.
16   Cheyenne, Wyoming: WDAI, Economic Analysis Division. 2011. <http://eadiv.state.wy.us/
17   pop/wyc&sc30.htm> (25 June 2012).
18
19   WDWS (Wyoming Department of Workforce Services Research and Planning). "Wyoming
20   Occupational Projections, 2011-2021." ML12243A386. Cheyenne, Wyoming: WDWS
21   Research and Planning, 2012. <http://doe.state.wy.us/lmi/projections/
22   WY_Occ_Proj_2011_2021.pdf> (25 June 2012).
23
24   Winham, R.P., L. Palmer, F. Sellet, and E.J. Lueck. "Intensive (Class III) Cultural Resouces
25   Inventory Survey of the Dacotah Cement Land Exchange Proposal with the Bureau of Land
26   Management in Southwestern Custer County, South Dakota." Volumes 1–6. Prepared by
27   Archeology Laboratory, Augustana College, Sioux Falls, South Dakota. 2001.
28
29   WYOGCC (Wyoming Oil and Gas Conservation Commission). "County Reports." Casper,
30   Wyoming: WYOGCC. 2012. <http://wogcc.state.wy.us/> (07 August 2012).

6 MITIGATION

6.1 Introduction

The Generic Environmental Impact Statement (GEIS) for *In-Situ* Leach Uranium Milling Facilities (NRC, 2009) described potential mitigation measures that a licensee or facility operator might use to reduce potential adverse impacts associated with construction, operation, aquifer restoration, and decommissioning of an *in-situ* recovery (ISR) milling facility. Under 40 CFR 1508.20, the Council on Environmental Quality defines mitigation to include activities that (i) avoid the impact altogether by not taking a certain action or parts of a certain action; (ii) minimize impacts by limiting the degree or magnitude of the action and its implementation; (iii) rectify the impact by repairing, rehabilitating, or restoring the affected environment; (iv) reduce or eliminate the impact over time by preservation and maintenance operations during the life of the action; and (v) compensate for the impact by replacing or providing substitute resources or environments.

Mitigation measures are those actions or processes that will be implemented to control and minimize potential adverse impacts from construction, operation, aquifer restoration, and decommissioning of the proposed Dewey-Burdock ISR Project. Potential mitigation measures can include general best management practices (BMPs) and more site-specific management actions.

BMPs are processes, techniques, procedures, or considerations that can be used to effectively avoid or reduce potential environmental impacts. While best management practices are not regulatory requirements, they can overlap and support such requirements. BMPs will not replace any U.S. Nuclear Regulatory Commission (NRC) requirements or other federal, state, or local regulations.

Management actions are active measures that a licensee or facility operator specifically implements to reduce potential adverse impacts to a specific resource area. These actions include compliance with applicable government agency stipulations or specific guidance, coordination with governmental agencies or interested parties, and monitoring of relevant ongoing and future activities. If appropriate, corrective actions could be implemented to limit the degree or magnitude of a specific action leading to an adverse impact (reducing or eliminating the impact over time by preservation and maintenance operations) and repairing, rehabilitating, or restoring the affected environment. The licensee may also minimize potential adverse impacts by implementing specific management actions such as programs, procedures, and controls for monitoring, measuring, and documenting specific goals or targets (for example, pollution prevention goals of reducing waste) and, if appropriate, instituting corrective actions. The management actions may be established through standard operating procedures that appropriate local, state, and federal agencies (including NRC) review and approve. NRC may also establish requirements for management actions by identifying license conditions. These conditions are written specifically into the NRC source and byproduct material license and then become commitments that are enforced through periodic NRC inspections.

The mitigation measures Powertech (USA) Inc. (Powertech) proposed to reduce and minimize adverse environmental impacts at the proposed Dewey-Burdock ISR Project are summarized in Section 6.2. Based on the potential impacts identified in Chapter 4 of this draft SEIS, the NRC staff have identified additional potential mitigation measures for the proposed Dewey-Burdock ISR Project. These mitigation measures are summarized in Section 6.3. The proposed

1   mitigation measures provided in this chapter do not include environmental monitoring activities.
2   Environmental monitoring activities are described in Chapter 7 of this draft SEIS.
3
4   ## 6.2   Mitigation Measures Proposed by Powertech
5
6   The applicant identified mitigation measures in its license application (Powertech, 2009a–c) as
7   well as in response to NRC staff requests for additional information (Powertech, 2010a–c,
8   2011). Table 6.2-1 lists the mitigation measures proposed for each resource area. Because
9   many of the applicant's proposed mitigation measures apply to all four phases of the ISR
10 process, they are listed together in the table.

Table 6.2-1. Summary of Mitigation Measures Proposed by Powertech

Resource Area	Activity	Proposed Mitigation Measures
Land Use	Land disturbance	Reclaim the surface and reestablish vegetation in areas disturbed by drilling, pipeline installation, and facility construction as soon as construction activities are completed.
		Minimize construction of new and secondary access roads.
		Restrict normal vehicular traffic to designated roads, and keep traffic in wellfields to a minimum.
		Develop wellfields sequentially, and restore and reclaim wellfields in interim steps to minimize land area impacted at any one time.
	Access restrictions	Construct fences and signage around processing facilities, radium settling and storage ponds, and potential land application areas.
		Construct temporary fencing around injection and production wellfield patterns (remove fencing after operations and reclamation of each wellfield is completed).
		Limit access to monitoring wells, Class V deep injection wells, and header houses by (i) covering each monitoring well with a locking device, (ii) securing the well head and pumping equipment for Class V injection wells within locked buildings, and (iii) securing header houses within the fenced area of the wellfield.
		Implement fencing construction techniques to minimize habitat alteration and impediments to large game migration.
		Work with BLM, SDGFP, and private landowners to limit recreational activities (primarily hunting) within the project area to the extent practicable.

Table 6.2-1. Summary of Mitigation Measures Proposed by Powertech (continued)

Resource Area	Activity	Proposed Mitigation Measures
Transportation	Transportation safety	Maintain access roads, and impose speed limits on unpaved roads to minimize or eliminate accidents.
		Comply with all applicable NRC and DOT packaging and transportation requirements for all shipments of yellowcake, process chemicals, ion-exchange resins, fuel, and radioactive materials to mitigate the potential impacts of a transportation accident.
		Use dedicated tanker trucks for transporting uranium-loaded or uranium-stripped resins between the central processing plant and satellite facilities.
		Survey the exterior and cab of the shipping truck for radiological contamination prior to each shipment of uranium-loaded or uranium-stripped resin or yellowcake.
		Equip both the transport vehicle and shipping facilities with communication devices that allow direct communication with Powertech (USA) personnel.
	Emergency response	Communicate with local and state authorities on transportation and emergency response procedures.
		Use standard operating procedures for transportation and emergency response.
		Require proper training for transport contractor personnel on transportation accident response based on the specific material(s) shipped. Written standard operating procedures would accompany all drivers to ensure proper response to accidents and spill containment.
		Supply both shipping and receiving facilities with emergency response kits.
		Ensure each resin or yellowcake transport vehicle carries an emergency spill kit that would help contain material in the event of a spill.
		Maintain shipping records (bill of lading) to identify the characteristics and quantity of material shipped.
		Notify NRC if a radiological accident occurs pursuant to requirements of 10 CFR Part 20 §2202 and §2203.

1

1

Table 6.2-1. Summary of Mitigation Measures Proposed by Powertech (continued)

Resource Area	Activity	Proposed Mitigation Measures
Geology and Soils	Soil disturbance and contamination	Salvage and stockpile soil from disturbed areas. Reestablish temporary or permanent native vegetation as soon as possible after disturbance utilizing the latest technologies in reseeding and sprigging, such as hydroseeding. Decrease runoff from disturbed areas by using structures to temporarily divert and/or dissipate surface runoff from undisturbed areas. Retain sediment within the disturbed areas by using silt fencing, retention ponds, and hay bales. Fill pipeline and cable trenches with appropriate material, and regrade surface soon after completion. Design drainages to minimize potential for erosion by creating slopes less than 4 to 1, and/or provide rip-rap or other soil stabilization controls. Construct roads using techniques that will minimize erosion, such as surfacing with a gravel road base, building stream crossings at right angles with adequate embankment protection and culvert installation. Use a spill prevention and cleanup plan to minimize soil contamination from vehicle accidents and/or wellfield spills or leaks. Collect and monitor soils and sediments for potential contamination including areas used for land application of treated wastewater, transport routes for yellowcake and ion exchange resins, and wellfield areas where spills or leaks are possible. Treat liquid wastes applied to land application areas to comply with release standards for radiological constituents in 10 CFR Part 20, Appendix B. Obtain an SDDENR groundwater discharge plan permit, and comply with applicable state discharge requirements for land application of treated liquid wastes.

2

1

Table 6.2-1. Summary of Mitigation Measures Proposed by Powertech (continued)

Resource Area	Activity	Proposed Mitigation Measures
Surface Water Resources	Erosion, runoff, and sedimentation	Refrain from consuming or discharging to surface waters. Obtain USACE permits and authorization from SDDENR when filling and crossing jurisdictional waters. Obtain construction and industrial NPDES permits in accordance with SDDENR regulations, and implement mitigation measures to control erosion, runoff, and sedimentation. Construct the Burdock central plant and Dewey satellite facility and their supporting buildings outside the 100-year floodplain of Pass and Beaver Creeks and away from their tributaries. Construct a system of structures such as straw bales, collector ditches, and engineered diversion structures or berms to protect facilities and infrastructures (e.g., storage ponds, access roads, plant-to-plant pipelines, wellfields) that will be located within the 100-year inundation boundary to protect them from flood damage. Implement a storm water management plan (SWMP) in accordance with SDDENR requirements to ensure that surface water runoff from disturbed areas meets NPDES permit limits. Avoid earthmoving activities at the proposed land-application sites. Divert potential runoff produced by snowmelt or precipitation in land application areas to adjacent catchment areas.
	Spills and leaks	Recontour land surface to restore surface drainage to blend with the natural terrain after completion of the proposed ISR project. Develop and implement emergency response procedures to correct and remediate accidental spills. Place liners, underdrains, and leak detection systems underneath settling and holding ponds. Bury pipelines to avoid freezing, and monitor pipeline pressures for leak detection.

2

1

Table 6.2-1. Summary of Mitigation Measures Proposed by Powertech (continued)

Resource Area	Activity	Proposed Mitigation Measures
Groundwater Resources	Water use	Obtain Class III UIC permit and aquifer exemption. Obtain Class V UIC permit for deep well disposal of treated liquid wastes, and monitor process effluents injected into Class V deep injections wells to comply with (i) release standards in 10 CFR Part 20, Subparts D and K and Appendix B and (ii) the drinking water standards if proposed injection zones are underground sources of drinking water (have total dissolved solids concentrations below 10,000 mg/L). Treat liquid wastes applied to land application areas to comply with release standards for radiological constituents in 10 CFR Part 20, Appendix B. Obtain an SDDENR groundwater discharge permit, and comply with applicable state discharge requirements for land application of treated liquid wastes. Obtain water appropriation permit to access groundwater from the Madison aquifer. Monitor private domestic, livestock, and agricultural wells as appropriate during operations, and provide alternative sources of water to landowners in the event of significant drawdown to wells within and adjacent to the proposed project area.
	Spills and leaks	Obtain construction and industrial NPDES permits from SDDENR, which require reporting of spills of petroleum products or hazardous chemicals. Implement a spill prevention and cleanup plan to minimize impacts to soils and groundwater, including rapid response cleanup and remediation. Place liners, underdrains, and leak detection systems underneath settling and holding ponds to prevent potential infiltration of liquid waste into soil and shallow aquifers. Bury pipelines to avoid freezing, and monitor pipeline pressures for leak detection.

2

1
2
3
4

Table 6.2-1. Summary of Mitigation Measures Proposed by Powertech (continued)

Resource Area	Activity	Proposed Mitigation Measures
	Excursions	Conduct periodic mechanical integrity testing at the injection, production, and monitoring wells to limit the likelihood of well integrity failure during operations.
		Collect detailed lithologic and hydrogeological data in each proposed wellfield prior to ISR operations to ensure hydraulic control of the production zone.
		Maintain production bleed rate at 0.5 to 3 percent to prevent lixiviant excursions.
		Conduct ISR operations only in confined portions of production aquifers.
		Install monitoring wells within and encircling the production zone for early detection of potential horizontal excursions.
		Install monitoring wells in aquifers above and below the production aquifer for early detection of potential vertical excursions.
		Implement corrective actions, and provide required notifications and reports to NRC in the event of an excursion.
	Restoration/reclamation	Submit wellfield operational plans including well layouts for NRC and EPA approval before conducting operations in wellfields.
		Return groundwater quality in the production zone to NRC-approved groundwater protection standards upon completion of ISR operations as required by 10 CFR Part 40, Appendix A, Criterion 5B(5).
		Plug and abandon all monitoring, injection, and production wells in accordance with applicable federal and state regulations, as part of decommissioning activities.

5

1

Table 6.2-1. Summary of Mitigation Measures Proposed by Powertech (continued)

Resource Area	Activity	Proposed Mitigation Measures
Ecology		Follow the Land Use mitigation measures for land disturbance activities and access restrictions, which will also minimize impacts to vegetation and wildlife.
	Restoration/reclamation	Minimize disturbance of surface areas and vegetation, where possible (also benefits wildlife).
		Construct new roads, power lines, and pipelines in the same corridors to the extent possible to reduce overall disturbance and minimize new surface disturbance (also benefits wildlife).
		Impose dust control measures as described under Air Quality to limit dust deposition on vegetation, both on- and offsite, affecting the forageability for obligate species.
		Implement weed control as needed to limit the spread of noxious, invasive, and nonnative species on disturbed areas.
		Reestablish temporary or permanent native vegetation as soon as possible after disturbance.
		Minimize the spread of undesirable, invasive, and nonnative species (weeds) in disturbed areas.
	Transmission Lines	Construct new overhead power lines using BMPs to reduce bird injuries and mortalities.
	Reduce Human Disturbances	Enforce speed limits to minimize collisions with wildlife.
		Use existing roads when possible, and limit construction of new primary and secondary roads to provide access to more than one drill site to minimize wildlife and habitat disturbance.
		Restore diverse landforms; direct topsoil replacement; and construct brush piles, snags, and/or rock piles to enhance habitat for wildlife.
		Prepare FWS-approved raptor monitoring and mitigation plan to minimize conflicts between active nest sites and project-related activities if direct impacts to raptors occur.

2

1

Table 6.2-1. Summary of Mitigation Measures Proposed by Powertech (continued)

Resource Area	Activity	Proposed Mitigation Measures
Air Quality	Fugitive dust and combustion emissions from construction equipment and vehicles	Use drill rigs with engines no larger than 300 horsepower (except for deep well drill rig) to limit combustion emissions. Use Tier 1 or higher drill rig engines and Tier 3 or higher construction equipment engines (see SEIS Section 4.7.1.1.1 for an explanation of "Tiers") to limit combustion emissions. Spray water to mitigate fugitive dust accounting for a 50 percent reduction in emissions generated from onsite unpaved roads. Impose speed limits for travel on unpaved roads and areas. Encourage carpooling. Restore or reseed disturbed areas promptly to limit the exposed/disturbed area at any given time. Coordinate construction and transportation activities to reduce maximum dust levels. Maintain vehicles to meet applicable EPA emission standards.
Noise	Exposure of workers and public to noise	Avoid construction activities during the night. Use sound abatement controls on operating equipment and facilities. Use personal hearing protection for workers in high noise areas. Adhere to FWS and SDGFP seasonal noise, vehicular traffic, and human proximity guidelines to limit noise impacts to raptors. Locate all planned facilities outside of BLM-=recommended buffer zones of raptor nests identified within the project area. Follow an FWS-approved raptor monitoring and mitigation plan to reduce conflicts between active raptor nests and project-related activities.

2

1

Table 6.2-1. Summary of Mitigation Measures Proposed by Powertech (continued)

Resource Area	Activity	Proposed Mitigation Measures
Cultural and Historic Resources	Disturbance of prehistoric archaeological sites and sites eligible for listing on the National Register of Historic Places (NRHP)	Conduct appropriate historic and cultural resource surveys as part of prelicensing application activities and eligibility evaluation of cultural resources for listing on the NRHP under criteria in 36 CFR 60.4(a)–(d). Conduct consultation under Section 106 of the National Historic Preservation Act (NHPA) with NRC, South Dakota State Historic Preservation Office (SD SHPO), other government agencies (e.g., FWS, EPA, and BLM), and Native American tribes.
Visual and Scenic	Potential visual intrusions in the existing landscape character	Cover wellheads with low structures that present low contrast with existing landscape. Reclaim disturbed areas, and remove debris after construction is complete. Remove and reclaim roads and structures after operations are complete. Select building materials and paint that complement the natural environment. Consider landscape topography to conceal wellheads, plant facilities, access roads, potential land application areas, and other areas of disturbance from public vantage points. Use standard dust control measures including water application, speed limits, and coordinating dust-producing activities to reduce fugitive dust impacts. Consider using exterior lighting only where needed, limiting the height of exterior lighting units, and using shielded or directional lighting to limit lighting to where it is needed.
Socioeconomics	Effects on surrounding communities	Preferentially source the labor force from the surrounding region to reduce any burden on public services and community infrastructure (e.g., housing, schools) in nearby towns.

2

1

Table 6.2-1. Summary of Mitigation Measures Proposed by Powertech (continued)

Resource Area	Activity	Proposed Mitigation Measures
Occupational and Public Health and Safety	Effects from facility construction	Implement standard dust control measures, such as water application and speed limits, to reduce and control fugitive dust emissions.
	Effects from facility operation	Comply with federal and state occupational safety regulations to limit nonradiological impacts of fugitive dust and diesel emissions to acceptable levels.
		Reduce radiological exposure to workers by (i) installing ventilation designed to limit worker exposure to radon; (ii) installing gamma exposure rate monitors, air particulate monitors, radon daughter product monitors to verify that expected radiation levels are not exceeded; and (iii) conducting work area radiation and contamination surveys.
		Use vacuum dryer technology during normal operations to limit radiological emissions other than radon gas.
		Comply with an NRC-approved Radiation Protection Program that would include routine radiation surveys, respiratory protection, standard operating procedures for spill response and cleanup, and worker training in radiological health and emergency response.
		Monitor radiation workers via use of dosimeters and area air sampling to ensure that radiological doses remain within regulatory limits and as low as reasonably achievable (ALARA).
		Implement engineering controls, such as concrete curbs and sumps, to contain process spills resulting from accidents.
		Comply with applicable EPA, OSHA, and SDDENR regulations concerning the use, inspection, and storage of hazardous and nonhazardous chemicals.
		Develop and implement standard operating procedures regarding receiving, storing, handling, and disposing of chemicals.

2

1

Table 6.2-1. Summary of Mitigation Measures Proposed by Powertech (continued)

Resource Area	Activity	Proposed Mitigation Measures
Waste Management	Disposal Capacity	Establish a solid byproduct material disposal agreement with a licensed facility prior to the start of operations.
	Waste Reduction	Recycle wastewater to reduce the amount of water needed for facilities and the amount of wastewater that could require disposal.
		Use decontamination techniques that reduce waste generation.
		Institute preventative maintenance and inventory management programs to minimize waste from breakdowns and overstocking.
		Recycle nonradioactive materials where appropriate.
		Salvage extra materials, and use them for other construction activities.
		Encourage the reuse of materials and use of recycled materials.
	Waste Storage and Containment	Avoid using hazardous materials when possible.
		Store and properly label solid byproduct material onsite to prevent any potential release. Isolate byproduct material inside a restricted area until a full shipment can be transferred to an NRC-approved disposal site.
		Install curbs or berms on all waste storage areas.
		Install leak detection and warning systems in all liquid waste facilities.
		Develop a spill prevention plan for petroleum products and other hazardous materials.
		Ensure that equipment is available to respond to spills, and identify the location of such equipment. Inspect and replace worn or damaged components.

2

3 ## 6.3   Potential Mitigation Measures Identified by NRC

4

5   The NRC staff have reviewed the mitigation measures the applicant proposed and have
6   identified additional mitigation measures that could potentially reduce impacts (Table 6.3-1).
7   NRC has the authority to address unique site-specific characteristics by identifying license
8   conditions based on conclusions reached in the safety and environmental reviews. These

Table 6.3-1. Summary of Mitigation Measures Identified by NRC

Resource Area	Activity	Proposed Mitigation Measures
Land Use	Land disturbance	Monitor and control potential irrigation areas, if used, to maintain levels of radioactive constituents in treated liquid wastes applied to land application areas to within allowable release limits to protect the agricultural and recreational integrity of the land. Use BMPs to control waste disposal, erosion, and runoff to limit the effect of facility operation on surrounding land use.
Transportation	Transportation safety	Use accepted industry codes and standards for handling and transporting hazardous chemicals. Implement safe driving training for personnel and truck drivers. Use check-in/check-out or global positioning satellite technology to track shipments. Construct turn lanes in both directions on Dewey Road for vehicles turning onto the main access roads to the central and satellite processing plants. Provide means of advance warning to oncoming traffic that large trucks are entering Dewey Road from site access roads (e.g., signage, flashing light, flagman).
Geology and Soils	Soils	Maintain a log of all spills occurring at the site whether or not these spills are reportable to NRC per 10 CFR 40.60. Implement alternatives or mitigation measures to manage drilling fluid during well drilling operations including (i) lining mud pits with an impermeable membrane, (ii) disposing of potentially contaminated drilling mud and other fluids offsite, and (iii) using portable tanks or tubs to contain drilling mud and other fluids.
Surface Water Resources	Water quality	Collect quarterly preoperational water quality samples from surface waters.
Groundwater Resources	Contamination and excursions	Locate all boreholes and wells within 305 m [1,000 ft] of a wellfield, if possible, and properly plug and abandon them. Submit results of the hydrogeological characterization and aquifer pump tests (hydrologic

1

1

Table 6.3-1. Summary of Mitigation Measures Identified by NRC (continued)

Resource Area	Activity	Proposed Mitigation Measures
		test data packages) for NRC review and approval prior to development of any proposed wellfields.
		Prior to ISR operations in partially saturated portions of the Chilson aquifer, require the applicant to demonstrate the ability to detect and remediate excursions in partially saturated production zones.
		Monitor potential mobilization and migration of contaminants from abandoned open pit mines into production zones during aquifer restoration.
Ecology	Restoration/reclamation	Use weed control techniques that incorporate BMPs approved by BLM and SDDNER.
	Fencing and screening	Cover vent pipes with either netting or other methods to prevent bats, birds, or small mammals from being trapped.
	Transmission lines	Follow the Avian Power Line Interaction Committee guidance to avoid impacts (electrocution and perching) to birds, especially prior to the fledging of young (Avian Power Line Interaction Committee, 2006).
		Bury transmission lines after (step-down) transforming to minimize risks to raptors and large birds.
	Reduce Human Disturbances	Adhere to BLM timing and distance restrictions provided in SEIS Table 4.6-3.
		Avoid drilling activity in Sections 29 and 30 T6S-R1E between February 1 and August 31 annually to prevent disruption of the active bald eagle nest and redtail hawk nest in the vicinity of these sections. Require the applicant to contact SDGFP if exploration activity is conducted between February 1 and August 31 in Sections 30 and 29 T6S-R1E so that additional distance and/or timing restrictions may be issued.
		Allow snakes and lizards that are encountered to retreat.

2
3
4
5

Table 6.3-1. Summary of Mitigation Measures Identified by NRC (continued)

Resource Area	Activity	Proposed Mitigation Measures
		Inform employees of applicable wildlife laws and penalties associated with unlawful taking and harassment of wildlife. Train employees on (i) the types of wildlife in the area susceptible to collisions with motor vehicles, (ii) the circumstances when collisions are most likely to occur, and (iii) measures that should be taken to avoid wildlife–vehicle collisions. Sign and gate as needed all new and improved roads related to the proposed project to minimize public traffic. Comply with applicable state and local requirements to design or treat mud pits and ponds to prevent the development of favorable mosquito habitat (to reduce possible transmission of West Nile virus).
Air Quality	Fugitive dust and combustion emissions from construction equipment and vehicles	Implement fuel saving practices such as minimizing vehicle and equipment idle time. Utilize fossil-fuel vehicles that meet the latest emission standards. Utilize newer, cleaner running equipment. Minimize unnecessary travel. Ensure that diesel-powered construction equipment and drill rigs are properly tuned and maintained. Limit access to construction sites, staging areas, and wellfields to authorized vehicles only, through designated treated roads. Pave or put gravel on dirt roads and parking lots if appropriate. Cover trucks carrying soil and debris to reduce dust emissions from the back of trucks. Burn low-sulfur fuels in all diesel engines and generators. Train workers to comply with the speed limit, use good engineering practices, minimize disturbed areas, and employ other BMPs as appropriate.

1

Table 6.3-1. Summary of Mitigation Measures Identified by NRC (continued)

Resource Area	Activity	Proposed Mitigation Measures
		To the extent practicable, avoid conducting soil-disturbing activities and travel on unpaved roads during periods of unfavorable meteorological conditions (e.g., high winds). Implement any permit conditions identified in the SDDENR air permit, if applicable. Limit the numbers of hours in a day that effluent-generating activities can be conducted. Perform road maintenance (i.e., promptly remove earthen material on paved roads). Apply erosion mitigation methods on disturbed lands.
Noise	Exposure of workers and the public to noise	Maintain noise levels in work areas to below OSHA regulatory limits. Reduce noise levels generated by irrigation equipment in potential land application areas by (i) installing exhaust and inlet silencers on engines, (ii) using electric motor drives instead of internal combustion engines, and (iii) erecting acoustic barriers to block the line of hearing from the exhaust engine and inlet toward human and wildlife receptors.
Cultural and Historic Resources	Disturbance of prehistoric archaeological sites and sites eligible for listing on the National Register of Historic Places (NRHP)	Stop work upon discovery of previously undocumented historic and cultural resources, and notify appropriate federal, tribal, and state agencies with regard to mitigation measures. Avoid historic properties within the project area that are currently listed or eligible for listing on the NRHP. Avoid identified sites within the project area with burial or cairn features. Develop an agreement between all interested parties outlining the mitigation process for each affected resource and why sites cannot be avoided, if required. Prior to construction, develop an Unexpected Discovery Plan that will outline the steps required in

1

1

Table 6.3-1. Summary of Mitigation Measures Identified by NRC (continued)

Resource Area	Activity	Proposed Mitigation Measures
		the event that unexpected historical and cultural resources are encountered at the site. Submit a decommissioning plan for NRC review to ensure compliance with Section 106 of NHPA during the decommissioning phase.
Visual and Scenic	Potential visual intrusions in the existing landscape character	Limit the number of drill rigs operating during wellfield construction. To the extent possible, use existing secondary roads within the project area to access wellfields, potential irrigation areas, and other facility infrastructure.
Socioeconomics	Effects on surrounding communities	Coordinate emergency response activities with local authorities, fire departments, medical facilities, and other emergency services before operations begin.
Occupational and Public Health and Safety	Effects from facility operation	Use high-efficiency particulate air filters or similar controls for particulates. Design task procedures to reduce potential accidents. Develop contingency plans with county and municipal governments to ensure adequate medical, fire, and emergency services are available in case of a major accident.
Waste Management	Disposal Capacity	Dispose of decommissioning nonhazardous solid waste at the Rapid City landfill in the event that the disposal capacities of local landfills are limited or otherwise unavailable at the time of decommissioning.

2
3   license conditions could include additional mitigation measures, such as modifications to
4   required monitoring programs. The NRC staff is conducting the safety review of the proposed
5   Dewey-Burdock ISR Project, which will be documented in a Safety Evaluation Report, and
6   license conditions resulting from the safety review will be included as part of the final SEIS.
7   While NRC cannot impose mitigation outside its regulatory authority under the Atomic Energy
8   Act, the NRC staff have identified mitigation measures in Table 6.3-1 that could potentially
9   reduce the impacts of the proposed Dewey-Burdock ISR Project. These additional mitigation
10 measures are not requirements being imposed upon the applicant. For the purposes of NEPA,
11 and consistent with 10 CFR 51.71(d) and 51.80(a), NRC is disclosing measures that could
12 potentially reduce or avoid environmental impacts of the proposed project.
13
14

6.4   References

10 CFR Part 20. Code of Federal Regulations, Title 10, *Energy*, Part 20. "*Standards for Protection Against Radiation.*" Washington, DC: U.S. Government Printing Office.

10 CFR Part 40. Code of Federal Regulations, Title 10, *Energy*, Part 40. "*Domestic Licensing of Source Material.*" Washington, DC: U.S. Government Printing Office.

10 CFR Part 40. Appendix A. Code of Federal Regulations, Title 10, *Energy*, Part 40. Appendix A. "*Criteria Relating to the Operation of Uranium Mills and to the Disposition of Tailings or Wastes Produced by the Extraction and Concentration of Source Material from Ores Processed Primarily from their Source Material Content.*" Washington, DC: U.S. Government Printing Office.

10 CFR Part 51. Code of Federal Regulations, Title 10, *Energy*, Part 51. "*Environmental Protection Regulations for Domestic Licensing and Related Regulatory Functions.*" Washington, DC: U.S. Government Printing Office.

36 CFR Part 60. Code of Federal Regulations, Title 36, *Parks, Forests, and Public Property*, Part 60. "*National Register of Historic Places.*" Washington, DC: U.S. Government Printing Office.

40 CFR Part 1508. Code of Federal Regulations, Title 40, *Protection of the Environment*, Part 1508. "*Terminology and Index.*" Washington, DC: U.S. Government Printing Office.

Avian Power Line Interaction Committee. "Suggested Practices for Avian Protection on Power Lines: The State of the Art in 2006." ML12243A391. Washington, DC: Edison Electric Institute and Sacramento, California: Avian Power Line Interaction Committee and the California Energy Commission. 2006. <http://www.aplic.org/uploads/files/2643/SuggestedPractices2006(LR-2).pdf> (13 October 2009).

NRC (U.S. Nuclear Regulatory Commission). NUREG–1910, "Generic Environmental Impact Statement for *In-Situ* Leach Uranium Milling Facilities." ML091480244, ML091480188. Washington, DC: NRC. May 2009.

Powertech [Powertech (USA) Inc.]. "Dewey-Burdock Project, Application for NRC Uranium Recovery License Fall River and Custer Counties, South Dakota, Technical Report RAI Responses, June, 2011." ML112071064. Greenwood Village, Colorado: Powertech. 2011.

Powertech. "Dewey-Burdock Project, Application for NRC Uranium Recovery License Fall River and Custer Counties, South Dakota ER_RAI Response August 11, 2010." ML102380516. Greenwood Village, Colorado: Powertech. 2010a.

1 Powertech. "Subject: Powertech (USA), Inc.'s Responses to the U.S. Nuclear Regulatory
2 Commission (NRC) Staff's Verbal and Email Requests for Clarification of Selected Issues
3 Related to the Dewey-Burdock Uranium Project Environmental Review Docket No. 40-9075;
4 TAC No. J 00533." Letter (November 4) to R. Burrows, Project Manager, Office of Federal and
5 State Materials and Environmental Management Programs, U.S. Nuclear Regulatory
6 Commission, from R. Blubaugh, Vice President-Environmental Health and Safety Resources.
7 ML110820582. Greenwood Village, Colorado: Powertech. 2010b.

9 Powertech. . "Subject: Powertech (USA), Inc.'s Responses to the U.S. Nuclear Regulatory
10 Commission (NRC) Staff's Verbal Request for Clarification of Response Regarding Inclusion of
11 Emissions from Drilling Disposal Wells; Dewey-Burdock Uranium Project Environmental Review
12 Docket No. 40-9075; TAC No. J 00533." Letter (November 17) to R. Burrows, Project Manager,
13 Office of Federal and State Materials and Environmental Management Programs, U.S. Nuclear
14 Regulatory Commission from R. Blubaugh, Vice President-Environmental Health and Safety
15 Resources. ML103220208. Greenwood Village, Colorado: Powertech. 2010c.

17 Powertech. "Dewey-Burdock Project, Application for NRC Uranium Recovery License Fall River
18 and Custer Counties, South Dakota—Environmental Report." Docket No. 040-09075.
19 ML092870160. Greenwood Village, Colorado: Powertech. August 2009a.

21 Powertech. "Dewey-Burdock Project, Application for NRC Uranium Recovery License Fall River
22 and Custer Counties, South Dakota—Technical Report." Docket No. 040-09075.
23 ML092870160. Greenwood Village, Colorado: Powertech. August 2009b.

25 Powertech. "Dewey-Burdock Project, Supplement to Application for NRC Uranium Recovery
26 License Dated February 2009." Docket No. 040-09075. ML092870160. Greenwood Village,
27 Colorado: Powertech. August 2009c.

7 ENVIRONMENTAL MEASURES AND MONITORING PROGRAMS

7.1   Introduction

As discussed in Section 8.0 of NUREG–1910, Generic Environmental Impact Statement for *In-Situ* Leach Uranium Milling Facilities (GEIS) (NRC, 2009), monitoring programs are developed for *in-situ* uranium recovery (ISR) facilities to verify compliance with standards for the protection of worker health and safety in operational areas and for protection of the public and environment beyond the facility boundary. Monitoring programs provide data on operational and environmental conditions so prompt corrective actions can be implemented when adverse conditions are detected. In this regard, these programs help to limit potential environmental impacts at ISR facilities and the surrounding areas.

Required monitoring programs can be modified to address unique site-specific characteristics by adding license conditions resulting from the conclusions of the U.S. Nuclear Regulatory Commission (NRC) safety and environmental reviews. The NRC staff are conducting the safety review of the proposed Dewey-Burdock ISR Project, which will be documented in a Safety Evaluation Report, and license conditions resulting from the safety review will be included as part of the final supplemental environmental impact statement (SEIS). The discussion of the proposed monitoring programs for the proposed Dewey-Burdock ISR Project is organized as follows:

- Radiological Monitoring (Section 7.2)
- Physiochemical Monitoring (Section 7.3)
- Ecological Monitoring (Section 7.4)
- Land Application Monitoring (Section 7.5)
- Class V Deep Injection Well Monitoring (Section 7.6)

The occurrence of spills and leaks at ISR facilities is considered in Section 2.11.2 of the GEIS (NRC, 2009), and the management of spills and leaks is not part of the routine environmental monitoring program described herein. Spills and leaks, including the design of the infrastructure to detect leaks, are described in the NRC safety evaluation.

7.2   Radiological Monitoring

This section describes Powertech (USA) Inc.'s (Powertech, referred to herein as the applicant) proposed radiological monitoring program as described in its license application, supporting documents for the proposed Dewey-Burdock ISR Project, and subsequent responses to NRC requests for additional information (Powertech, 2009a–c, 2010, 2011). The purpose of the monitoring program is to (i) characterize and evaluate the radiological environment, (ii) provide data on measurable levels of radiation and radioactivity, and (iii) provide data on the principal pathways of radiological exposure to the public (NRC, 2003). Although not a requirement, NRC Regulatory Guide 4.14 (NRC, 1980) provides guidance for establishing a radioactive effluent and environmental monitoring program for uranium mills, which includes ISR facilities. In accordance with NRC regulations in 10 CFR Part 40, Appendix A, Criterion 7, a preoperational monitoring program is required to establish facility baseline conditions. After establishing the baseline program, ISR facility operators must conduct an operational monitoring program to measure or evaluate compliance with standards and to evaluate environmental impacts of an ISR facility under operational conditions. In accordance with 10 CFR 40.65, the applicant must

1   submit to NRC a semiannual effluent and environmental monitoring report (Powertech, 2009b).
2   This report would specify the quantity of each of the principal radionuclides released to
3   unrestricted areas in liquid and in gaseous effluents during the previous 6 months of operation.
4   This report would also provide other NRC required information to estimate the maximum
5   potential annual radiation doses to the public resulting from effluent releases.
6
7   The results of the applicant's baseline radiological monitoring program are presented in SEIS
8   Section 3.12.1. The following sections briefly describe the applicant's proposed operational
9   monitoring program.
10

7.2.1   Airborne Radiation Monitoring

13   The applicant proposes to conduct continuous air particulate sampling at five locations identified
14   in Figure 7.2-1 (Powertech, 2011). The filters from air samplers will be analyzed biweekly, or
15   more frequently if required for dust loading, for natural uranium, Th-230, Ra-226, and Pb-210 in
16   accordance with Regulatory Guide 4.14 (NRC, 1980; Powertech, 2011). Samplers will be
17   equipped with sensors to measure total air flow within a sampling period and detect changes in
18   air flow due to dust loading, barometric pressure, and temperature (Powertech, 2011).
19
20   Passive track-etch detectors will be deployed at each air monitoring station for monitoring
21   Rn-222 on a monthly basis, consistent with Regulatory Guide 4.14 and NUREG–1569 (NRC,
22   1980, 2003; Powertech, 2011). Thermoluminescent dosimeters (TLDs) will be located with air
23   particulate samplers at each station (Powertech, 2011). The TLDs will be exchanged quarterly
24   and used to assess gamma exposure rates at each air monitoring station. Additionally, effluents
25   from the yellowcake dryer and packaging stacks will be sampled quarterly. The effluent
26   samples will be isokinetic in nature and would be analyzed for natural uranium, Th-230, Ra-226,
27   and Pb-210 (Powertech, 2009a).
28

7.2.2   Soils and Sediment Monitoring

31   Samples of surface soil from a 0–5 cm [0–2 in] depth will be collected annually at each of the air
32   monitoring stations shown in Figure 7.2-1. The samples will be analyzed for natural uranium,
33   Ra-226, and Pb-210 (Powertech, 2009a). Sediments will also be collected annually at each of
34   the 24 impoundments and 10 stream sampling sites proposed for operational surface water
35   monitoring (see SEIS Sections 7.2.4 and 7.3.3). The sediment samples will be analyzed for
36   natural uranium, Th-230, Ra-226, and Pb-210 (Powertech, 2011). The maximum lower limits of
37   detection for the analyses will be consistent with the recommendations of Regulatory
38   Guide 4.14 (NRC, 1980) unless matrix interferences prohibit attainment of these low detection
39   limit goals.
40

7.2.3   Vegetation, Food, and Fish Monitoring

43   The applicant plans to annually collect samples of livestock raised within 3.2 km [2 mi] of the
44   project area, consistent with the recommendations of Regulatory Guide 4.14 (NRC, 1980). The
45   samples will include cattle, pigs, and other livestock present at the time of sampling. Currently,
46   cattle and pigs are the only livestock within 3.2 km [2 mi] of the proposed project area. If other
47   livestock are found during annual land surveys, the applicant will seek the livestock owner's
48   approval to collect tissue samples at the time of slaughter (Powertech, 2011). Consistent with
49   Regulatory Guide 4.14 (NRC, 1980), fish will be collected semiannually provided they exist in

1

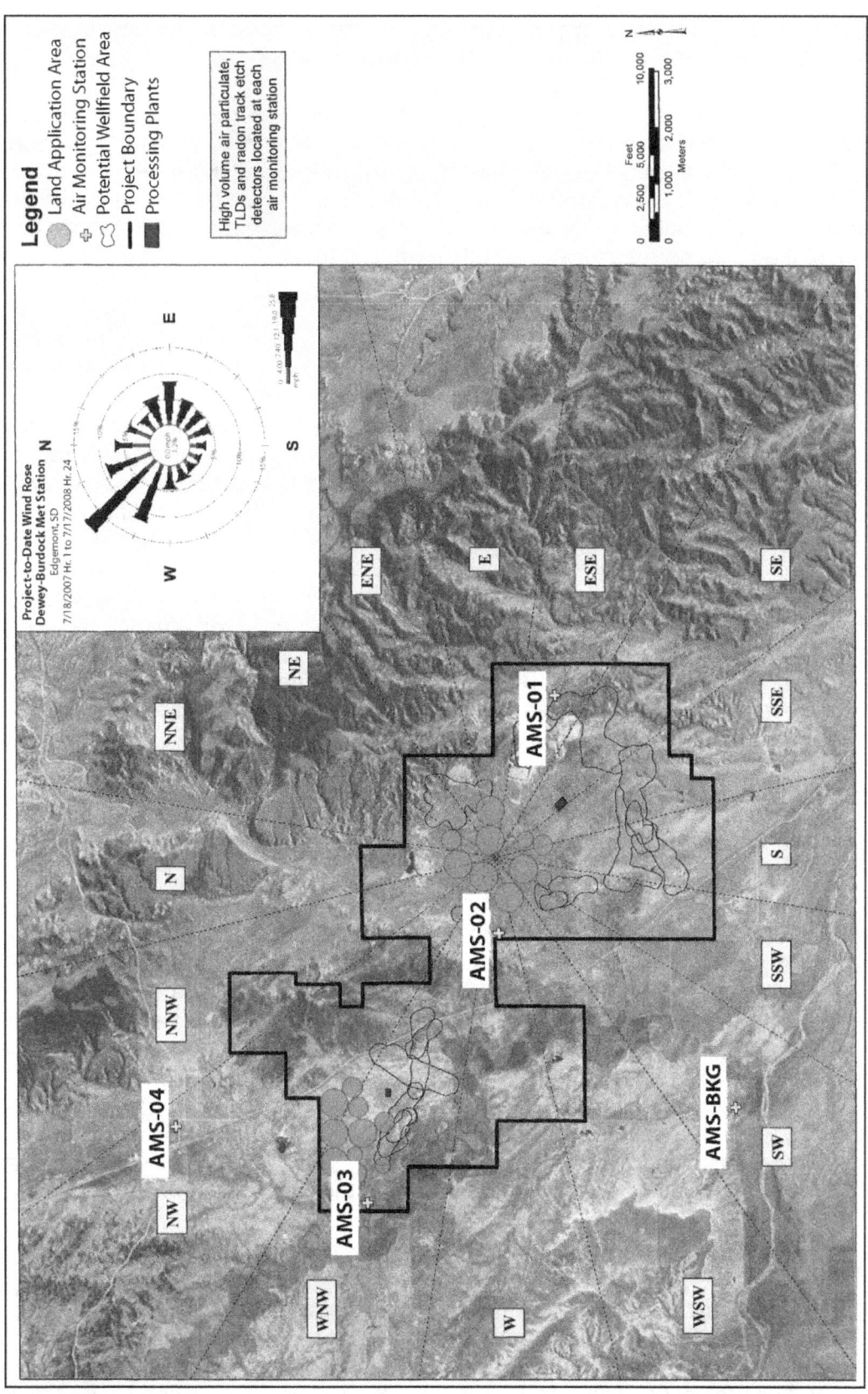

Figure 7.2-1. Locations of Operational Air Monitoring Stations at the Proposed Dewey-Burdock ISR Project Site. Source: Modified from Powertech (2011).

1   water bodies that may be affected by seepage or surface drainage from potentially
2   contaminated areas (Powertech, 2011). Livestock and fish samples will be analyzed for natural
3   uranium, Th-230, Ra-226, Pb-210, and Po-210 (Powertech, 2011).

5   The applicant plans to collect samples of vegetation three times during the grazing season.
6   The applicant will collect samples in the vicinity of each operational air monitoring station
7   (Figure 7.2-1). The samples of vegetation will be analyzed for Ra-226 and Pb-210 (Powertech,
8   2009b). The maximum lower limits of detection for the analyses will be consistent with the
9   recommendations of Regulatory Guide 4.14 (NRC, 1980) unless matrix interferences prohibit
10 attainment of these low detection limit goals (Powertech, 2009b).

12 **7.2.4   Surface Water Monitoring**

14 Operational surface water sampling will be conducted on (i) all surface impoundments located
15 downgradient of proposed ISR facilities and activities and (ii) perennial and ephemeral streams
16 passing through the site or located downgradient of proposed ISR activities (Powertech, 2011).
17 The applicant plans to monitor 24 impoundments and 10 stream sampling sites as part of
18 operational monitoring (Figure 7.2-2). Consistent with recommendations in Regulatory Guide
19 4.14 (NRC, 1980), grab samples will be collected quarterly from the impoundments and
20 analyzed for dissolved and suspended natural uranium, Ra-226, Th-230, Pb-210, and Po-210.
21 A grab sample is a sample of water, rock, or sediment taken more or less indiscriminately. Grab
22 samples will also be collected quarterly from perennial stream sampling locations on Beaver
23 Creek (BVC11 and BVC14) and the Cheyenne River (CHR01 and CHR05) (see Figure 7.2-2).
24 Passive samplers will be installed at the six remaining stream sampling sites, which are located
25 on ephemeral drainages (Pass Creek, Bennett Canyon, and unnamed tributaries), to
26 automatically sample during flow events. All stream samples will be analyzed for dissolved and
27 suspended uranium, Ra-226, Th-230, Pb-210, and Po-210 (Powertech, 2011).

29 **7.2.5   Groundwater Monitoring**

31 The operational groundwater monitoring program at the proposed Dewey-Burdock ISR Project
32 site will sample domestic wells, stock wells, and monitoring wells located hydrologically
33 upgradient and downgradient of proposed ISR facilities and wellfields (Powertech, 2011).
34 Consistent with Regulatory Guide 4.14 (NRC, 1980), the applicant proposes to collect annual
35 groundwater samples from all domestic wells within 2 km [1.2 mi] of the project boundary
36 (Figure 7.2-3) (Powertech, 2011). Quarterly groundwater samples will be collected from stock
37 wells within the project area (Figure 7.2-3) and from monitoring wells located hydrologically
38 upgradient and downgradient of proposed ISR facilities and wellfields (Figure 7.2-4). The
39 monitoring wells will be situated in the alluvium, Fall River Formation, Chilson Member of the
40 Lakota Formation, and the Unkpapa Formation. Water samples collected from the domestic
41 and monitoring wells will be analyzed for uranium and other radiological parameters, including
42 gross alpha, gross beta, and Ra-226 (Powertech, 2011). SEIS Section 7.3.4 further details the
43 applicant's preoperational and operational groundwater monitoring programs.

45 **7.3   Physiochemical Monitoring**

47 This section describes the applicant's proposed physiochemical monitoring program as
48 detailed in its license application and supporting documents (Powertech, 2009a–c, 2011). The
49 purpose of this monitoring program is to (i) provide data on operational and environmental
50 conditions so that prompt corrective actions can be taken when adverse conditions are detected

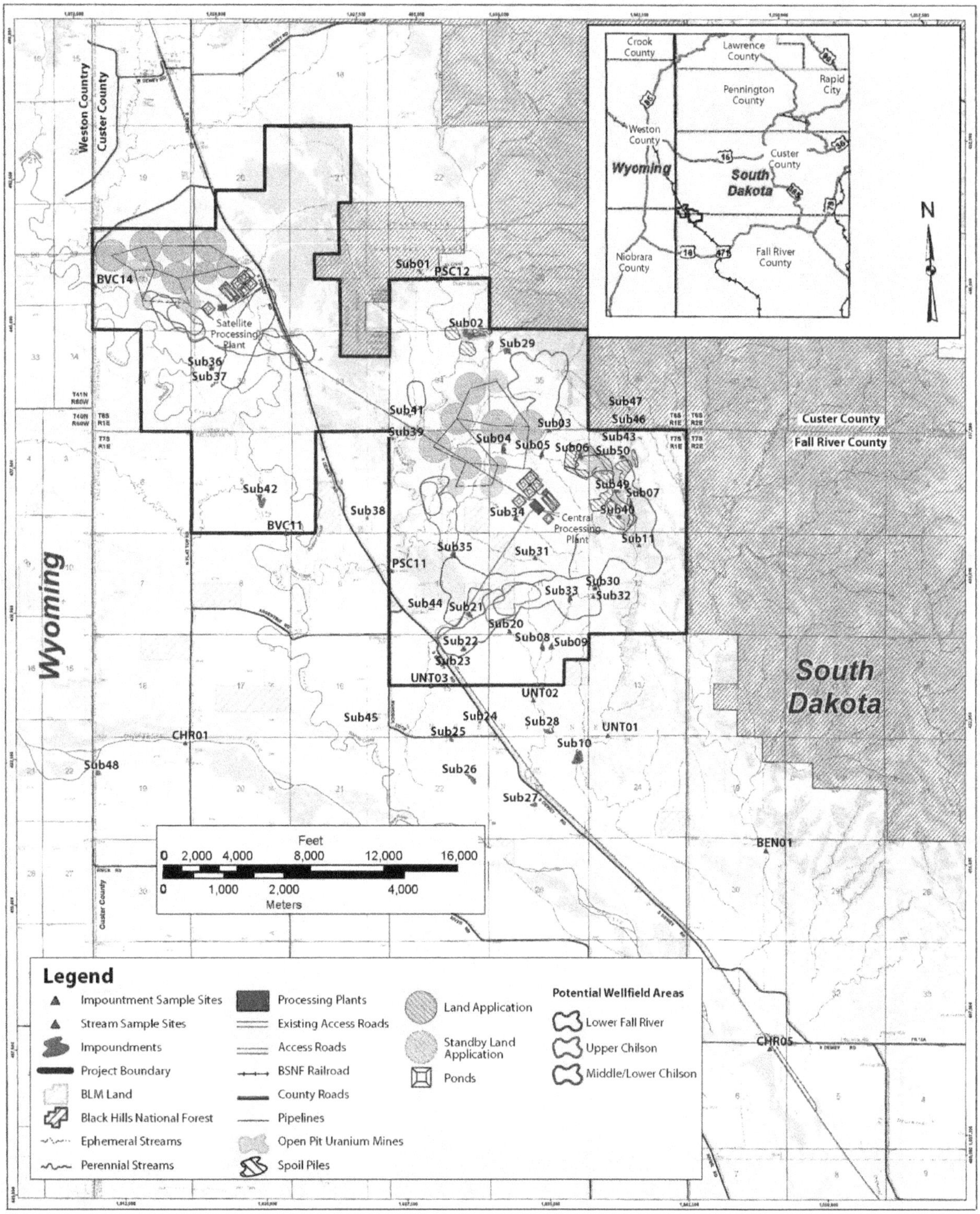

1

Figure 7.2-2. Locations of Operational Surface Water Monitoring Sites.
Source: Modified from Powertech (2011).

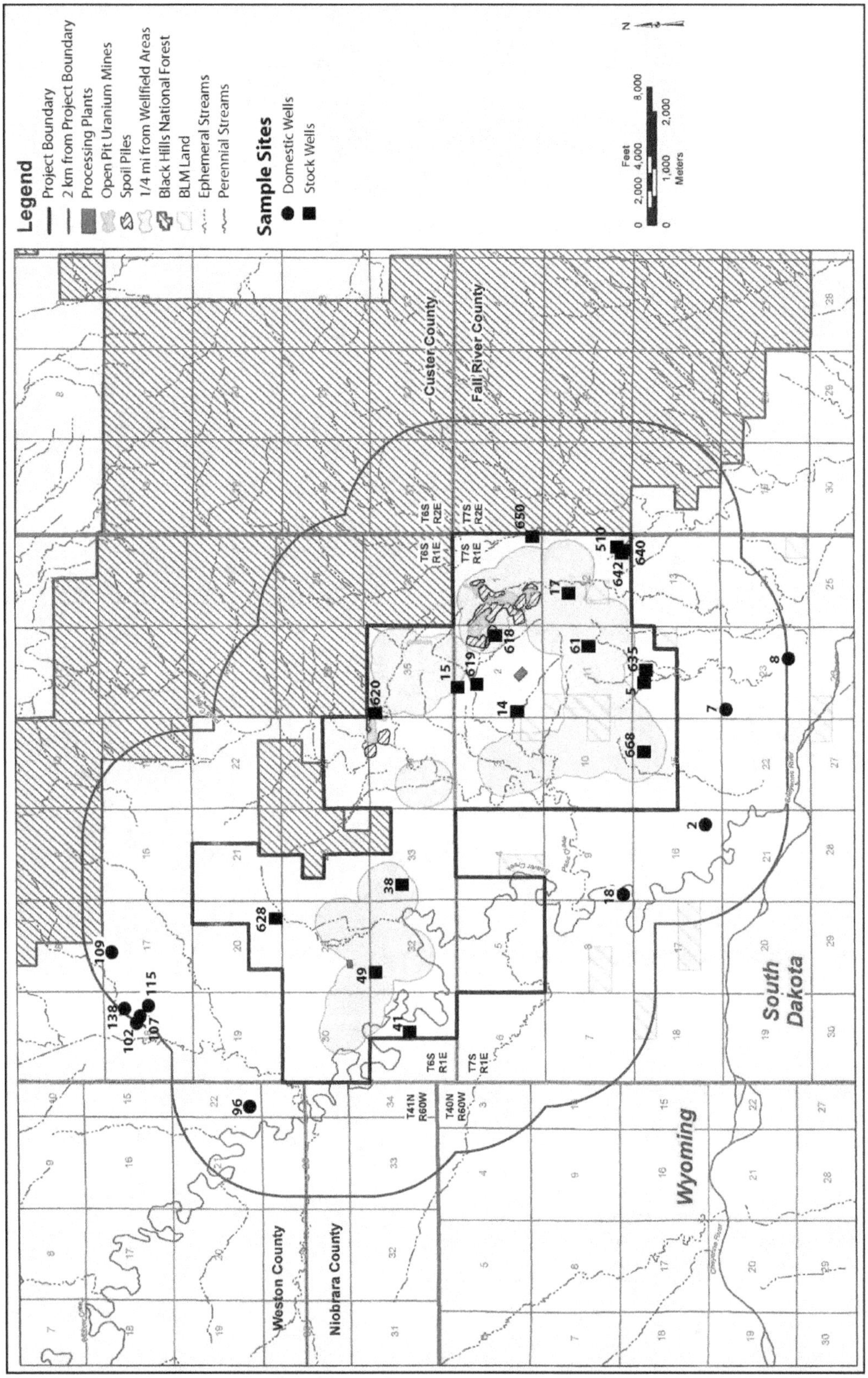

1

Figure 7.2-3. Locations of Operational Domestic and Stock Monitoring Wells. Source: Modified from Powertech (2011).

2

1

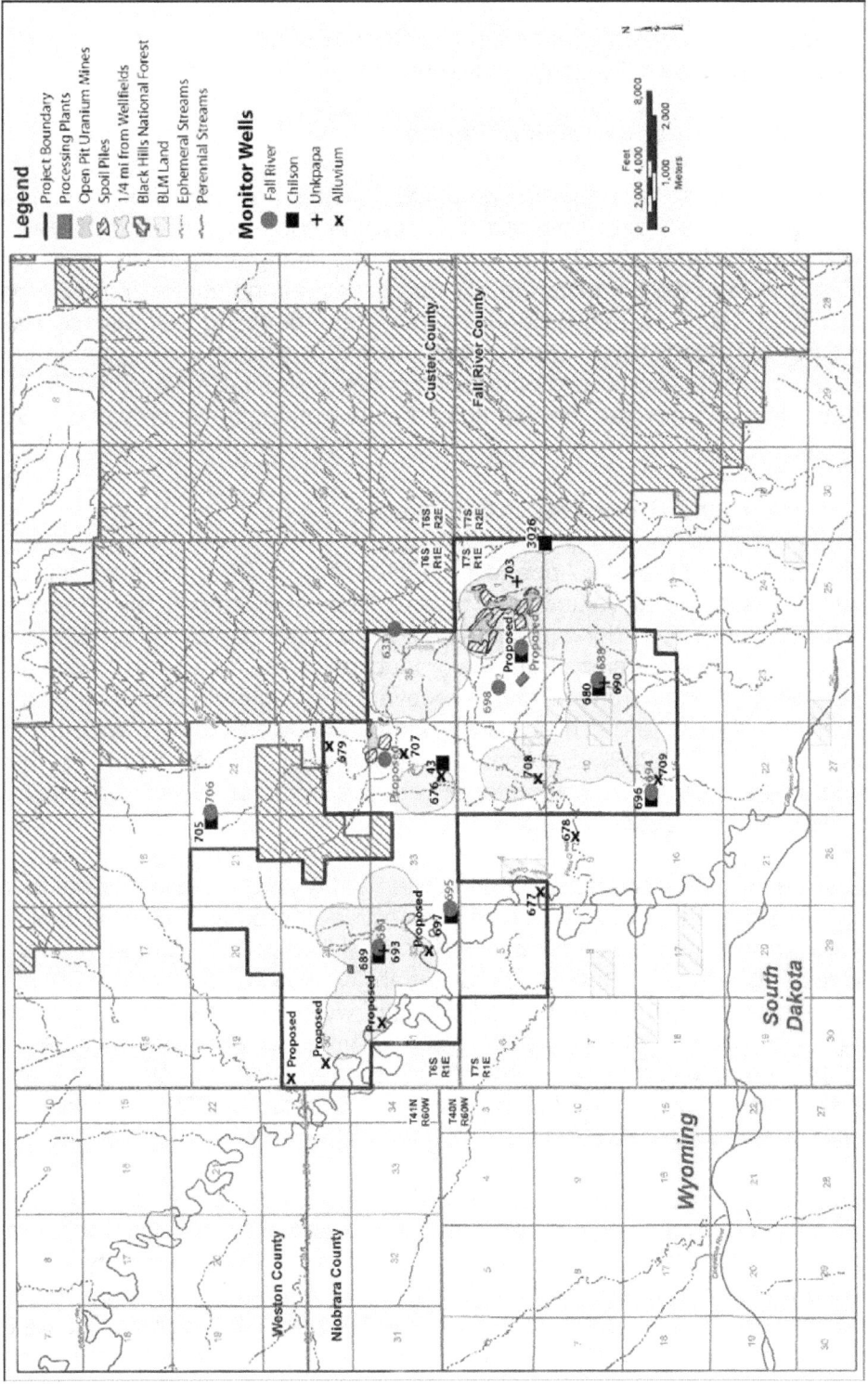

Figure 7.2-4. Locations of Operational Groundwater Monitoring Wells.
Source: Modified from Powertech (2011).

1
2 and (ii) comply with environmental requirements or license conditions. In this regard, this
3 monitoring program helps to limit potential environmental impacts at an ISR facility.
4
5 **7.3.1   Wellfield Groundwater Monitoring**
6
7 As discussed in GEIS Section 8.3, the ISR production process directly affects the groundwater
8 near the operating wellfield. For this reason, groundwater conditions are extensively monitored
9 both before and during operations. The groundwater monitoring program includes production
10 zone monitoring wells and wells monitoring aquifers overlying and underlying the production
11 aquifer zone (NRC, 2009). The background groundwater monitoring that will occur as part of
12 the proposed Dewey-Burdock ISR Project is discussed in Section 7.3.1.1. The groundwater
13 quality monitoring that will occur during operations is discussed in Section 7.3.1.2. The
14 applicant's restoration groundwater monitoring and stabilization plan is provided in SEIS
15 Section 2.1.1.1.4.2.
16
17 **7.3.1.1   Background Groundwater Sampling**
18
19 In accordance with 10 CFR Part 40, Appendix A, Criterion 5B(5), Commission-approved
20 background groundwater quality values must be established before beginning uranium
21 production in a wellfield. This is done to characterize the water quality in monitoring wells that
22 are used to detect lixiviant excursions from the production zone. This is also done to establish
23 standards for aquifer restoration after uranium recovery is complete. The requirements and
24 details of sampling programs to establish background groundwater quality are described in
25 GEIS Section 8.3.1.1 (NRC, 2009). Background water quality can be established through
26 examining records and reports for existing local water wells and/or by sampling wells developed
27 for the ISR project before production begins.
28
29 The applicant will establish background groundwater quality before beginning operations by
30 sampling a subset of wells that will later serve as injection or production wells installed in the
31 uranium mineralization zones. The applicant will sample these wells at least four times over a
32 sufficiently spaced interval to indicate seasonal variability (Powertech, 2011). The subset of
33 wells will include at least one well per 1.6 ha [4.0 ac] of wellfield pattern area, or six wells,
34 whichever is greater. These wells will be sampled four times for baseline characterization, with
35 a minimum of 14 days between sampling events. The water level in each well will also be
36 measured and recorded prior to each sampling event (Powertech, 2009a). Samples will be
37 analyzed for the parameters shown in Table 7.3-1.
38
39 Prior to calculating background water quality statistics, the water quality data will be examined
40 for differences between hydrogeologic units within each wellfield using visual screening, such as
41 trilinear diagrams, and statistical analyses (Powertech, 2011). If heterogeneity exists in the
42 data, then background water quality will be established for each hydrogeologic unit; otherwise,
43 background water quality will be established for the entire production zone of the wellfield. After
44 grouping the water quality data into hydrogeologic units and removing outliers (i.e., anomalously
45 high or low values relative to other values) if necessary, the applicant will calculate background
46 water quality as the arithmetic average for each sample parameter. Target restoration goals,
47 which will be used to assess the effectiveness of groundwater restoration activities, will be
48 established as a function of the average background water quality and the variability in each
49 parameter based on statistical methods. Before wellfield background evaluation, the applicant
50

Table 7.3-1. Background Water Quality Parameters and Indicators for Operational Groundwater Monitoring*

Test Analyte/Parameter	
BULK PROPERTIES	pH Total Dissolved Solids (TDS) Conductivity
CATIONS/ANIONS	Bicarbonate Alkalinity (as $CaCO_3$) Calcium, Ca Carbonate Alkalinity (as $CaCO_3$) Chloride, Cl Magnesium, Mg Nitrate, NO_3^- (as Nitrogen) Potassium, K Sodium, Na Sulfate, SO_4 Total Alkalinity (as $CaCO_3$)
TRACE METALS	Arsenic, As Barium, Ba Boron, B Cadmium, Cd Chromium, Cr Copper, Cu Fluoride, F Iron, Fe Lead, Pb Manganese, Mn Mercury, Hg Molybdenum, Mo Nickel, Ni Selenium, Se Silver, Ag Uranium, U Vanadium, V Zinc, Zn
RADIONUCLIDES	Gross Alpha=Alpha Particles Gross Beta=Beta Particles and Photons Radium, Ra-226

*All metals analyses are for dissolved metals.
Source: NRC (2003); Powertech (2011).

1
2 will consult with NRC for approval of the statistical methods used to determine target restoration
3 goals (Powertech, 2011). NRC will consult with EPA before establishing water quality standards
4 at the Dewey-Burdock site.
5
6 **7.3.1.2   Groundwater Quality Monitoring**
7
8 As discussed in GEIS Section 8.3.1.2, monitoring wells are situated around the wellfields, in the
9 aquifers overlying and underlying the ore-bearing production aquifers, and within the wellfields.
10 Wells are placed in these locations to ensure the early detection of potential horizontal and

1   vertical excursions of lixiviants. Monitoring well placement is based on what is known about the
2   nature and extent of the confining layer and the presence of drill holes, hydraulic gradient and
3   aquifer transmissivity, and well abandonment procedures used in the region. The ability of a
4   monitoring well to detect groundwater excursions is influenced by several factors, such as the
5   thickness of the aquifer, the distance between the monitoring wells and the wellfield, the
6   distance between the adjacent monitoring wells, the frequency of groundwater sampling, and
7   the magnitude of changes in lixiviant migration indicator parameters. As a result, the spacing,
8   distribution, and number of monitoring wells at a given ISR facility are site specific. The factors
9   that control the spacing, distribution, and number of monitoring wells are detailed in GEIS
10   Section 8.3.1.2 (NRC, 2009). The applicant's monitoring well design is described in SEIS
11   Section 2.1.1.1.2.3.2 and summarized next.
12
13   The applicant proposes to install production and nonproduction zone monitoring wells to detect
14   any horizontal and vertical lixiviant excursions at the proposed project site (Powertech, 2009a).
15   The production zone monitoring wells will be located in the ore zone, in a ring around the
16   perimeter of the production wellfields. They will be spaced at a maximum of 122 m [400 ft]
17   outside the production wellfield and evenly spaced around the perimeter of the wellfield with a
18   minimum spacing of either 122 m [400 ft] or the spacing that will ensure that no greater than a
19   70 degree angle between adjacent production zone monitoring wells and the nearest injection
20   well (Mackin, et al., 2001; NRC, 2009, 2003; Powertech, 2009a, 2011). The applicant
21   conducted numerical simulations using site-specific hydrologic data and proposed production
22   flow rates to support the proposed spacing of monitoring wells (Powertech, 2011). Simulation
23   results indicated that the proposed maximum monitoring well spacing of 122 m [400 ft] would be
24   adequate to detect potential excursions (Powertech, 2011).
25
26   Nonproduction monitoring wells may consist of two types of monitoring wells: overlying and
27   underlying (Mackin, et al., 2001; NRC, 2003, 2009). The screened intervals of overlying wells
28   will be located in the sand unit or aquifer immediately above the ore-bearing stratum. The
29   overlying nonproduction monitoring wells are designed to monitor any upward movement of
30   leach fluids that may occur from the production zone and to guard against potential leakage
31   from production and injection well casings into any overlying aquifer (Mackin, et al., 2001; NRC,
32   2003, 2009). The overlying wells are used to obtain background water quality data and to
33   develop upper control limits (UCLs) for the overlying zones that will be used to determine
34   whether vertical migration of leach fluids is occurring.
35
36   Vertical monitoring is generally set up with a density of wells ranging from one every 1.2 to 2 ha
37   [3 to 5 ac]. However, where confining layers are very thick and permeabilities are negligible,
38   requirements for vertical excursion monitoring can be relaxed or eliminated (Mackin, et al.,
39   2001). The screened zone for the overlying wells is determined from electric logs by qualified
40   geologists or hydrogeologists.
41
42   The applicant's nonproduction zone monitoring plan is described in SEIS Section 2.1.1.1.2.3.2.
43   Following the previously outlined guidance, the applicant plans to design and install both
44   overlying and underlying monitoring wells. The first layer of overlying nonproduction zone
45   monitoring wells will be evenly distributed through the production area with a minimum of one
46   well for every 1.6 ha [4.0 ac] of production area (Powertech, 2009a). The overlying wells will be
47   placed within the geology just above the proposed project's upper confining layer (the Skull
48   Creek Shale), which has a thickness of approximately 61 m [200 ft]. Core samples collected
49   from the lower Skull Creek Shale demonstrate that the Skull Creek clays have extremely low
50   vertical permeabilities. The thicknesses of the upper confining layer {approximately 61 m

1   [200 ft]} and the lower confining layer {approximately 30 m [100 ft]} minimize concerns about
2   vertical excursions of lixiviant.
3
4   The monitoring ring and overlying and underlying monitoring wells will be designed for each
5   wellfield according to site-specific lithology and processes of the production zone(s) of
6   each wellfield. To ensure administrative approval, the applicant would present each monitoring
7   well program to NRC and the U.S. Environmental Protection Agency (EPA) before proposed
8   wells are installed (Powertech, 2009a). After the required hydrologic tests are complete, it may
9   be necessary to revise the location and/or number of wells proposed. Each wellfield will be
10   handled on a case-by-case basis in consultation with NRC and EPA.
11
12   UCLs are selected and set for chemical constituents or parameters that will be indicative of
13   lixiviant migration from the wellfield (Mackin, et al., 2001; NRC, 2003, 2009). The constituents
14   and parameters selected as lixiviant migration indicators and for which UCLs will be set at the
15   proposed Dewey-Burdock ISR Project are chloride, conductivity, and total alkalinity (Powertech,
16   2011). Chloride is measured because the ion exchange process increases concentrations in
17   the lixiviant. In addition, chloride is highly mobile in groundwater and is not influenced by pH
18   changes and oxidation-reduction reactions that occur in the production zone (Powertech, 2011).
19   Conductivity is evaluated because it indicates changes in groundwater quality and is more
20   easily measured than parameters such as total dissolved solids. Total alkalinity will be
21   examined because its concentration significantly increases during the ISR process and,
22   therefore, provides a conservative indicator (Powertech, 2011).
23
24   The applicant followed guidance in NUREG–1569 (NRC, 2003) to establish and set UCLs in
25   wellfields. All monitoring wells in the production zone aquifer and nonproduction zone aquifers
26   (i.e., underlying and overlying aquifers) will be sampled 4 times with a minimum of 14 days
27   between sampling events (Powertech, 2011). All samples will be analyzed for the parameters in
28   Table 7.3-1. The mean concentration and standard deviation of the constituents or parameters
29   selected as UCLs (i.e., chloride, conductivity, and total alkalinity) will be calculated for samples
30   taken from the production zone aquifer and nonproduction zone aquifers. UCLs for each
31   production zone monitoring well in a wellfield will be set at the mean concentration of the
32   production zone aquifer plus five standard deviations for each excursion indicator. UCLs for
33   each nonproduction zone monitoring well will be set at the mean concentration of the
34   nonproduction zones aquifers plus five standard deviations for each excursion indicator. Some
35   aquifers exhibit a low chloride concentration with an insignificant standard deviation (i.e., a
36   narrow concentration range). Consistent with NUREG–1569 (NRC, 2003), when setting the
37   UCL for chloride the applicant will use either the mean plus five standard deviations or the mean
38   plus 15 mg/L [15 ppm], whichever is greater (Powertech, 2011).
39
40   The applicant proposes to sample monitoring wells at the proposed Dewey-Burdock ISR Project
41   at approximately 2-week intervals (at least 10 days apart) (Powertech, 2009a). The samples
42   will be analyzed for and compared against the excursion parameter UCL values. The water
43   level in each monitoring well will also be measured and recorded prior to each sampling event
44   (Powertech, 2009a). Water level and analytical monitoring data for the UCL parameters will be
45   reported to EPA quarterly and retained onsite for NRC review.
46
47   After operations are complete, the wellfields will be restored. As described in SEIS
48   Section 2.1.1.1.4.2, as part of aquifer restoration the applicant will sample the same horizontal
49   perimeter and overlying/underlying monitoring wells used during production. During restoration,
50   lixiviant injection ceases, thereby reducing the potential for an excursion. The applicant will

1 implement a reduced groundwater monitoring program during aquifer restoration because
2 lixiviant injection will have ceased. During the aquifer restoration phase, wells located in the
3 perimeter monitoring ring and completed in the overlying and underlying aquifers will be
4 sampled every 60 days for chloride, alkalinity, and conductivity excursion parameters. An
5 excursion will be defined in the same manner as during operations and subject to the same
6 corrective action requirements.
7
8 ### 7.3.2   Wellfield and Pipeline Flow and Pressure Monitoring
9
10 As indicated in GEIS Section 8.3.2, the operator typically monitors injection and production well
11 flow rates to manage water balance for the entire wellfield. Additionally, the pressure of each
12 production well and the production trunk line in each wellfield header house is monitored.
13 Unexpected losses of pressure may indicate equipment failure, a leak, or a problem with
14 well integrity (NRC, 2009).
15
16 The applicant's program will include monitoring of the injection well and production well flow
17 rates and pressures at each header house. Individual well flow readings will be recorded during
18 each shift, and the overall wellfield flow rates will be balanced daily (Powertech, 2009a,b). Flow
19 and total volume data will be transferred to and checked automatically at the Burdock central
20 processing plant and Dewey satellite facility. The recovery and injection trunk lines will have
21 electronic pressure gauges. Information from these gauges will be monitored from each unit's
22 control room. The control system will have both high and low alarms for pressure and flow. If
23 the pressure and/or flow are out of range, the alarms will sound, alerting personnel to make
24 adjustments. Certain high or low readings will signal automatic shutoffs or shutdowns.
25 Activation of the flow alarms will prompt the applicant to take corrective actions, which include
26 inspections for leaks and spills.
27
28 ### 7.3.3   Surface Water Monitoring
29
30 The applicant will conduct surface water monitoring on all surface impoundments located
31 downgradient from ISR activities. The applicant will also monitor surface waters passing
32 through the site or located downgradient of ISR activities (Powertech, 2011). As described in
33 SEIS Section 7.2.4, the applicant plans to monitor 24 impoundments and 10 stream sampling
34 sites as part of the operational surface water monitoring program. The operational surface
35 water sampling sites are shown in Figure 7.2-2 and listed in Table 7.3-2.
36

Table 7.3-2. Impoundments and Stream Sampling Locations Proposed for Operational Monitoring

Site ID	Type/Name
Impoundments	
Sub02	Triangle Mine Pit
Sub03	Mine Dam
Sub04	Stock Pond
Sub05	Mine Dam
Sub06	Darrow Mine Pit Northwest
Sub07	Stock Dam
Sub08	Stock Pond
Sub09	Stock Pond
Sub10	Stock Pond
Sub11	Stock Pond

Table 7.3-2. Impoundments and Stream Sampling Locations Proposed for Operational Monitoring (continued)

Site ID	Type/Name
Impoundments (continued)	
Sub20	Stock Pond
Sub21	Stock Pond
Sub22	Stock Pond
Sub29	Stock Pond
Sub30	Stock Pond
Sub31	Stock Pond
Sub32	Stock Pond
Sub33	Stock Pond
Sub34	Stock Pond
Sub35	Stock Pond
Sub36	Stock Pond
Sub40	Darrow Mine Pit Southeast
Sub49	Darrow Mine Pit
Sub50	Darrow Mine Pit
Streams	
BVC11	Beaver Creek Downstream
BVC14	Beaver Creek Upstream
CHR01	Cheyenne River Upstream
CHR05	Cheyenne River Downstream
PSC11	Pass Creek Downstream
PSC12	Pass Creek Upstream
BEN01	Bennett Canyon
UNT01	Unnamed Tributary
UNT02	Unnamed Tributary
UNT03	Unnamed Tributary
Source: Powertech, 2011.	

1
2   Prior to ISR operations, the applicant plans to sample each impoundment sampling site 4 times
3   and each stream sampling site monthly for 12 consecutive months in accordance with
4   preoperational monitoring recommendations in Regulatory Guide 4.14 (NRC, 1980). Water
5   samples will be collected from the impoundments, when available, and analyzed for the
6   constituents in Table 7.3-1. Grab samples will be collected from perennial stream sampling
7   locations on Beaver Creek (BVC11 and BVC14) and the Cheyenne River (CHR01 and CHR05).
8   Passive samplers will be installed at the remaining sites to collect samples during ephemeral
9   flow events. All stream samples will be analyzed for the constituents listed in Table 7.3-1.
10
11 During ISR operations, water samples collected from the impoundment and stream sampling
12 sites will be analyzed for the constituents listed in Table 7.3-1 along with dissolved and
13 suspended natural uranium, Ra-226, Th-230, Pb-210, and Po-210. In addition, the samples
14 would be analyzed in the field for pH, conductivity, and temperature (Powertech, 2011).
15
16 **7.3.4   Groundwater Monitoring (Project-Wide)**
17
18 The groundwater monitoring program will include domestic wells, stock wells, and monitoring
19 wells located hydrologically upgradient and downgradient of proposed ISR activities

1   (Powertech, 2011). Consistent with Regulatory Guide 4.14 (NRC, 1980), all domestic and stock
2   wells within 2 km [1.2 mi] of the project area and all monitoring wells will be sampled quarterly
3   over a 1-year period to establish baseline water quality before operations begin. All the
4   preoperational groundwater samples will be analyzed for the constituents listed in Table 7.3-1.
5
6   Prior to operations, all domestic wells within the proposed project boundary will be removed
7   from private use (Powertech, 2011). The applicant will work with the well owner to provide an
8   alternative water source such as a replacement well or alternate water supply for domestic use
9   (Powertech, 2011). Depending on well construction, location, and screen interval, the applicant
10   could continue to use the well for monitoring or plug and abandon the well. During operations,
11   the applicant will monitor all domestic wells within 2 km [1.2 mi] of the project boundary
12   (Figure 7.2-3). Samples will be collected annually and analyzed for the constituents listed in
13   Table 7.3-1.
14
15   Prior to operation of nearby wellfields, all stock wells within 0.4 km [0.25 mi] of wellfields will be
16   removed from private use (Powertech, 2011). In addition, all nearby stock wells that have the
17   potential to be adversely affected by ISR operations or to adversely affect ISR operations will be
18   removed from private use (Powertech, 2011). Depending on well construction, location, and
19   screen interval, the applicant could continue to use the stock well for monitoring or plug and
20   abandon the well. During operations, the applicant must monitor all stock wells within the
21   project area (Figure 7.2-3). Water samples will be collected quarterly and analyzed for three
22   excursion indicators: chloride, total alkalinity, and conductivity (Powertech, 2011).
23
24   During operations, the monitoring wells located hydrologically upgradient and downgradient of
25   ISR activities will be sampled quarterly and analyzed for the constituents listed in Table 7.3-1.
26   The operational monitoring wells proposed will be in the alluvium, Fall River Formation, Chilson
27   Member of the Lakota Formation, and the Unkpapa Formation. The position of each well
28   relative to site facilities and features is shown in Figure 7.2-4 and listed in Table 7.3-3.
29
30   **7.3.5   Meteorological Monitoring**
31
32   The applicant does not specify a plan for continued meteorological monitoring at the proposed
33   site (Powertech, 2009a,b). As part of the site characterization process, the applicant installed
34   a weather station near the center of the proposed action area. This weather station was
35   monitored from July 2007 through July 2008 to analyze and describe the long-term and
36   site-specific meteorological conditions and trends. In addition, data sets from several regional
37   weather stations were reviewed (see SEIS Section 3.7).
38

Table 7.3-3. Monitoring Wells Proposed for Operational Monitoring

Well ID	Aquifer	Relative Position
676	Alluvium	Downgradient of Land Application
677	Alluvium	Downgradient
678	Alluvium	Downgradient
679	Alluvium	Upgradient
707	Alluvium	Downgradient of Triangle Pit
708	Alluvium	Downgradient of Land Application
Proposed	Alluvium	Downgradient of Wellfield
Proposed	Alluvium	Downgradient of Wellfield
Proposed	Alluvium	Downgradient of Land Application
631	Fall River	Upgradient

Table 7.3-3. Monitoring Wells Proposed for Operational Monitoring (continued)

Well ID	Aquifer	Relative Position
709	Alluvium	Downgradient of Wellfield
Proposed	Alluvium	Upgradient
681	Fall River	Production Zone
688	Fall River	Overlying Production Zone
694	Fall River	Upgradient
695	Fall River	Downgradient
698	Fall River	Downgradient
706	Fall River	Upgradient
Proposed	Fall River	Downgradient of Triangle Pit
Proposed	Fall River	Downgradient of Darrow Pit
43	Chilson	Downgradient of Triangle Pit
680	Chilson	Production Zone
689	Chilson	Production Zone
696	Chilson	Downgradient
697	Chilson	Downgradient
705	Chilson	Upgradient
3026	Chilson	Upgradient
Proposed	Chilson	Downgradient of Darrow Pit
690	Unkpapa	Production Zone
693	Unkpapa	Production Zone
703	Unkpapa	Production Zone
Source: Powertech, 2011		

7.4   Ecological Monitoring

This section describes the applicant's proposed ecological monitoring program as described in its license application (Powertech, 2009a–c). As discussed in GEIS Section 8.4, ecological monitoring may include surveys of habitat, species counts, or other measures of the health of endangered, threatened, and sensitive species (NRC, 2009). Records of all sampling activities and analyses will be maintained onsite for NRC review, and periodic reports of all sampling and analyses will be submitted to NRC.

7.4.1   Vegetation Monitoring

Site characterization studies (Powertech, 2009a) indicate the proposed project area consists of five vegetation communities: Big Sagebrush Shrubland, Greasewood Shrubland, Ponderosa Pine Woodland, Upland Grassland, and Cottonwood Gallery. Each community was investigated for baseline vegetation information in support of an NRC Source Materials License and SDDENR Regular Mine Permit Application. No threatened or endangered species were encountered within the proposed project area. The applicant noted the presence of the state-designated weed Canada thistle (*Cirsium avense*) within the Cottonwood Gallery community and the presence of the Fall River County-designated weed field bindweed (*Convolvulus arvensis*) within the Greasewood Shrubland vegetation community. The applicant proposes weed control to mitigate further intrusion of invasive species in disturbed areas.

1   **7.4.2   Wildlife Monitoring**
2
3   The applicant will conduct annual wildlife monitoring at the project site during the lifespan of the
4   project (Powertech, 2009a). The annual wildlife monitoring surveys will follow the same
5   regimen as other ISR operations in the region (NRC, 2009). This will facilitate comparisons
6   among survey results and impact assessments. As described in SEIS Section 3.6, no federally
7   listed threatened and endangered species were documented within the project area during the
8   baseline study. However, eight raptor nests were identified within the proposed project area,
9   including one active bald eagle nest. The bald eagle is currently listed as threatened and
10   endangered by the South Dakota Department of Game, Fish, and Parks (SDGFP). The
11   applicant's annual monitoring surveys will include the following:
12
13   (1)   Early spring surveys for, and monitoring of, Greater sage-grouse leks {no sage-grouse
14   leks were identified within 10 km [6 mi] of the proposed action area}; new and/or
15   occupied raptor territories and/or nests; threatened and endangered species (federal
16   and state); and species tracked by the South Dakota Natural Heritage Program, as
17   directed, on and within 1.6 km [1 mi] of the proposed project area
18
19   (2)   Late spring and summer surveys for raptor production at occupied nests, and
20   opportunistic observations of all wildlife species, including threatened and endangered
21   species, and other species of management concern
22
23   (3)   Other surveys regulating agencies require
24
25   The applicant will employ a number of possible mitigation strategies to reduce the impact of its
26   activities on raptors in the project area (Powertech, 2009a). These strategies include possible
27   relocation of raptor nests. In the unlikely event that the applicant determines it necessary to
28   disturb a raptor nest, the applicant will develop a mitigation plan and consult with SDGFP and
29   the U.S. Fish and Wildlife Service, at which time any applicable permits will be obtained from
30   the appropriate agencies (Powertech, 2009a).
31
32   The applicant does not plan to sample aquatic species Powertech, 2009a). As described in
33   SEIS Section 3.6.2, aquatic species are limited within the proposed project area due to a lack of
34   persistent aquatic resources (i.e., surface waters) and poor habitat conditions.
35
36   Because the proposed project area does not include any critical big game habitats (see SEIS
37   Section 3.6) and is already included in SDGFP big game surveys, SDGFP did not require big
38   game surveys for the applicant's baseline wildlife surveys. Consequently, no long-term big
39   game monitoring requirements are planned (Powertech, 2009a). A similar approach has been
40   applied to other baseline projects (uranium, coal, bentonite, gold) in South Dakota and
41   Wyoming and is the current policy of both states for annual monitoring at surface mines in the
42   two-state region.
43
44   **7.5   Land Application Monitoring**
45
46   This section describes the applicant's proposed land application monitoring program as
47   described in the applicant's Groundwater Discharge Plan submitted to SDDENR (Powertech,
48   2012). As described in SEIS Section 2.1.1.1.2.4, the applicant is proposing options for liquid
49   waste disposal at the proposed Dewey-Burdock ISR Project that include deep well disposal,
50   land application, or combined deep well disposal and land application. If land application is
51   used for liquid waste disposal at the proposed project, the applicant will implement this program

1   in a manner that ensures beneficial uses will not be impaired and there will be no hazard to
2   human health and the environment (Powertech, 2012). Records of all sampling activities and
3   analyses will be maintained onsite for NRC review, and periodic reports of all sampling and
4   analyses will be submitted to SDDENR (Powertech, 2012).
5
6   **7.5.1   Groundwater**
7
8   The land application groundwater monitoring program will include alluvial monitoring wells within
9   and hydrologically upgradient and downgradient of proposed land application systems. In
10   addition, the shallowest bedrock aquifer, the Fall River Formation, will be monitored and suction
11   lysimeters will be installed to monitor the vadose groundwater quality beneath the land
12   application systems. The groundwater monitoring program is designed to provide a
13   comprehensive evaluation of potentially affected groundwater quality within and near the
14   proposed perimeter of operational pollution (POP) for proposed land application areas.
15   Proposed POP zones in the Dewey and Burdock land application areas are shown in
16   Figures 7.5-1 and 7.5-2, respectively.
17
18   **7.5.1.1   Alluvial Monitoring Wells**
19
20   Three types of alluvial monitoring wells are proposed to assess baseline conditions and impacts
21   to alluvial water quality during operations: compliance wells, interior wells, and other wells.
22   Proposed alluvial monitoring wells in the Dewey area are presented in Table 7.5-1 and depicted
23   in Figure 7.5-1. Proposed alluvial monitoring wells in the Burdock area are presented in
24   Table 7.5-2 and depicted in Figure 7.5-2. Compliance wells will be hydrologically downgradient
25   from land application systems at the POP zone boundaries and will serve as compliance
26   locations for potential impacts to alluvial water quality outside of the POP zone. Interior wells
27   will be within each POP zone and will measure potential changes in alluvial water quality within
28   the POP zones. Other wells are proposed to measure ambient alluvial water quality within the
29   project area (see SEIS Section 7.2.5). These wells are outside of the POP zones both
30   upgradient and downgradient of proposed land application systems.
31
32   Prior to operations of land application systems, all compliance, interior, and other wells will be
33   sampled to determine baseline water quality. Each compliance and interior well will be sampled
34   a minimum of four times within a 6-month period with no two samples taken in the same month.
35   During operations of land application systems, compliance, interior, and other wells will be
36   sampled quarterly. All baseline and operational water samples will be analyzed for the
37   parameters in Table 7.3-1.
38
39   For each compliance and interior well, baseline water quality for each parameter will be
40   established as an arithmetic mean of baseline water samples plus one standard deviation of the
41   sample data. Compliance limits for constituents in compliance wells will be established on a
42   well-by-well basis as the human health standards in Administrative Rules of South Dakota
43   (ARSD) 74:54:01:04 or baseline water quality, whichever is greater. Out-of-compliance status
44   will be defined in accordance with ARSD 74:54:02:28 as two consecutive samples that exceed
45   the permitted allowable limit by two standard deviations. Interior wells will not have established
46   compliance limits, but a contingency plan will be implemented if the monitored constituent
47   concentrations increase (Powertech, 2012).
48

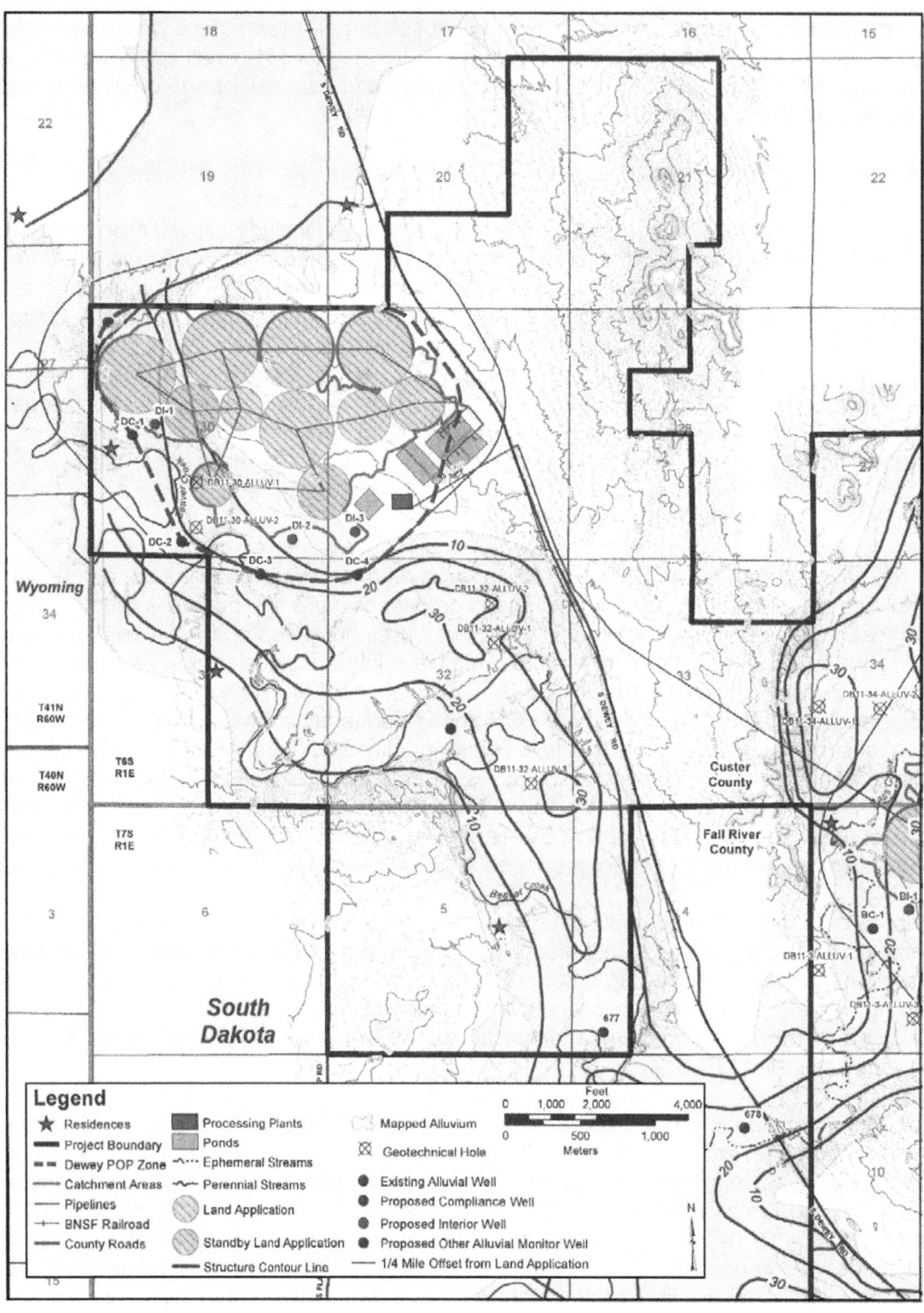

Figure 7.5-1. Map of Dewey Land Application Areas Showing the Perimeter of Operational Pollution (POP) and Proposed Alluvial Monitoring Wells. Source: Powertech (2012).

1
2

1

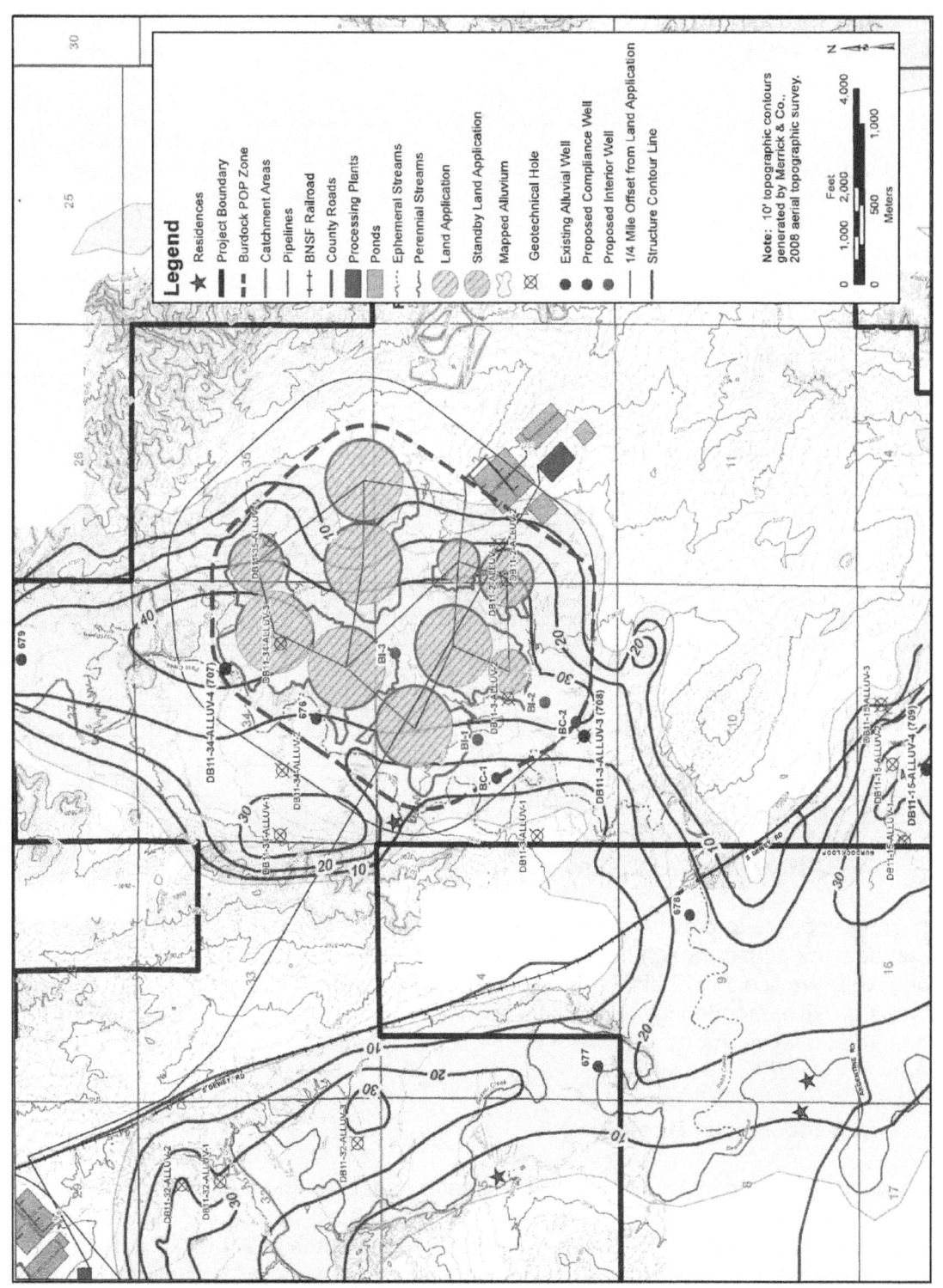

Figure 7.5-2. Map of Burdock Land Application Areas Showing the Perimeter of Operational Pollution (POP) and Proposed Alluvial Monitoring Wells.
Source: Powertech (2012).

1

Table 7.5-1. Proposed Alluvial Monitoring Wells in the Dewey Area

Monitoring Well Type	Well ID	Status
Compliance Wells	DC-1	Proposed
	DC-2	Proposed
	DC-3	Proposed
	DC-4	Proposed
Interior Wells	DI-1	Proposed
	DI-2	Proposed
	DI-3	Proposed
Other Wells	TBD	Proposed
	TBD	Proposed
	677	Existing
Source: Powertech, 2012		

2
3

Table 7.5-2. Proposed Alluvial Monitoring Wells in the Burdock Area

Monitoring Well Type	Well ID	Status
Compliance Wells	BC-1	Proposed
	BC-2	Proposed
Interior Wells	BI-1	Proposed
	BI-2	Proposed
	BI-3	Proposed
Other Wells	676	Existing
	678	Existing
	679	Existing
	707	Existing
	708	Existing
Source: Powertech, 2012		

4
5   **7.5.1.2   Bedrock Aquifer Monitoring**
6
7   The applicant proposes to provide monitoring results from operational monitoring wells in the
8   shallowest bedrock aquifer, which occurs in the Fall River Formation. These Fall River
9   monitoring wells are listed in Table 7.3-3 and depicted in Figure 7.2-4. Prior to ISR operations,
10   each of the Fall River monitoring wells will be sampled quarterly for 1 year. During ISR
11   operations, the Fall River monitoring wells will be sampled quarterly and analyzed for the
12   parameters in Table 7.3-1.
13
14   **7.5.1.3   Vadose Zone Monitoring**
15
16   The applicant proposes to install one suction lysimeter in each of the center pivot circles and
17   catchment areas at both the Dewey and Burdock areas to obtain pore water samples from
18   unsaturated soil. The suction lysimeters will be installed at depths of 2.4 to 3.7 m [8 to 12 ft].
19   Prior to operations of land application systems, pore water samples will be collected a minimum
20   of four times within a 6-month period with no two samples taken in the same month. During
21   operations, pore water samples will be collected once prior to each irrigation season, once
22   during each irrigation season, and once after each irrigation season. Samples will be analyzed
23   for the parameters in Table 7.3-1.

1   ### 7.5.2   Surface Water
2
3   The locations of stream sampling sites on Beaver and Pass Creeks are BVC11, BVC14,
4   PSC11, and PSC12. These sites are listed in Table 7.3-2 and depicted in Figure 7.2-2. The
5   upstream sites on Beaver Creek (BVC14) and Pass Creek (PSC12) are approximately at the
6   boundary of the permit area and will represent ambient water quality. The downstream site on
7   Beaver Creek (BVC11) is downstream of the Dewey land application area, and the downstream
8   site on Pass Creek (PSC11) is downstream of the Burdock land application area. Samples for
9   each sampling site will be collected monthly for 12 consecutive months prior to ISR operations.
10 Grab samples will be collected from sites BVC11 and BVC14. Passive samplers will be
11 installed at sites PSC11 and PSC12 to collect samples during ephemeral flow events. Water
12 samples will be analyzed for the constituents listed in Table 7.3-1. During ISR operations,
13 including operation of land application systems, grab samples will be collected quarterly from
14 perennial stream sampling locations on Beaver Creek and passive samplers installed on Pass
15 Creek will automatically collect samples following runoff events from April through October.
16 Grab samples will be analyzed in the field for pH, conductivity, and temperature. All stream
17 samples will be analyzed for the constituents listed in Table 7.3-1 along with dissolved and
18 suspended uranium, Ra-226, Th-230, Pb-210, and Po-210 to monitor for impacts to surface
19 water from uranium ISR operations.
20
21 The applicant has proposed operational monitoring of all impoundments within and adjacent to
22 the project area downgradient of proposed ISR facilities (e.g., wellfields, plants, pipelines, and
23 land application areas). Impoundments downstream of land application areas in the Dewey and
24 Burdock areas are listed in Table 7.3-2 and depicted in Figure 7.2-2. Prior to operations,
25 ambient water samples will be collected, when available, from the impoundments four times
26 and analyzed for the constituents listed in Table 7.3-1. All the impoundments will be sampled
27 on a quarterly basis throughout construction and operations and analyzed for the constituents
28 listed in Table 7.3-1 along with dissolved and suspended uranium, Ra-226, Th-230, Pb-210,
29 and Po-210.
30
31 ### 7.5.3   Process-Related Liquid Waste
32
33 Grab samples of process-related liquid wastewater will be collected monthly during operation of
34 each land application system and analyzed for the parameters listed in Table 7.3-1. In addition
35 to the parameters in Table 7.3-1, monthly liquid wastewater will be analyzed for compliance with
36 the 10 CFR Part 20, Appendix B radionuclide effluent discharge limits in Table 7.5-3.
37
38 ### 7.5.4   Soil
39
40 Two baseline soil samples will be collected from each quadrant of each center pivot (eight total
41 samples per pivot) prior to operation of land application systems. During operations, a minimum
42

Table 7.5-3. NRC Radionuclide Discharge Limits for Land Application

Radionuclide	µCi/ml	pCi/L
Pb-210	1E-8	10
Ra-226	6E-8	60
Uranium-natural	3E-7	300
Th-230	1E-7	100
Source: 10 CFR Part 20, Appendix B, Table 2, Column 2		

43

1   of two soil samples will be collected each year for each land application pivot active during
2   the year. Both the baseline and operational samples will be collected at depths of 0–46 and
3   46–91 cm [0–18 and 18–36 in] and analyzed for the parameters in Table 7.5-4.
4
5   **7.5.5   Biomass**
6
7   Samples of crops grown on three land application areas from each of the Dewey and Burdock
8   sites will be collected at the end of each irrigation season during operations. If crops are not
9   grown, samples of existing vegetation will be collected. Samples will be analyzed for the
10   parameters in Table 7.5-5.
11
12   Livestock samples will be collected during operation of land application systems if livestock
13   graze or consume crops grown on land application areas. The applicant will collect one grab
14   sample per year taken at the time of slaughter and have it analyzed for the parameters in
15   Table 7.5-5.

Table 7.5-4. Soil Sampling Parameters

Parameter
Conductivity, paste extract
pH, paste extract
Chloride, soluble
Chloride
Sulfate
Arsenic
Barium
Boron
Cadmium
Chromium
Lead
Mercury
Selenium
Silver
Vanadium
Nitrate as N, KCl extract
Uranium-natural
Ra-226
Th-230
Pb-210
Po-210
Source: Powertech, 2012

16

Table 7.5-5. Biomass Sampling Parameters

Constituent
Uranium-natural
Ra-226
Th-230
Pb-210
Po-210
Selenium
Arsenic
Source: Powertech, 2012

17

7.6   Class V Deep Injection Well Monitoring

This section describes the Class V deep injection well monitoring program the applicant proposed in its Class V UIC permit application submitted to EPA (Powertech, 2011, Appendix 2.7-L). The proposed injection zones for the Class V deep injection wells are the Minnelusa Formation and the Deadwood Formation (Figure 3.5-5). The applicant estimates the need for disposal capacity of 1,135 Lpm [300 gpm] {about 1,635,120 L [432,000 gal] per day per well assuming 24 hour/7 day injection}. Two Class V injection wells are proposed in the Dewey area: one injecting into the Deadwood and one injecting into the Minnelusa. Two deep Class V injection wells are also proposed in the Burdock area: one injecting into the Deadwood and one injecting into the Minnelusa. In all, this totals four deep injection wells. If the disposal capacity for either the Deadwood Formation or the Minnelusa Formation is not as great as anticipated, the EPA UIC Class V permit will allow up to four Class V wells each at the Dewey and the Burdock sites to increase the disposal capacity. The applicant's preference is to utilize the deep injection wells for the disposal of all process waste fluids, but if the deep injection wells cannot accommodate the total volume of waste fluids, land application will be used to dispose of the volume of waste fluids unable to be accommodated by the deep injection wells. EPA will not authorize injection into the Class V deep injection wells unless the permittee demonstrates the wells are properly sited, such that confinement zones and proper well construction minimize the potential for migration of fluids outside of the approved injection zone.

The deep injection wells are Class V wells because (i) Class I disposal wells are prohibited in South Dakota by state statute and (ii) the deep injection wells proposed for injection into the Minnelusa Formation would be injecting into or above an underground source of drinking water. (The definition for underground source of drinking water is found at 40 CFR Part 144.3 and p. 2-15 of this SEIS.) Although the deep injection wells are Class V wells, many of the protective requirements found at 40 CFR Part 146 Subpart B, Criteria and Standards Applicable to Class I Wells, will be included in the EPA UIC Class V Permit. Because Class V deep injection wells are being used for disposal rather than Class I wells, the injectate will have to be treated to remove radioactive constituents to below the radioactive waste standards at 10 CFR Part 20, Appendix B, Table II. The injectate will not need to be treated for injection into a Class I well. If the Total Dissolved Solids concentration in the proposed injection zone is below 10,000 mg/L [10,000 ppm], the injection zone is an underground source of drinking water. In that case, to be injected into an underground source of drinking water, the injectate will need to be treated to meet drinking water standards, or contaminant-specific background concentrations for constituents regulated under the Safe Drinking Water Act, whichever is greater.

A variety of data will be collected to monitor the deep injection well operations. This monitoring will use both periodic and continuous techniques. The EPA UIC Class V permit will require the annulus between the tubing and the long string of casings to be filled with a fluid and adequate pressure maintained on the annulus. The EPA UIC Class V permit will require installation and use of continuous recording devices to monitor injection pressure, flow rate and volume, and the pressure on the annulus between the tubing and the long string of casing as required under 40 CFR 146.13(b)(2). The continuous monitoring of the pressurized fluid-filled annulus will provide the necessary information for the internal mechanical integrity test required under 40 CFR 146.8(a)(1), which determines whether there is any significant fluid leak in the casing tubing an packer. The permit will also require a demonstration of external mechanical integrity pursuant to 40 CFR 146.8(a)(2) at least once every 5 years during the life of the well as required under 40 CFR 146.13(b)(3).

7.6.1   Injection Pressure Monitoring

As required by 40 CFR 146.13(a)(1), injection pressure at the wellhead shall not exceed a maximum value, which shall be calculated so as to assure that the pressure in the injection zone during injection does not initiate new fractures or propagate existing fractures in the injection zone. In no case shall injection pressure initiate fractures in the confining zone or cause the movement of injection or formation fluids into an underground source of drinking water. A data acquisition system will be used to monitor injection rate, injection pressure, annulus pressure, and simultaneous differential pressure. Maximum, minimum, and average values for each of the four parameters, along with total volume, will be recorded at least once every 15 minutes. Pressure transducers located near the wellhead and downstream of any pumping devices will be used to measure pressures. Flow rate is to be measured utilizing an inline turbine meter and totalizer or equivalent. In the case of a manned operation, well operators will be required to visually inspect the recorder and computer on a weekly basis when injection occurs to verify proper operation.

A backup power source (battery) will be used to ensure continuous collection of operating and well alarm data for up to a minimum of 30 minutes should power failure occur. If a power failure persists past the ability of the battery systems to allow power, the wells will be shut in. Upon discovery of the shut in, readings will be recorded a minimum of once every day until power is restored to the monitoring equipment.

If any of the permit conditions are exceeded, including injection pressure or differential pressure between the annulus pressure and the injection pressure, a visual alarm light will be illuminated at the well building. In addition, the computerized data acquisition system will be coupled to a telephone autodialer that will send a page to the operator to ensure that the condition is communicated. Upon an alarm condition, the operator will stop injection until the problem is identified and corrected and the system manually restarted.

7.6.2   Annulus Monitoring System

The permittee plans to fill the annulus area between the protective casings and injection tubing strings with fresh water containing an approved corrosion inhibitor. Annulus pressure will be continuously monitored to detect any potential leaks in the tubing or casing strings, and annulus pressures will be maintained at more than 100 psi above the tubing pressure.

The proposed annulus monitoring system will consist of an annulus fluid tank with a level indicator or site glass, pressure transducers and gauges, a nitrogen regulator, and a nitrogen supply cylinder. Annulus pressure in this system will be maintained with a nitrogen blanket supplied from pressurized nitrogen cylinders. In the event of power failure, positive pressure can still be maintained on the annulus.

The annulus tank will have sufficient reservoir capacity to accommodate double the anticipated volume fluctuations due to temperature and pressure limitations. The pressurized nitrogen cylinders will be replaced and recharged as required. The annulus tank is to be equipped with a level indicator or a full length armored reflex sight glass, a pressure relief valve, and an independent liquid fill nozzle. Well operators will record the annulus tank level and any annulus fluid added to the system.

The annulus pressure will be recorded continuously for each well. Electronic pressure transducers will be placed in pressure taps on the annulus system and injection flow lines. A

1   signal will be sent from these transducers to a digital recorder and/or a chart recorder. The
2   automated control system data will be visually inspected a minimum of once daily for anomalies
3   when the well is operating. As part of the process and controls, the monitoring system will
4   record maximum, minimum, and average information. Differential pressures (the difference
5   between the pressure applied to the annulus and the injection pressure) are to be obtained by
6   comparison of simultaneous readings of the annulus and injection pressure transducer readings
7   obtained for the wells.
8
9   In addition to the annulus pressure operating and monitoring requirements, an interlock system
10 will be installed to prevent the well from being operated if permit conditions are exceeded or if
11 unsafe conditions exist.
12
13 ### 7.6.3   Mechanical Integrity Demonstration
14
15 Under 40 CFR Part 146.8, periodic monitoring must be performed on both the internal and
16 external mechanical integrity of the deep disposal wells to demonstrate (i) there is no
17 significant leak in the casing, tubing, or packer and (ii) there is no significant fluid movement
18 into an underground source of drinking water through vertical channels adjacent to the injection
19 well bore.
20
21 ### 7.6.3.1   Internal Mechanical Integrity Demonstration
22
23 To demonstrate mechanical integrity for the casing, tubing and packer, the EPA UIC Class V
24 permit will require monitoring of the tubing–casing annulus pressure with sufficient frequency to
25 be representative while maintaining an annulus pressure different from atmospheric pressure
26 measured at the surface. Monitoring the pressure changes in the sealed annulus space is a
27 means of verifying the continued mechanical integrity of the well. The annulus pressure is to be
28 continually monitored to detect any leaks in the tubing or casing.
29
30 ### 7.6.3.2   External Mechanical Integrity Demonstration
31
32 To demonstrate that there is no significant fluid movement into an underground source of
33 drinking water through vertical channels adjacent to the injection well bore, the EPA UIC Class
34 V permit will require one of the following logs to be recorded once each fifth calendar year:
35 temperature, noise, or oxygen activation. If determined necessary because of operational or
36 regulatory concerns, casing inspection logs may be conducted to investigate corrosion when
37 tubing is already removed from the borehole during a workover or stimulation.
38
39 ### 7.6.4   Injection Zone Pressure Monitoring
40
41 The EPA UIC Class V permit will require monitoring of the pressure buildup in the injection zone
42 annually, including shutting down the well for a time sufficient to conduct a valid observation of
43 the pressure fall off as described under 40 CFR 146.13(d).
44
45 ### 7.6.5   Injectate Monitoring
46
47 The EPA UIC Class V permit will require the analysis of the injected fluids with sufficient
48 frequency to yield representative data of their characteristics. If the proposed injection zones
49 are demonstrated not to be underground sources of drinking water, the permit will require the
50 injectate to be treated to meet radioactive waste standards at 10 CFR Part 20, Appendix B,

1   Table II. If the proposed injection zones are underground sources of drinking water, the permit
2   will require the injectate to be treated to meet drinking water standards. Injectate characteristics
3   will be monitored by collecting samples following procedures of a permittee-proposed waste
4   analysis plan, which is reviewed and approved by EPA and becomes part of the permit
5   requirements. At a minimum, the composition parameters listed in Table 7.6-1 will be monitored
6   once quarterly for any quarterly period that fluid is injected.
7

8   ## 7.7   References

9
10   10 CFR Part 20. *Code of Federal Regulations*, Title 10, *Energy*, Part 20. "Standards for
11   Protection Against Radiation." Washington, DC: U.S. Government Printing Office.
12
13   10 CFR Part 20, Appendix B. *Code of Federal Regulations*, Title 10, *Energy*, Part 20. "Annual
14   Limits on Intake (ALIs) and Derived Air Concentrations (DACs) of Radionuclides for
15   Occupational Exposure; Effluent Concentrations; Concentrations for Release to Sewerage."
16   Washington, DC: U.S. Government Printing Office.
17
18   10 CFR Part 40. *Code of Federal Regulations*, Title 10, *Energy*, Part 40, "Domestic Licensing of
19   Source Material." Washington, DC: U.S. Government Printing Office.
20
21

Table 7.6-1. Composition Parameters for Class V Injectate Monitoring

Test Analyte/Parameter*
pH
total dissolved solids
total suspended solids
specific gravity
arsenic
barium
bicarbonate alkalinity
calcium
chloride
iron
lead
mercury
Ra-226
selenium
sodium
sulfate
Th-230
uranium
vanadium
*All metal analyses under the EPA UIC Class V permit are for total metals.

22
23

1   10 CFR Part 40, Appendix A. *Code of Federal Regulations*, Title 10, *Energy*,
2   Part 40 Appendix A. "Criteria Relating to the Operation of Uranium Mills and to the Disposition
3   of Tailings or Wastes Produced by the Extraction or Concentration of Source Material from Ores
4   Processed Primarily from their Source Material Content." Washington, DC: U.S. Government
5   Printing Office.
6
7   40 CFR Part 144. *Code of Federal Regulations*, Title 40, *Protection of the Environment*,
8   Part 144. "Underground Injection Control Program." Washington, DC: U.S. Government
9   Printing Office.
10
11 40 CFR Part 146. *Code of Federal Regulations*, Title 40, *Protection of the Environment*,
12 Part 146. "Underground Injection Control Program: Criteria and Standards." Washington, DC:
13 U.S. Government Printing Office.
14
15 ARSD Section 74:54:01:04. "Standards for Groundwater of 10,000 mg/L TDS Concentration or
16 Less." South Dakota Legislature Administrative Rules.
17
18 ARSD Section 74:54:02:28. "Out-of-Compliance Status." South Dakota Legislature
19 Administrative Rules.
20
21 Mackin, P.C., D. Daruwalla, J. Winterle, M. Smith, and D.A. Pickett. NUREG/CR–6733,
22 "A Baseline Risk-Informed Performance-Based Approach for *In-Situ* Leach Uranium Extraction
23 Licensees." Washington, DC: NRC. September 2001.
24
25 NRC (U.S. Nuclear Regulatory Commission). NUREG–1910, "Generic Environmental Impact
26 Statement for *In-Situ* Leach Uranium Milling Facilities." ML091480244, ML091480188.
27 Washington, DC: NRC. May 2009.
28
29 NRC. NUREG–1569, "Standard Review Plan for *In-Situ* Leach Uranium Extraction License
30 Applications—Final Report." Washington, DC: NRC. June 2003.
31
32 NRC. "Regulatory Guide 4.14, Radiological Effluent and Environmental Monitoring at Uranium
33 Mills, Rev. 1." Washington, DC: NRC. 1980.
34
35 Powertech [Powertech (USA) Inc.]. "Dewey-Burdock Project Groundwater Discharge Plan
36 Custer and Fall River Counties, South Dakota." ML12195A039, ML12195A040. Edgemont,
37 South Dakota: Powertech. March 2012.
38
39 Powertech. "Dewey-Burdock Project, Application for NRC Uranium Recovery License Fall River
40 and Custer Counties, South Dakota, Technical Report RAI Responses, June 2011."
41 ML112071064. Greenwood Village, Colorado: Powertech. 2011.
42
43 Powertech. "Dewey-Burdock Project, Application for NRC Uranium Recovery License Fall River
44 and Custer Counties, South Dakota ER_RAI Response August 11, 2010." ML102380516.
45 Greenwood Village, Colorado: Powertech. 2010.
46
47 Powertech. "Dewey-Burdock Project, Application for NRC Uranium Recovery License Fall River
48 and Custer Counties, South Dakota—Environmental Report." Docket No. 040-09075.
49 ML092870160. Greenwood Village, Colorado: Powertech. 2009a.
50

1   Powertech. "Dewey-Burdock Project, Application for NRC Uranium Recovery License Fall River
2   and Custer Counties, South Dakota—Technical Report." Docket No. 040-09075.
3   ML092870160. Greenwood Village, Colorado: Powertech. 2009b.
4
5   Powertech. "Dewey-Burdock Project, Supplement to Application for NRC Uranium Recovery
6   License Dated February 2009." Docket No. 040-09075. ML092870160. Greenwood Village,
7   Colorado: Powertech. 2009c.

8 COST-BENEFIT ANALYSIS

8.1   Introduction

This chapter summarizes benefits and costs associated with the proposed action and the
No-Action alternative. The proposed action is to issue the applicant, Powertech (USA) Inc., an
NRC license. The applicant will use the license for the construction, operation, aquifer
restoration, and decommissioning of the proposed Dewey-Burdock *in-situ* uranium recovery
(ISR) project. Section 4.11 of this Supplemental Environmental Impact Statement (SEIS)
discusses the potential socioeconomic impacts of the proposed action.

Implementation of the proposed action will generate regional and local benefits and costs. The
regional and local benefits of constructing and operating the proposed Dewey-Burdock ISR
Project include increases in employment, economic activity, and tax revenues. The benefits of
increased tax revenues will accrue primarily to Fall River and Custer Counties, South Dakota,
and the surrounding towns of Edgemont, Hot Springs, and Custer. Increases in economic
activity and employment may extend to Rapid City in neighboring Pennington County and the
city of Newcastle in Weston County, Wyoming. Costs associated with the proposed
Dewey-Burdock ISR Project will be, for the most part, limited to the area surrounding the site.
Examples of these costs include changes to current land and water use, and increased
road traffic.

8.2   Proposed Action (Alternative 1)

Under the proposed action, the NRC will issue the applicant an NRC license. With this license,
the applicant will construct, operate, restore the aquifer, and decommission the proposed
Dewey-Burdock ISR Project. Under the proposed action, the applicant is also seeking BLM
approval of its modified Plan of Operations subject to mitigation included in the license
application and this draft SEIS. Following 2 years of site development and facility construction,
there will be 8 years of wellfield and uranium recovery operations (see Figure 2.1-1). During the
8-year operations phase of the project, wellfield construction will continue as additional
wellfields are sequentially developed along the uranium roll fronts in both the Dewey and
Burdock areas. Wellfield restoration at the Dewey-Burdock site will begin immediately after
production activities in the wellfields end. The applicant projects that restoration activities in the
first wellfields will begin 2 years after production activities commence. Aquifer restoration
activities, including restoration construction, stability monitoring, and regulatory approval of
restoration, will continue for 11 years.

Some overlap between wellfield decommissioning and groundwater restoration activities is
expected. Wellfield decommissioning is estimated to continue for 8 years. Decommissioning of
the Burdock central processing plant and Dewey satellite facility will begin after aquifer
restoration and wellfield decommissioning activities are complete. It is anticipated that these
activities will take 2 years to complete (Powertech, 2009).

8.2.1   Benefits of the Proposed Action

The principal socioeconomic benefit expected to result from the Dewey-Burdock ISR Project is
an increase in employment opportunities in the region. The applicant expects to directly employ
86 workers during construction and 84 workers during operations of the proposed project

1   (Powertech, 2009). Fewer workers will be involved in aquifer restoration and decommissioning
2   activities (Powertech, 2010). The applicant expects nine workers will be directly involved in
3   aquifer restoration activities and nine workers will be directly involved in decommissioning
4   activities. As discussed in SEIS Section 4.11.1, the construction workforce will most likely not
5   relocate permanently to the area because of the short duration (1 to 2 years) of these activities.
6   Workers are expected to be more likely to relocate near the facility during the operations,
7   aquifer restoration, and decommissioning phases of the proposed project.
8
9   The majority of jobs are expected to be filled by workers from outside the region. A standard
10 employment multiplier of 0.7[1] was used to calculate the expected influx of approximately
11 60 jobs (i.e., 86 jobs × 0.7 = 60) during construction, 59 jobs (i.e., 84 jobs × 0.7 = 59) during
12 operations, 6 jobs during aquifer restoration (i.e., 9 jobs × 0.7 = 6), and 6 jobs during
13 decommissioning (i.e., 9 jobs × 0.7 = 6) activities.[1]
14
15 The town nearest to the proposed project is Edgemont, with a population of 774 (USCB, 2012).
16 However, employees supporting project activities might prefer to reside in larger surrounding
17 communities such as Hot Springs, Custer, and Newcastle, which have populations of 3,711,
18 2,067, and 3,532, respectively (USCB, 2012). The influx of jobs created by the Dewey-Burdock
19 ISR Project and the expected reduction in unemployment are expected to have a MODERATE
20 beneficial impact to the businesses of Edgemont and a SMALL beneficial impact to the
21 businesses of larger towns surrounding the proposed site, such as Hot Springs, Custer,
22 and Newcastle.
23
24 In addition to job creation, the proposed project's operations and the addition of regionally
25 based employees are expected to contribute to local, regional, and state revenues. Revenues
26 are expected to increase through the purchase of goods and services and through the taxes
27 levied on goods and services. Overall, the project is expected to generate $13.54 million in total
28 indirect business tax revenue over the lifetime of construction, operation, restoration, and
29 decommissioning activities (Powertech, 2009). Sources of indirect business tax revenue
30 include property taxes, sales taxes, and motor vehicle license charges.
31
32 The Special Tax Division of the Department of Revenue and Regulation of South Dakota levies
33 a severance tax of 4.5 percent (South Dakota Codified Law 10-39A-1), as well as a 0.24 percent
34 conservation tax (South Dakota Codified Law 10-39B-2), on the taxable value of the uranium
35 produced from uranium milling and mining. The applicant's estimate of uranium resources to be
36 recovered at the Dewey-Burdock ISR Project is 3.45 million kg [7.6 million lb] of uranium
37 (as U_3O_8) (Powertech, 2009). If the applicant fully recovers this quantity of uranium and sells it
38 at market prices of approximately $49.25 per pound (August 6, 2012, quoted price), the
39 severance tax is expected to yield $16,843,500 and the conservation tax is expected to yield
40 $898,320 in economic benefits over the life of the project. Fall River and Custer Counties would
41 collect 50 percent of the severance tax. The State of South Dakota collects the remainder of
42 the severance tax and the conservation tax.
43
44 In addition, the proposed Dewey-Burdock ISR Project is expected to generate
45 $186,700,000 in value-added benefits over the life of the project (Powertech, 2009). These

---

[1]The economic multiplier provides a statistical estimate of the total impact that is expected from a regional change in a given economic activity. The multiplier is a ratio of total change to initial change. The multiplier of 0.7 is used in these calculations because it is the standard employment multiplier for the milling/mining industry (Economic Policy Institute, 2003).

1   include employee wages and benefits; payments to self-employed individuals;
2   payments from interest, rents, royalties, dividends, and profits; and excise and sales taxes
3   paid on retail and commercial transactions.
4
5   **8.2.2   Benefits From Uranium Production**
6
7   The taxes to be generated by operations at the proposed Dewey-Burdock ISR Project will be
8   dependent on yellowcake production levels and the number of persons employed in facility
9   operations. The applicant projects 3.45 million kg [7.6 million lb] of uranium will be recovered.
10 However, production of yellowcake will depend on the market price for yellowcake (as uranium)
11 and production costs. Since 2007, the spot market price for uranium has fluctuated significantly,
12 from a high of more than $130 per pound in 2007 to a low of $40 per pound in 2009. As of
13 August 6, 2012, the price was $49.25 per pound (UXC, 2012).
14
15 The project's potential benefits to the local community depend on the applicant's operating costs
16 being lower than the future price of uranium. If the price of uranium falls below the costs of
17 operation, then operations would likely be suspended or discontinued.
18
19 **8.2.3   Costs to the Local Communities**
20
21 Table 8.2-1 lists the towns within an 80-km [50-mi] radius of the proposed project. These towns
22 are expected to provide the majority of the workers for the proposed project. The table also lists
23 the population of the towns and the distances to the proposed project site. As stated in
24 Section 8.2.1, the construction of the proposed project is expected to employ 86 workers, and if
25 it is assumed that the majority of the construction employment requirements are filled by a
26 workforce from outside the region, there could be an influx of 60 jobs (86 jobs × 0.7^2 = 60).
27 Because of the short duration of construction (1 to 2 years) and small size of the construction
28 force, the impact to housing demand would be SMALL (see SEIS Section 4.11.1.1). Workers
29 would not be expected to bring families and school-aged children with them; therefore, there
30 would be a SMALL impact on education services and on health and social services (see SEIS
31 Section 4.11.1.1).
32
33 As mentioned in SEIS Section 8.2.1, the proposed project is expected to employ
34 84 workers during the period of operations, 9 workers during the period of aquifer restoration,
35 and 9 workers during the period of site decommissioning. As described in SEIS
36 Section 4.11.1.2, employment types are expected to be more technical during operations, and
37

Table 8.2-1. Towns Near the Proposed Dewey-Burdock ISR Project

Town	Population (2010 Estimate)	Distance From Project in km [mi]
Edgemont, SD	774	21 [13]
Custer, SD	2,067	80 [50]
Hot Springs, SD	3,711	64 [40]
Newcastle, WY	3,532	64 [40]
Source: USCB (2012)		

38
39

---

[2]The multiplier of 0.7 is used in these calculations because it is the standard employment multiplier for the milling/mining industry (Economic Policy Institute, 2003).

1   as a result, the majority of the operational workforce is expected to be staffed from outside the
2   region. Therefore, it is anticipated that there will be an influx of workers into the towns closest
3   to the project area. Specifically, it is anticipated that there will be an influx of 59 workers
4   (84 jobs × 0.7^3 = 59) during operations, 6 jobs during aquifer restoration (i.e., 9 jobs × 0.7 = 6),
5   and 6 jobs during decommissioning (i.e., 9 jobs × 0.7 = 6) activities.
6
7   It is also expected that workers moving from outside the region to communities within
8   commuting distance of the Dewey-Burdock project site for employment opportunities will arrive
9   with their families. The average household size in the State of South Dakota is 2.42 persons
10   (USCB, 2012). Therefore, newly created jobs have the potential to increase the local population
11   by as many as 172 persons (59 + 6 + 6 = 71 workers from outside the region × 2.42 persons per
12   household = 172 persons). The influx of workers and their families will increase the demand for
13   housing and may spur an increase in the construction of new homes in towns surrounding the
14   proposed site. It is anticipated that the impact of increased housing demand and construction
15   may be MODERATE for small towns such as Edgemont. For larger towns such as Hot Springs,
16   Custer, and Newcastle, which have more available housing, the impact will be SMALL.
17
18   The projected population growth from the proposed project will have a SMALL impact on
19   education infrastructure and health and social services. As assessed in SEIS Section 4.11.1,
20   the impact on schools and education-related services during operations, aquifer restoration, and
21   decommissioning will be SMALL. As presented in SEIS Section 3.11.7, towns surrounding the
22   proposed project have adequate medical facilities, social services, and police, fire, and
23   emergency medical services to accommodate the projected project workforce and their families.
24   NRC staff discussions with city and county planners indicate that current and planned upgrades
25   to health care facilities and hospitals in the region will accommodate projected increases in
26   population (NRC, 2009). Furthermore, as discussed in Section 4.11.1, local governments are
27   expected to have the capacity to effectively plan for and manage increased demand for
28   health and social services from workers and their families relocating to towns near the
29   proposed project.
30

31   ## 8.3   Evaluation of Findings of the Proposed Dewey-Burdock Project
32
33   If NRC issues the applicant a license, it is anticipated that the Dewey-Burdock ISR Project will
34   have a SMALL to MODERATE overall economic impact on the region of influence and will
35   generate primarily regional and local benefits and costs. As discussed earlier, the regional
36   benefits of the project are increased employment opportunities and increased economic activity
37   that will add to tax revenues in the region. Increases in tax revenues are expected to bring the
38   largest benefit to Fall River and Custer Counties, although economic benefits will most likely be
39   shared by neighboring counties and communities in South Dakota and Wyoming. Social and
40   economic costs associated with the Dewey-Burdock project will, for the most part, be limited to
41   communities within commuting distance of the site. Table 8.3-1 summarizes the costs and
42   benefits of the proposed Dewey-Burdock ISR Project.
43

44   ## 8.4   No Action (Alternative 2)
45
46   Under the No-Action alternative, NRC will not approve the license application for the proposed
47   Dewey-Burdock ISR Project and the Bureau of Land Management (BLM) will not approve the
48   applicant's modified Plan of Operations. The No-Action alternative will result in the applicant not

---

[3]Ibid.

1

Table 8.3-1. Summary of Costs and Benefits of the Proposed Dewey-Burdock ISR Project

Cost-Benefit Category	Proposed Action
Benefits	
Production Capacity	7.6 million pounds of yellowcake (as uranium)
Other Monetary: Severance and conservation taxes Indirect business tax revenues	 $17.7 million (estimated) $13.54 million (estimated)
Nonmonetary benefits (50% of jobs would be from Custer and Fall River Counties)	86 jobs—during construction 60 jobs—local jobs from economic multiplier during construction 84 jobs—during operations 59 jobs—local jobs from economic multiplier during operations 9 jobs—during aquifer restoration 6 jobs—local jobs from economic multiplier during aquifer restoration 9 jobs—during decommissioning 6 jobs—local jobs from economic multiplier during decommissioning
Costs	
Education Infrastructure	SMALL
Health and Social Services	SMALL
Housing Demand	SMALL for larger towns (Hot Springs, Custer, Newcastle) MODERATE for Edgemont
Emergency Response	SMALL
Source: Powertech (2009, 2010)	

2
3 constructing and operating the proposed project. No facilities, roads, or wellfields will be built,
4 and no pipelines will be laid as described in SEIS Section 2.1.2. No uranium will be recovered
5 from the subsurface ore body; therefore, injection, production, and monitoring wells will not be
6 installed to operate the facility. No lixiviant will be introduced in the subsurface, and no
7 buildings will be constructed to process extracted uranium or store chemicals involved in
8 that process. Because no uranium will be recovered, neither aquifer restoration nor
9 decommissioning activities will occur. No liquid or solid effluents will be generated. As a result,
10 the proposed site will not be disturbed by proposed project activities and ecological, natural, and
11 socioeconomic resources will remain unaffected. All potential environmental impacts from the
12 proposed action will be avoided. Similarly, all project-specific socioeconomic impacts
13 (e.g., employment, economic activity, population, housing, and local finance) will also
14 be avoided.
15
16

8.5   References

Economic Policy Institute. "Updated Employment Multipliers for the U.S. Economy." ML12243A398. Washington, DC: Economic Policy Institute. 2003. <http://www.epi.org/page/-/old/workingpapers/epi_wp_268.pdf> (30 August 2010).

NRC (U.S. Nuclear Regulatory Commission). "Site Visit to the Proposed Dewey-Burdock Uranium Project, Fall River and Custer Counties, South Dakota, and Meetings with Federal, State, and County Agencies, and Local Organizations, November 30–December 4, 2009." ML093631627. Washington, DC: NRC. 2009.

Powertech (Powertech (USA) Inc.). "Response to U.S. Nuclear Regulatory Commission Request for Additional Information Powertech (USA) Inc. Dewey-Burdock Project Environmental Review of Application for a U.S. Nuclear Regulatory Commission Source Material License, August 2010." Docket No. 040-09075. ML102238516. Greenwood Village, Colorado: Powertech. 2010.

Powertech. "Dewey-Burdock Project, Application for NRC Uranium Recovery License Fall River and Custer Counties, South Dakota—Environmental Report." Docket No. 040-09075. ML092870160. Greenwood Village, Colorado: Powertech. 2009.

South Dakota Codified Law 10-39A-1. Severance Tax Imposed on Energy Minerals—Rate. South Dakota Legislature.

South Dakota Codified Law 10-39B-2. Imposition of Tax—Rate—Payment—Disposition–Collection. South Dakota Legislature.

USCB (United States Census Bureau). "American Factfinder, Census 2000 and 2010, 2006–2010 American Community Survey 5-Year Estimate, State and County Quickfacts." ML12248A240. 2012. <http://factfinder.census.gov, http://quickfacts.census.gov> (17 April 2012).

UXC (The Ux Consulting Company). "Ux U_3O_8 Prices." 2012. <http://www.uxc.com/> (27 July 2012).

9 SUMMARY OF ENVIRONMENTAL IMPACTS

This chapter summarizes the potential environmental impacts of the proposed action and the No-Action alternative. The potential impacts of the proposed action are discussed in terms of (i) unavoidable adverse environmental impacts, (ii) irreversible and irretrievable commitments of resources, (iii) short-term impacts and uses of the environment, and (iv) long-term impacts and the maintenance and enhancement of productivity. The information is presented for each of the 13 resource areas that may be affected by the proposed Dewey-Burdock ISR Project. This information addresses the impacts during each phase of the project (i.e., construction, operation, aquifer restoration, and decommissioning). The specific impacts are described in Table 9-1.

The following terms are defined in NUREG–1748 (NRC, 2003).

- Unavoidable adverse environmental impacts: applies to impacts that cannot be avoided and for which no practical means of mitigation are available

- Irreversible: involves commitments of environmental resources that cannot be restored

- Irretrievable: applies to material resources and will involve commitments of materials that, when used, cannot be recycled or restored for other uses by practical means

- Short-term: represents the period from preconstruction to the end of the decommissioning activities and, therefore, generally affects the present quality of life for the public

- Long-term: represents the period of time following the termination of the site license, with the potential to affect the quality of life for future generations

As discussed in Chapter 4, the significance of potential environmental impacts is categorized as follows:

SMALL: The environmental effects are not detectable or are so minor that they will neither destabilize nor noticeably alter any important attribute of the resource

MODERATE: The environmental effects are sufficient to alter noticeably, but not to destabilize, important attributes of the resource

LARGE: The environmental effects are clearly noticeable and are sufficient to destabilize important attributes of the resource

The alternatives and their environmental impacts are summarized in the following sections. Section 9.1 describes the environmental impacts from implementing the proposed action, and Section 9.2 describes the environmental impacts from implementing the No-Action alternative.

9.1   Proposed Action (Alternative 1)

Powertech (USA) Inc. (Powertech, referred to herein as the applicant) is seeking an NRC source material license for the construction, operation, aquifer restoration, and decommissioning of the proposed Dewey-Burdock ISR Project (Powertech, 2009a–c). Under

1   the proposed action, NRC would grant Powertech's license request. The proposed project will
2   consist of processing facilities and sequentially developed wellfields sited in two contiguous
3   areas: the Burdock area and the Dewey area.
4
5   Construction of the Dewey-Burdock ISR Project is expected to last about 2 years (see
6   Figure 2.1-1). During this phase, the applicant will construct buildings, access roads, wellfields,
7   pipelines, Class V injection wells, and potential land application areas to be used for liquid
8   waste disposal. Operations are expected to last 8 years. Construction and operations activities
9   would disturb approximately 98 ha [243 ac] if deep well disposal via Class V injection wells is
10   used to dispose of treated wastewater and approximately 566 ha [1,398 ac] if land application is
11   used to dispose of treated wastewater (Powertech, 2010).
12
13   During the operations phase, injection wells will be used to inject lixiviant (recovery) solutions
14   into the orebody to recover uranium. Production wells will be used to recover the dissolved
15   uranium, which then will be processed through the central plant. Finally, monitoring wells will be
16   installed to monitor the performance of the wellfields and to mitigate potential excursions from
17   the production zone.
18
19   Approximately 0.45 million kg [1 million lb] of U_3O_8 (triuranium octoxide) would be produced per
20   year. After operations at a wellfield cease, the applicant will have to begin aquifer restoration,
21   which will ensure that water quality and groundwater use from surrounding aquifers is not
22   impacted by the proposed action.
23
24   The aquifer restoration process is expected to last about 9 years. The methods selected for
25   aquifer restoration will depend on the liquid waste disposal option. For the Class V deep
26   injection well disposal option, groundwater treatment using reverse osmosis (RO) with permeate
27   injection will be the primary restoration method (Powertech, 2011). If land application is used
28   for liquid waste disposal, then groundwater sweep with injection of clean makeup water from the
29   Madison Formation will be used to restore the aquifer. During wellfield and facility
30   decommissioning (expected to last 10 years), disturbed lands will be returned to their prior uses.
31   Wells will be plugged and abandoned, and the land surface will be reclaimed.
32
33   The potential environmental impacts from the proposed action are summarized in Table 9-1.
34

9.2   No Action (Alternative 2)

37   Under the No-Action alternative, NRC would not issue a license. The applicant will neither
38   construct buildings, roads, or wellfields nor will the facility be operated at the proposed
39   Dewey-Burdock ISR Project. Uranium ore will not be recovered from the site, and the applicant
40   will not receive a license. Under the No-Action alternative, there will be no impact to any of the
41   13 resource areas from the proposed licensing action. There will be no unavoidable adverse
42   environmental impacts attributable to the proposed action and no relationship between local
43   short-term or long-term uses of the environment. Therefore, there will be no irreversible and
44   irretrievable commitment of resources.
45

9.3   References

48   10 CFR Part 20. *Code of Federal Regulations*, Title 10, *Energy*, Part 20. "Standards for
49   Protection Against Radiation." Washington, DC: U.S. Government Printing Office.

1   NRC (U.S. Nuclear Regulatory Commission). NUREG–1748, "Environmental Review Guidance
2   for Licensing Actions Associated With NMSS Programs." Washington, DC: NRC. August
3   2003.
4
5   Powertech. "Dewey-Burdock Project, Application for NRC Uranium Recovery License Fall River
6   and Custer Counties, South Dakota, Technical Report RAI Responses, June, 2011."
7   ML112071064. Greenwood Village, Colorado: Powertech (USA) Inc. 2011.
8
9   Powertech. "Dewey-Burdock Project, Application for NRC Uranium Recovery License Fall River
10 and Custer Counties, South Dakota ER_RAI Response August 11, 2010." ML102380516.
11 Greenwood Village, Colorado: Powertech (USA) Inc. 2010.
12
13 Powertech. "Dewey-Burdock Project, Application for NRC Uranium Recovery License Fall River
14 and Custer Counties, South Dakota—Environmental Report." Docket No. 040-09075.
15 ML092870160. Greenwood Village, Colorado: Powertech. August 2009a.
16
17 Powertech. "Dewey-Burdock Project, Application for NRC Uranium Recovery License Fall River
18 and Custer Counties, South Dakota—Technical Report." Docket No. 040-09075.
19 ML092870160. Greenwood Village, Colorado: Powertech. August 2009b.
20
21 Powertech. "Dewey-Burdock Project, Supplement to Application for NRC Uranium Recovery
22 License Dated February 2009." Docket No. 040-09075. ML092870160. Greenwood Village,
23 Colorado: Powertech. August 2009c.

Table 9-1. Summary of Environmental Impacts of the Proposed Action

Impact Category	Unavoidable Adverse Environmental Impacts	Irreversible and Irretrievable Commitment of Resources	Short-Term Impacts and Uses of the Environment	Long-Term Impacts and the Maintenance and Enhancement of Productivity
Land Use (SEIS Section 4.2.1)	There will be a SMALL impact to land use. During construction and operation, the total amount of land affected by earthmoving activities to construct surface facilities, wellfields and associated infrastructure, and to build access roads will depend on the option used to dispose of liquid wastes. For Class V well injection, approximately 98 ha [243 ac] or 2 percent	No impact. There will be no irreversible and irretrievable commitment of land resources from implementing the proposed action. The duration of the project will be approximately 17 years after which time the land could be reclaimed and made available for other uses.	There will be a SMALL impact to land use from implementing the proposed action. Depending on the option used to dispose of liquid wastes, approximately 98 ha [243 ac] (Class V well injection) or 566 ha [1.398 ac] (land application) of the proposed license area will be unavailable for other uses such as grazing and recreation; oil and	There will be no long-term impact to land resources from implementing the proposed action. The land will be available for other uses at the end of the license period.

24

Table 9-1. Summary of Environmental Impacts of the Proposed Action (continued)

Impact Category	Unavoidable Adverse Environmental Impacts	Irreversible and Irretrievable Commitment of Resources	Short-Term Impacts and Uses of the Environment	Long-Term Impacts and the Maintenance and Enhancement of Productivity
	of the proposed license area will be disturbed. For land application, approximately 566 ha [1,398 ac] or 13 percent of the proposed license area will be disturbed. During decommissioning, land will be impacted by earthmoving activities to reclaim and reseed the affected areas.		gas exploration could coexist with the applicant's proposed action.	
Transportation (SEIS Section 4.3.1)	During the construction and operation phases, there will be a MODERATE increase in local traffic counts associated with project-related traffic on Dewey Road, the nearest road to the proposed project. Increased traffic will degrade the road surface, increase dust generation, and increase the potential for traffic accidents and wildlife and livestock kills. During all phases, there will be a SMALL increase in traffic on the more well-traveled regional roads.	There will be an irreversible and irretrievable commitment of fuel for vehicle and equipment operation, heating, commuter traffic, and regional transport.	During construction and operations, there will be a MODERATE impact due to increased traffic on Dewey Road, which will degrade the road surface, increase dust generation, and increase the potential for traffic accidents and wildlife and livestock kills. During operation, aquifer restoration, and decommissioning, there will be a SMALL increased accident risk from transporting yellowcake, ion-exchange resin, byproduct material, and hazardous chemicals. During construction, no	There will be no long-term impacts to transportation following license termination.

Table 9-1. Summary of Environmental Impacts of the Proposed Action (continued)

Impact Category	Unavoidable Adverse Environmental Impacts	Irreversible and Irretrievable Commitment of Resources	Short-Term Impacts and Uses of the Environment	Long-Term Impacts and the Maintenance and Enhancement of Productivity
			short-term hazardous material transportation impacts will occur because no chemical or radioactive material will be transported.	
Geology and Soil (SEIS Section 4.4.1)	There will be a SMALL impact on geology and soils. The construction, operations, and decommissioning phases will disturb surface soils during construction of the central and satellite plants, development of the wellfields, laying of pipelines, and construction of new access roads. These impacts will be temporary, and at the end of the decommissioning phase topsoil will be replaced and reseeded.	Soil layers will be irreversibly disturbed by the proposed action; however, topsoil salvaged during the construction phase will be stored and replaced during decommissioning. Therefore, the potential impact will be SMALL. Reseeding and recontouring will mitigate the impact to topsoil.	There will be a SMALL impact to geology and soils. No significant matrix compression or ground subsidence is expected because the net withdrawal of fluid from the production zone aquifers will be about 3 percent or less. Approximately 5.3 ha [13 ac] of topsoil will be stripped. Topsoil salvaged during the construction phase of the project will be replaced during the reclamation and reseeding processes.	There will be no long-term impacts to geology and soils following license termination.
Surface Waters and Wetlands (SEIS Section 4.5.1.1)	There will be a SMALL impact to surface water and wetlands from the proposed action. The occurrence of surface water is limited, and surface water flow in channels is intermittent. U.S. Army Corps of Engineers permits under Section 404 of	There will be no irreversible and irretrievable commitment of either surface water or wetlands from implementing the proposed action. No drainage or body of water will be significantly altered by the proposed action. The impact	There will be a SMALL impact to surface waters and wetlands. The proposed action will not discharge to perennial or ephemeral surface water drainages.	No impact. The proposed action will not discharge to perennial or ephemeral surface water drainages.

Table 9-1. Summary of Environmental Impacts of the Proposed Action (continued)

Impact Category	Unavoidable Adverse Environmental Impacts	Irreversible and Irretrievable Commitment of Resources	Short-Term Impacts and Uses of the Environment	Long-Term Impacts and the Maintenance and Enhancement of Productivity
	the Clean Water Act will be required before conducting work in jurisdictional wetlands. The applicant will use best management practices and implement a storm water pollution management plan to ensure surface water runoff from disturbed areas meets NPDES permit limits.	to wetlands will be SMALL because stream flow is intermittent and the applicant will implement best management practices to control erosion, runoff, and sedimentation.		
Groundwater (SEIS Section 4.5.2.1)	There will be a SMALL impact on groundwater from implementing the proposed action by consumption of groundwater, degradation of water quality in the ore production zone, and the drawdown in water levels in wells located outside the project boundaries that are drilled into the ore-bearing aquifer(s). The applicant will provide alternative water sources in the event of significant drawdown to private wells adjacent to the proposed project area. The establishment of an inward hydraulic gradient, as well as an applicant-installed groundwater monitoring network to detect potential	There will be a SMALL impact on groundwater resources. Between 97 and 99.5 percent of groundwater used during the ISR process at the proposed project will be treated and reinjected into the subsurface and/or applied to land irrigation areas. Between 0.5 and 3 percent of groundwater will be consumed.	Short-term impacts to groundwater will include degradation of water quality in production zones and the potential to draw down the water level in neighboring private wells. These impacts will be SMALL. The applicant will provide alternative water sources if water-level drawdowns affect water yields in domestic and livestock wells within and adjacent to the proposed project area.	There will be no long-term impacts to groundwater resources. Both the State of South Dakota and NRC require restoration of affected groundwater following operations. The groundwater quality will be restored to ensure that aquifers will not be affected. Although water levels will be affected in the short term, the water levels will eventually recover after operations and aquifer restoration are completed.

Table 9-1. Summary of Environmental Impacts of the Proposed Action (continued)

Impact Category	Unavoidable Adverse Environmental Impacts	Irreversible and Irretrievable Commitment of Resources	Short-Term Impacts and Uses of the Environment	Long-Term Impacts and the Maintenance and Enhancement of Productivity
	vertical and horizontal excursions, will limit the potential for undetected groundwater excursions that could degrade groundwater quality.			
Ecological Resources (SEIS Section 4.6.1)	There will be SMALL to MODERATE impacts until vegetation has been reestablished, and then the impact will be SMALL. Construction and decommissioning of the proposed Dewey-Burdock Project will result in short-term loss (over the ISR facility lifecycle) of vegetation on approximately 98 ha [243 ac] if deep Class V well injection is used to dispose of liquid wastes and approximately 566 ha [1,398 ac] if land application is used to dispose of liquid wastes. The short-term loss of vegetation could stimulate the introduction and spread of undesirable and invasive, nonnative species, and displacement of wildlife species. During operations and aquifer restoration, use of fences will limit.	Vegetative communities directly impacted by earthmoving activities and wildlife injuries and mortalities will be irreversible. However, the implementation of mitigation measures, such as the use of fencing to limit wildlife movement and the applicant's enforcement of speed limits, will reduce potential impacts to wildlife. Furthermore, areas impacted by earthmoving activities will be reclaimed and reseeded.	During any of the ISR phases, SMALL direct impacts to ecological resources could include injuries and fatalities to wildlife caused by either collisions with project-related traffic or habitat removal actions involving the removal of topsoil. Habitat disruption will consist of scattered, confined drill sites for the deep Class V injection well option. Large transformation of the existing habitat would be a MODERATE impact during the decommissioning phase of the deep Class V injection well disposal option and during all facility lifecycle phases of the land application option. Wildlife could be temporarily displaced by increased noise and traffic during either waste	Some of the vegetative communities that exist within the proposed Dewey-Burdock Project could be difficult to reestablish through artificial plantings, and natural seeding could take many years resulting in MODERATE long-term impacts. Wildlife species associated with those communities could experience SMALL to MODERATE long-term impacts if animal populations are reduced in number or replaced by other species with broader habitat requirements.

Table 9-1. Summary of Environmental Impacts of the Proposed Action (continued)

Impact Category	Unavoidable Adverse Environmental Impacts	Irreversible and Irretrievable Commitment of Resources	Short-Term Impacts and Uses of the Environment	Long-Term Impacts and the Maintenance and Enhancement of Productivity
	wildlife ingress and egress to wellfields		disposal option. The applicant has committed to implement mitigation measures to reduce the potential impact to SMALL for wildlife species.	
Meteorology, Climatology, and Air Quality (SEIS Section 4.7.1)	There will be a SMALL to MODERATE impact to air quality. During all four phases the generation of air pollutants results in the degradation of air quality. Combustion emissions in and around the proposed site will be lower than NAAQS and PSD Class II regulatory thresholds. Fugitive dust emissions will also be lower than these regulatory thresholds. However, due to the level and nature of fugitive emissions, there is potential for intermittent impacts to localized areas in and around the proposed site. Fugitive emission will also contribute to visibility impacts at Wind Cave National Park, but the impact from the proposed action will be minimal.	There will be no irreversible or irretrievable commitment of air resources from the proposed action.	There will be SMALL to MODERATE impacts. Fugitive dust generated from all four phases has the potential to result in short-term, intermittent impacts in and around the site particularly when vehicles travel on unpaved roads. The effect will be localized and temporary. Use of mitigation measures, such as applying water for dust suppression, will limit fugitive dust emissions.	No impact. There will be no long-term effect on air quality either from the proposed project or following license termination.

1

Table 9-1. Summary of Environmental Impacts of the Proposed Action (continued)

Impact Category	Unavoidable Adverse Environmental Impacts	Irreversible and Irretrievable Commitment of Resources	Short-Term Impacts and Uses of the Environment	Long-Term Impacts and the Maintenance and Enhancement of Productivity
Noise (SEIS Section 4.8.1)	There will be a SMALL impact. Two onsite dwellings (Daniels residence and Beaver Creek Ranch Headquarters) will experience noise above background levels due to their proximity to wellfields and land application areas. However, noise impacts at these residences will be short term, intermittent, and mitigated by sound abatement controls on operating equipment. Noise impacts to raptors will be mitigated by adhering to timing and spatial restrictions within specified distances of active raptor nests as determined by appropriate regulatory agencies (e.g., BLM, FWS, and SDGFP).	Not applicable.	There will be a SMALL impact on two onsite dwellings (Daniels residence and Beaver Creek Ranch Headquarters) due to their proximity to wellfields and land application areas. However, noise impacts at these residences will be short-term, intermittent, and mitigated by sound abatement controls on operating equipment.	No impact. There will be no noise impact following license termination.
Historical and Cultural Resources (SEIS Section 4.9.1)	Impact on historic and cultural resources during the ISR construction phase will be SMALL to LARGE. To mitigate the impact, NRC, BLM, SD SHPO, tribes, and the applicant will develop and execute an agreement that will formalize treatment plans for adversely impacted resources	If archaeological and historic sites cannot be avoided, or the impacts to these sites cannot be mitigated, this could result in an irreversible and irretrievable loss of cultural resources.	There will be a SMALL to LARGE impact on historic and cultural resources during the ISR construction phase. The development of an agreement between NRC, BLM, SD SHPO, tribes, and the applicant will address adverse impacts to cultural	If potential impacts from implementation of the proposed action are not mitigated, then long-term impacts to cultural and historic resources will result.

Table 9-1. Summary of Environmental Impacts of the Proposed Action (continued)

Impact Category	Unavoidable Adverse Environmental Impacts	Irreversible and Irretrievable Commitment of Resources	Short-Term Impacts and Uses of the Environment	Long-Term Impacts and the Maintenance and Enhancement of Productivity
	during construction. If other NRHP-eligible sites cannot be avoided, then treatment plans will be developed. If other historic and cultural resources are encountered during the ISR lifecycle, the applicant will notify the appropriate authorities per an unexpected discovery plan.		and historic sites and historic properties of traditional religious and cultural importance to Native American tribes. If any unidentified historic or cultural resources are encountered, work will stop and appropriate authorities will be notified per the unexpected discovery plan.	
Visual and Scenic Resources (SEIS Section 4.10.1)	There would be a SMALL impact on the visual landscape. Visual impacts from drilling and earthmoving activities that generate fugitive dust will be short term. Mitigation measures will be implemented to reduce fugitive dust and visual impacts from buildings. Center pivot irrigation systems in proposed land application areas in the Dewey area will be visible to travelers on Dewey Road; however, Dewey Road is lightly traveled with few residences. Proposed activities will be consistent with the BLM VRM Class III and IV designation for the area.	No impact.	There will be a SMALL short-term impact to the visual landscape from implementing the proposed action. The activities will be consistent with the BLM VRM Class III and IV designation of the area and the existing natural resource exploration activities in the area.	No impact. There will be no impact on the visual landscape following license termination.

Table 9-1. Summary of Environmental Impacts of the Proposed Action (continued)

Impact Category	Unavoidable Adverse Environmental Impacts	Irreversible and Irretrievable Commitment of Resources	Short-Term Impacts and Uses of the Environment	Long-Term Impacts and the Maintenance and Enhancement of Productivity
Socioeconomics (SEIS Section 4.11.1)	Implementing the proposed action will have a SMALL socioeconomic impact over the life of the project.	Not applicable.	Implementing the proposed action will have a SMALL impact on local communities.	Following license termination, workers who supported activities at the Dewey-Burdock site will need to find other employment. There will be a loss of revenue to nearby communities, Fall River and Custer Counties, and the state following license termination.
Environmental Justice (SEIS Section 4.12.1)	There will be no disproportionately high and adverse impacts to minority or low-income populations from the construction, operation, aquifer restoration, and decommissioning of the proposed Dewey-Burdock ISR Project. While certain Native Americans may have a heightened interest in cultural resources potentially affected by the proposed action, the impacts to Native Americans in this and other areas is not expected to be disproportionately high or adverse.	Not applicable.	Implementing the proposed action will have a SMALL impact on environmental justice. There will be no disproportionately high and adverse impacts to minority or low-income populations from the construction, operation, aquifer restoration, and decommissioning of the proposed Dewey-Burdock ISR Project.	There will be no long-term environmental justice impacts following license termination. While certain Native Americans have a heightened interest in cultural resources potentially affected by the proposed action, the impacts to Native Americans in this and other areas is not expected to be disproportionately high or adverse. To the extent there might be adverse impacts to historic and cultural sites of interest to Native Americans, these impacts will be mitigated by an agreement that will formalize treatment plans during construction. If NRHP-eligible

1

Table 9-1. Summary of Environmental Impacts of the Proposed Action (continued)

Impact Category	Unavoidable Adverse Environmental Impacts	Irreversible and Irretrievable Commitment of Resources	Short-Term Impacts and Uses of the Environment	Long-Term Impacts and the Maintenance and Enhancement of Productivity
				sites cannot be avoided, treatment plans will be developed. If other historic and cultural resources are encountered during the ISR lifecycle, the applicant will notify appropriate authorities per an unexpected discovery plan.
Public and Occupational Health (SEIS Section 4.13.1)	There will be a SMALL impact on public and occupational health. Construction and decommissioning will generate fugitive dust emissions that will not result in a significant dose to the public or site workers. The emissions from construction equipment will be of short duration and readily dispersed into the atmosphere.	Not applicable.	There will be a SMALL impact from radiological exposure. Dose calculations under normal operations showed that the highest potential dose within the proposed project area is 6 percent of the 1 mSv [100 mrem] per year public dose limit specified in NRC regulations. The radiological impacts from accidents will be SMALL for workers if procedures to deal with accident scenarios are followed, and SMALL for the public because of the facility's remote location. The nonradiological public and occupational health impacts from normal operations, accidents, and	No impact. There will be no long-term impact to public and occupational health following license termination.

1

Table 9-1. Summary of Environmental Impacts of the Proposed Action (continued)

Impact Category	Unavoidable Adverse Environmental Impacts	Irreversible and Irretrievable Commitment of Resources	Short-Term Impacts and Uses of the Environment	Long-Term Impacts and the Maintenance and Enhancement of Productivity
			chemical exposures will be SMALL if handling and storage procedures are followed.	
Waste Management (SEIS Section 4.14.1)	Solid byproduct material generation and disposal from activities implemented during all postconstruction phases of the Dewey-Burdock ISR Project will result in SMALL impacts on available disposal capacity, because permitted facilities are available to accept the wastes. Disposal of treated liquid byproduct material using Class V injection, land application, or a combination of both will be conducted in accordance with NRC effluent discharge limits in 10 CFR Part 20, Appendix B and EPA (Class V well) or state (land application) permit conditions, and impacts will be SMALL. During decommissioning, the amount of nonhazardous solid waste will exceed available local landfill capacity and will result in MODERATE impacts unless local capacity is expanded prior to decommissioning or waste is shipped to a	The energy consumed during the ISR phases, the construction materials used that could not be reused or recycled, and the space used to properly handle and dispose of all waste types (i.e., wells for liquid wastes and permitted disposal space of solid wastes) will represent an irretrievable commitment of resources, resulting in a SMALL to MODERATE impact.	During all phases, hazards associated with handling and transport of wastes will represent a short-term and SMALL impact.	During all phases, permanent disposal of liquid wastes in onsite injection wells will represent a SMALL impact on the long-term productivity of the land allocated for these wells. Buildup of constituents in soil from potential land application of treated liquid wastes could affect productivity of irrigated land, but proposed monitoring is expected to detect potential problems early, resulting in a SMALL impact.

Table 9-1. Summary of Environmental Impacts of the Proposed Action (continued)

Impact Category	Unavoidable Adverse Environmental Impacts	Irreversible and Irretrievable Commitment of Resources	Short-Term Impacts and Uses of the Environment	Long-Term Impacts and the Maintenance and Enhancement of Productivity
	larger regional landfill; then impacts will be SMALL.			

1

10 LIST OF PREPARERS

This section documents all individuals who were involved with the preparation of this Supplemental Environmental Impact Statement. Contributors include staff from the U.S. Nuclear Regulatory Commission and consultants. Each individual's role, education, and experience are outlined next.

10.1   U.S. Nuclear Regulatory Commission Contributors

Haimanot Yilma: SEIS Project Manager
 M.B.A , University of Maryland, College Park, 2010
 B.S., Chemical Engineering, University of Maryland, College Park, 1998
 Year of Experience: 12

Kellee Jamerson: SEIS Co-Project Manager
 B.S., Environmental Science, Tuskegee University, 2006
 Years of Experience: 3

Jennifer A. Davis: Cultural Resources Reviewer
 B.A., Historic Preservation and Classical Civilization (Archaeology), Mary Washington College, 1996
 Years of Experience: 12

Nathan Goodman: Ecology Reviewer
 M.S., Environmental Science, Johns Hopkins University, 2000
 B.S., Biology, Muhlenberg College, 1998
 Years of experience: 11

Asimios Malliakos: Socioeconomics and Cost Benefit Reviewer
 Ph.D., Nuclear Engineering, University of Missouri-Columbia, 1980
 M.S., Nuclear Engineering, Polytechnic Institute of New York, 1977
 B.S., Physics, University of Thesealoniki, Greece, 1975
 Years of Experience: 32

Johari Moore: Health Physics Reviewer
 M.S., Nuclear Engineering and Radiological Sciences, University of Michigan, 2005
 B.S., Physics, Florida Agricultural and Mechanical University, 2003
 Years of Experience: 7

Stephen J. Cohen: Team Leader, Hydrogeologist
 Registered Professional Geologist, PA—1994
 M.S., Geological Engineering, University of Idaho, 2004
 Certificate of Continuing Engineering Studies, Johns Hopkins University, 1998
 B.S., Geology, University of Maryland, 1986
 Years of experience: 25

1   Ronald A. Burrows: Safety Project Manager
2   Certified Health Physicist, 1999
3   Registered Radiation Protection Technologist, 1997
4   M.S., Health Physics, Texas A&M University, 1995
5   MBA, Southern New Hampshire University, 1991
6   B.S., Nuclear Engineering, University of New Mexico, 1988
7   Years of experience: 23
8
9 **10.2   Center for Nuclear Waste Regulatory Analyses**
10 **(CNWRA®) Contributors**
11
12   Hakan Basagaoglu: Water Resources
13   Ph.D., Civil and Environmental Engineering, University of California, Davis 2000
14   M.S., Geological Engineering, Middle East Technical University, Turkey 1993
15   B.S., Geological Engineering, Middle East Technical University, Turkey 1991
16   Years of Experience: 19
17
18   Paul Bertetti: Environmental Measurements and Monitoring, Public and Occupational Health
19   and Safety, Water Resources
20   M.S., Geology, University of Texas at San Antonio, 1999
21   B.S., Geology, University of Texas at San Antonio, 1991
22   Years of Experience: 20
23
24   James Durham: Public and Occupational Health and Safety
25   Ph.D., Nuclear Engineering, University of Illinois, Urbana, 1987
26   M.S., Nuclear Engineering, University of Illinois, Urbana, 1984
27   B.S., Nuclear Engineering, University of Illinois, Urbana, 1980
28   Years of Experience: 31
29
30   Amy Glovan: Ecological Resources, Socioeconomics
31   B.A., Environmental Studies, University of Kansas, 1998
32   Years of Experience: 13
33
34   Patrick LaPlante: Transportation, Waste Management
35   M.S., Biostatistics and Epidemiology, Georgetown University, 1994
36   B.S., Environmental Studies, Western Washington University, 1988
37   Years of Experience: 24
38
39   Robert Lenhard: Program Manager
40   Ph.D., Soil Physics, Oregon State University, 1984
41   M.S., Forest Soils, University of Idaho, 1978
42   B.S., Forest Science, Humboldt State University, 1976
43   Years of Experience: 31
44
45   Robert Pauline: Cumulative Impacts
46   M.S., Biology, Environmental Science and Policy, George Mason University, 1999
47   B.S., Biology, Bates College, 1989
48   Years of Experience: 23

1   James Prikryl: Principal Investigator; Land Use, Noise, Visual/Scenic Resources, Cost/Benefit
2   Analysis, Cumulative Impacts
3   M.A., Geology, University of Texas at Austin, 1989
4   B.S., Geology, University of Texas at Austin, 1984
5   Years of Experience: 28
6
7   Marla Roberts: Geology and Soils
8   M.S., Geology, University of Texas at San Antonio, 2007
9   B.A., Geology, Vanderbilt University, 2001
10   Years of Experience: 11
11
12 John Stamatakos: Program Director
13   Ph.D., Geology, Lehigh University, 1990
14   M.S., Geology, Lehigh University, 1988
15   B.S., Geology, Franklin and Marshall College, 1981
16   Years of Experience: 31
17
18 Deborah Waiting: GIS Analyst
19   B.S., Geology, University of Texas at San Antonio, 1999
20   Years of Experience: 14
21
22 Bradley Werling: Meteorology, Climatology, Air Quality
23   M.S., Environmental Science, University of Texas at San Antonio, 2000
24   B.S., Chemistry, Southwest Texas State University, 1999
25   B.A., Engineering Physics, Westmont College, Santa Barbara, 1985
26   Years of Experience: 27
27

28   10.3   CNWRA Consultants and Subcontractors

29
30 Pollyanna Clark: Cultural and Historic Resources
31   M.A., Anthropology, University of Mississippi, Oxford, 2004
32   A.B., Anthropology, Princeton University, 1992
33   Years of Experience: 19
34
35 Randall Withrow: National Historic Preservation Act Section 106 Support
36   M.A., Anthropology, University of Minnesota, 1983
37   B.A., History, University of Wisconsin-LaCrosse, 1980
38   Years of Experience: 28

11   DISTRIBUTION LIST

The U.S. Nuclear Regulatory Commission (NRC) is providing copies of this draft Supplemental
Environmental Impact Statement (SEIS) to the organizations listed as follows.

11.1   Federal Agency Officials

Kenneth Distler
    U.S. Environmental Protection Agency
    Region 8
    Denver, CO

Marian Atkins
    Bureau of Land Management
    South Dakota Field Office
    Belle Fourche, SD

Gregory Fesko
    Bureau of Land Management
    Montana State Office
    Billings, MT

Mark Sant
    Bureau of Land Management
    Montana State Office
    Billings, MT

Brenda Shierts
    Bureau of Land Management
    South Dakota Field Office
    Belle Fourche, SD

Gary Smith
    Bureau of Land Management
    Montana State Office
    Billings, MT

Janet Carter
    U.S. Geological Survey
    Rapid City, SD

1 Lynn Kolud
2   U.S. Forest Service
3   Black Hills National Forest
4   Custer, SD
5
6 Mike McNeill
7   U.S. Forest Service
8   Buffalo Gap National Grassland
9   Hot Springs, SD
10
11 Steve Naylor
12   U.S. Army Corps of Engineers
13   Hot Springs, SD
14
15 ## 11.2   Tribal Government Officials
16
17 Leroy Spang
18   Northern Cheyenne
19   Tribal President
20   Lame Deer, MT
21
22 Conrad Fisher
23   Northern Cheyenne
24   Tribal Historic Preservation Office
25   Lame Deer, MT
26
27 John Yellow Bird Steele
28   Oglala Sioux
29   Tribal President
30   Pine Ridge, SD
31
32 Wilmer Mesteth
33   Oglala Sioux
34   Tribal Historic Preservation Office
35   Pine Ridge, SD
36
37 Charles Murphy
38   Standing Rock Sioux
39   Tribal Chairman
40   Fort Yates, ND
41

1   Waste'Win Young
2   Standing Rock Sioux
3   Tribal Historic Preservation Office
4   Fort Yates, ND
5
6   Kevin Keckler
7   Cheyenne River Sioux
8   Tribal Chairman
9   Eagle Butte, SD
10
11   Steve Vance
12   Cheyenne River Sioux
13   Tribal Historic Preservation Office
14   Eagle Butte, SD
15
16   Rodney Bordeaux
17   Rosebud Sioux
18   Tribal Chairman
19   Rosebud, SD
20
21   Russell Eagle Bear
22   Rosebud Sioux
23   Tribal Historic Preservation Office
24   Rosebud, SD
25
26   Thurman Cournoyer, Sr.
27   Yankton Sioux
28   Tribal Chairperson
29   Wagner, SD
30
31   Lana Gravatt
32   Yankton Sioux
33   Tribal Historic Preservation Office
34   Marty, SD
35
36   Brandon Sauze, Sr.
37   Crow Creek Sioux
38   Tribal Chairman
39   Ft. Thompson, SD
40

1	Wanda Wells
2	Crow Creek Sioux
3	Tribal Historic Preservation Office
4	Ft. Thompson, SD
5	
6	Jim Shakespeare
7	Northern Arapaho
8	Tribal Chairman
9	Fort Washakie, WY
10	
11	Darlene Conrad
12	Northern Arapaho
13	Tribal Historic Preservation Office
14	Fort Washakie, WY
15	
16	A.T. "Rusty" Stafne
17	Fort Peck Assiniboine Sioux
18	Tribal Chairman
19	Poplar, MT
20	
21	Darrell "Curley" Youpee
22	Fort Peck Assiniboine Sioux
23	Tribal Historic Preservation Office
24	Poplar, MT
25	
26	Mike LaJeunesse
27	Eastern Shoshone Tribe
28	Tribal Chairman
29	Fort Washakie, WY
30	
31	Wilfred Ferris
32	Eastern Shoshone
33	Tribal Historic Preservation Office
34	Fort Washakie, WY
35	
36	Cedric Black Eagle
37	Crow Tribe of Montana
38	Tribal Chairman
39	Crow Agency, MT
40	
41	Hubert B. Two Leggings
42	Crow Tribe of Montana
43	Tribal Historic Preservation Office
44	Crow Agency, MT

1   Michael Jandreau
2   Lower Brule Sioux
3   Tribal Chairman
4   Lower Brule, SD
5
6   Claire Green
7   Lower Brule Sioux
8   Tribal Historic Preservation Office
9   Lower Brule, SD
10
11 Roger Trudell
12   Santee Sioux Tribe of Nebraska
13   Tribal Chairman
14   Niobrara, NE
15
16 Rick Thomas
17   Santee Sioux Tribe of Nebraska
18   Tribal Historic Preservation Office
19   Niobrara, NE
20
21 Robert Shepherd
22   Sisseton-Wahpeton Tribe
23   Tribal Chairman
24   Agency Village, SD
25
26 Dianne Desrosiers
27   Sisseton-Wahpeton Tribe
28   Tribal Historic Preservation Office
29   Agency Village, SD
30
31 Roger Yankton
32   Spirit Lake Tribe
33   Tribal Chairperson
34   Fort Totten, ND
35
36 Darrell Smith
37   Spirit Lake Tribe
38   Tribal Historic Preservation Office
39   Fort Totten, ND
40
41

1   Anthony Reider
2   Flandreau–Santee Sioux
3   Tribal Chairman
4   Flandreau, SD
5
6   James B. Weston
7   Flandreau–Santee Sioux
8   Tribal Historic Preservation Office
9   Flandreau, SD
10
11   Gabe Prescott
12   Lower Sioux Tribe
13   Tribal President
14   Morton, MN
15
16   Anthony Morse
17   Lower Sioux Tribe
18   Tribal Historic Preservation Office
19   Morton, MN
20
21   G. Tex Hall
22   Three Affiliated Tribes
23   Tribal Chairman
24   New Town, ND
25
26   Elgin Crows Breast
27   Three Affiliated Tribes
28   Tribal Historic Preservation Office
29   New Town, ND
30
31   Rebecca White
32   Ponca Tribe of Nebraska
33   Tribal Chairwoman
34   Niobrara, NE
35
36   Gloria Hamilton
37   Ponca Tribe of Nebraska
38   Tribal Historic Preservation Office
39   Niobrara, NE
40
41   Merle St. Claire
42   Turtle Mountain Chippewa Tribe
43   Tribal Chairman
44   Belcourt, ND

1   Kade Ferris
2   Turtle Mountain Chippewa Tribe
3   Tribal Historic Preservation Office
4   Belcourt, ND
5
6   **11.3   State Agency Officials**
7
8   Matt Hicks
9   South Dakota Department of Environment and Natural Resources
10   Pierre, SD
11
12   Paige Olson
13   State Historic Preservation Office
14   Pierre, SD
15
16   Stan Michals
17   South Dakota Game, Fish, & Parks Department
18   Pierre, SD
19
20   Mike Fosha
21   South Dakota State Historical Society, Archaeological Research Center
22   Rapid City, SD
23
24   Roger Campbell
25   Office of Tribal Government Relations
26   Pierre, SD
27
28   Mary Cerney
29   Governor's Office of Economic Development
30   Pierre, SD
31
32   **11.4   Local Agency Officials**
33
34   David Green
35   Custer County Planning and Economic Development
36   Custer, SD
37
38   Bill Curran
39   Edgemont Area Chamber of Commerce
40   Edgemont, SD
41
42   Lisa Scheinost
43   Edgemont Area Chamber of Commerce
44   Edgemont, SD

11.5   Other Organizations and Individuals

Richard Blubaugh
        Powertech USA, Inc.
        Greenwood Village, CO

Mark Hollenbeck
        Powertech USA, Inc.
        Edgemont, SD

Eric Jantz
        New Mexico Environmental Law Center
        Santa Fe, NM

Christopher S. Pugsley, Esq.
        Thompson & Simmons, PLLC
        Washington, DC

Anthony Thompson
        Thompson & Simmons, PLLC
        Washington, DC

Jeffrey C. Parsons
        Western Mining Action Project
        Lysons, CO

Travis E. Stills
        Energy Minerals Law Center
        Durango, CO

Thomas J. Ballanco, Esq.
        Attorney for Dayton Hyde
        Pine Ridge, SD

David Frankel, Esq.
        Aligning for Responsible Mining
        Pine Ridge, SD

Bruce Ellison
        Law Office of Bruce Ellison
        Rapid City, SD

1　　Charmaine White Face
2　　　　　Defenders of the Black Hills
3　　　　　Rapid City, SD
4
5　　Lillias Jarding
6　　　　　The Lakota People's Law Project
7　　　　　Rapid City, SD
8
9　　Cindy Gillis
10　　　　Counsel for the Oglala Sioux Tribe
11　　　　Gonzalez Law Firm
12　　　　Rapid City, SD
13
14　　Martha Graham
15　　　　SRI Foundation
16　　　　Rio Rancho, NM
17
18　　Edgemont Public Library
19　　　　Edgemont, SD
20
21　　Rapid City Public Library
22　　　　Rapid City, SD
23
24　　Custer County Library
25　　　　Custer, SD
26
27　　Weston County Library
28　　　　Newcastle, WY
29
30　　Hot Springs Public Library
31　　　　Hot Springs, SD
32
33　　Susan Henderson
34　　　　Edgemont, SD
35
36　　Dayton Hyde
37　　　　The Black Hills Wild Horse Sanctuary
38　　　　Hot Springs, SD

APPENDIX A

CONSULTATION CORRESPONDENCE

CONSULTATION CORRESPONDENCE

The Endangered Species Act of 1973, as amended, and the National Historic Preservation Act of 1966, as amended, require that Federal agencies consult with applicable State and Federal agencies and groups prior to taking action that may affect threatened and endangered species, essential fish habitat, or historical and archaeological resources. This appendix contains consultation documentation related to these federal acts.

Table A–1. Chronology of Consultation Correspondence

Author	Recipient	Date of Letter	ADAMS Accession Number
U.S. Nuclear Regulatory Commission (K. Hsueh)	Fish and Wildlife Conservation Office (P. Gober)	March 15, 2010	ML100331503
U.S. Nuclear Regulatory Commission (K. Hsueh)	Cheyenne River Sioux Tribe (J. Brings Plenty)	March 19, 2010*	ML100331999
U.S. Fish and Wildlife Service (S. Larson)	U.S. Nuclear Regulatory Commission (K. Hsueh)	March 29, 2010	ML100970556
U.S. Nuclear Regulatory Commission (K. Hsueh)	Oglala Sioux Tribe (T. Two Bulls)	September 8, 2010	ML102450647
Turtle Mountain Band of Chippewa	U.S. Nuclear Regulatory Commission	April 7, 2010	ML101100137
U.S. Nuclear Regulatory Commission (K. Hsueh)	Standing Rock Sioux Tribe (R. His Horse is Thunder)	September 10, 2010*	ML102520308
Three Affiliated Tribes, Mandan Hidatsa Arikara (P. "No Tears" Brady)	U.S. Nuclear Regulatory Commission (K. Hsueh)	September 20, 2010	ML102780369
Sisseton Wahpeton Oyate (D. Desrosiers)	U.S. Nuclear Regulatory Commission (H. Yilma)	October 1, 2010	ML103050026
Sisseton Wahpeton Oyate (D. Desrosiers)	U.S. Nuclear Regulatory Commission (K. Hsueh)	November 2, 2010	ML103200287
Rosebud Sioux Tribe (R. Eagle Bear)	U.S. Nuclear Regulatory Commission (K. Hsueh)	November 7, 2010	ML103270443
U.S. Nuclear Regulatory Commission (H. Yilma)	Lower Brule Sioux Tribe (C. Green)	November 12, 2010	ML103330215
Lower Brule Sioux Tribe (M. Jandreau)	U.S. Nuclear Regulatory Commission (K. Hsueh)	November 15, 2010	ML103340146
U.S. Nuclear Regulatory Commission (H. Yilma)	Yankton Sioux Tribe (L. Gravatt)	November 22, 2010	ML103330220
Yankton Sioux Tribe (L. Gravatt)	U.S. Nuclear Regulatory Commission (K. Hsueh)	December 3, 2010	ML110030430
Standing Rock Sioux Tribe (A. Swallow)	U.S. Nuclear Regulatory Commission (K. Hsueh)	December 8, 2010	ML110030700
U.S. Nuclear Regulatory Commission (K. Hsueh)	Advisory Council on Historic Preservation (J. Fowler	December 15, 2010	ML103270171

Table A–1. Chronology of Consultation Correspondence (continued)

Author	Recipient	Date of Letter	ADAMS Accession Number
Oglala Sioux Tribe (M. Catches Enemy and W. Mesteth)	U.S. Nuclear Regulatory Commission (K. Hsueh)	January 31, 2011	ML110340107
U.S. Nuclear Regulatory Commission (L. Camper)	Crow Tribe of Montana (C. Black Eagle)	March 4, 2011*	ML110550535
Crow Tribe (H.B. Two Leggins)	U.S. Nuclear Regulatory Commission (H. Yilma)	March 10, 2011	ML110690166
U.S. Nuclear Regulatory Commission (L. Camper)	Yankton Sioux Tribe (L. Gravatt)	May 12, 2011*	ML111320395
U.S. Nuclear Regulatory Commission (K. Hsueh)	Powertech (USA) Inc. (R. Blubaugh)	August 12, 2011	ML112170237
Powertech (USA) Inc. (R. Blubaugh)	U.S. Nuclear Regulatory Commission (K. Hsueh)	August 31, 2011	ML112700464
U.S. Nuclear Regulatory Commission (K. Hsueh)	Oglala Sioux Tribe (Mr. James Laysbad)	October 20, 2011*	ML112440097
U.S. Nuclear Regulatory Commission (K. Hsueh)	Oglala Sioux Tribe (Mr. James Laysbad)	October 28, 2011*	ML112980555
U.S. Bureau of Land Management (M. Atkins)	U.S. Nuclear Regulatory Commission (L. Camper)	November 22, 2011	ML113340322
U.S. Nuclear Regulatory Commission (K. Hsueh)	Tribal Historic Preservation Officers	January 19, 2012†	ML120330066
Sisseton Wahpeton Oyate (D. Desrosiers)	U.S. Nuclear Regulatory Commission (H. Yilma)	January 24, 2012	ML12031A279
U.S. Nuclear Regulatory Commission (K. Hsueh)	Tribal Historic Preservation Officers	March 6, 2012†	ML120670079
U.S. Nuclear Regulatory Commission (K. Hsueh)	Tribal Historic Preservation Officers	March 9, 2012†	ML120730509
U.S. Nuclear Regulatory Commission (L. Camper)	Apache Tribe of Oklahoma (Mr. L. Maynahonah)	March 19, 2012*	ML120600178
U.S. Nuclear Regulatory Commission (L. Camper)	Apache Tribe of Oklahoma (Mr. L. Maynahonah)	March 26, 2012*	ML120670319
U.S. Nuclear Regulatory Commission (K. Hsueh	Tribal Historic Preservation Officers	April 5, 2012†	ML12130A067
U.S. Nuclear Regulatory Commission (H. Yilma)	Tribal Historic Preservation Officers	April 20, 2012‡	ML121180264
U.S. Nuclear Regulatory Commission (L. Camper)	Crow Creek Sioux Tribe (Mr. D. Big Eagle)	May 7, 2012*	ML121250102
U.S. Nuclear Regulatory Commission (L. Camper)	Oglala Sioux Tribe (Mr. J. Yellow Bird Steele)	May 23, 2012*	ML12143A185
U.S. Nuclear Regulatory Commission (K. Hsueh)	Northern Cheyenne Tribe (C. Fisher)	June 20, 2012*	ML12172A356
U.S. Nuclear Regulatory Commission (K. Hsueh)	Standing Rock Sioux Tribe (W. Young)	June 26, 2012*	ML12177A319

Table A–1. Chronology of Consultation Correspondence (continued)

Author	Recipient	Date of Letter	ADAMS Accession Number
U.S. Nuclear Regulatory Commission (L. Camper)	Northern Arapaho Tribe (Mr. J. Shakespeare)	June 29, 2012*	ML12181A324
Powertech (USA) Inc. (R. Blubaugh)	U.S. Nuclear Regulatory Commission (H. Yilma)	July 20, 2012	ML12213A694
U.S. Nuclear Regulatory Commission (H. Yilma)	Tribal Historic Preservation Officers	August 7, 2012‡	ML12261A375
U.S. Nuclear Regulatory Commission (H. Yilma)	Tribal Historic Preservation Officers	August 9, 2012‡	ML12261A429
U.S. Nuclear Regulatory Commission (H. Yilma)	Tribal Historic Preservation Officers	August 20, 2012‡	ML12261A463
U.S. Nuclear Regulatory Commission (H. Yilma)	Tribal Historic Preservation Officers	August 21, 2012‡	ML12261A454
Powertech (USA) Inc. (R. Blubaugh)	U.S. Nuclear Regulatory Commission (K. Hsueh)	August 29, 2012	ML12243A158
U.S. Nuclear Regulatory Commission (K. Hsueh)	Tribal Historic Preservation Officers	August 30, 2012‡	ML12261A470
U.S. Nuclear Regulatory Commission (K. Hsueh	Tribal Historic Preservation Officers	September 18, 2012†	ML12264A594
U.S. Nuclear Regulatory Commission (K. Hsueh)	Powertech (USA) Inc. (R. Blubaugh)	October 4, 2012	ML12278A185
U.S. Nuclear Regulatory Commission (L. Camper)	Crow Tribe of Montana (C. Black Eagle)	October 11, 2012*	ML12283A156
U.S. Nuclear Regulatory Commission (K. Hsueh)	Tribal Historic Preservation Officers	October 12, 2012†	ML12286A310
*Similar letters were sent to tribes listed in SEIS Section 1.7.3.5. †Letter sent via email to tribes listed in SEIS Section 1.7.3.5. ‡Email sent to tribes listed in SEIS Section 1.7.3.5.			

March 15, 2010

Pete Gober
Fish and Wildlife Conservation Office
420 South Garfield Avenue, Suite 400
Pierre, SD 57501-5408

SUBJECT: REQUEST FOR INFORMATION REGARDING ENDANGERED OR
THREATENED SPECIES AND CRITICAL HABITAT FOR THE POWERTECH
INC. PROPOSED DEWEY-BURDOCK IN-SITU RECOVERY FACILITY NEAR
EDGEMONT, SOUTH DAKOTA (Docket 040-09075)

Dear Mr. Gober:

The U.S. Nuclear Regulatory Commission (NRC) has received an application from Powertech
Inc. (Powertech) for a new radioactive source materials license to develop and operate the
Dewey-Burdock Project located near Edgemont, South Dakota in Fall River and Custer
Counties. The facility, if licensed, would use an *in-situ* recovery methodology to extract uranium
at the Dewey-Burdock site. The proposed project area consists approximately of 10,580 acres
(4,282 ha) located on both sides of Dewey Road (County Road 6463) and portions of Sections
1-5, 10-12, 14, and 15, Township 7 South, Range 1 East and Sections 20, 21, 27, 28, 29, and
30-35, Township 6 South, Range 1 East, Black Hill Meridian. A map showing the proposed
project boundary is enclosed (Powertech Figure 1.4-1).

As established in Title 10 *Code of Federal Regulations* Part 51 (10 CFR 51), the NRC regulation
that implements the National Environmental Policy Act of 1969, as amended, the agency is
preparing a Supplemental Environmental Impact Statement (SEIS). In accordance with Section
7 of the Endangered Species Act, the SEIS will include an analysis of potential impacts to
endangered or threatened species or critical habitat in the proposed project area. To support
the environmental review, the NRC is requesting information from the U.S. Fish and Wildlife
Service to facilitate the identification of endangered or threatened species or critical habitat that
may be affected by the proposed project. After a careful review and assessment of all the
comments received, the NRC will determine what additional actions are necessary to comply
with Section 7 of the Endangered Species Act.

The Powertech Dewey-Burdock Project license application is publicly available in the NRC
Public Document Room located at One White Flint North, 11555 Rockville Pike, Rockville,
Maryland 20852, or from the NRC's Agency Wide Documents Access and Management System
(ADAMS). The ADAMS Public Electronic Reading Room is accessible at
http://www.nrc.gov/reading-rm/adams.html. The accession numbers for the Powertech
application including the environmental report is ML092870160.

P. Gober   2

Please submit any information that you may have regarding this environmental review within 30 days of the receipt of this letter to NRC, Attention: Mr. Kevin Hsueh, Mail Stop T8F05, Washington, DC 20555. If you have any questions, please contact Ms. Haimanot Yilma of my staff by telephone at 301-415-8029 or by email at Haimanot.Yilma@nrc.gov. Thank you for your assistance.

Sincerely,

/RA/

Kevin Hsueh, Chief
Environmental Review Branch-B
Environmental Protection and
 Performance Assessment Directorate
Division of Waste Management
 and Environmental Protection
Office of Federal and State Materials
 and Environmental Management Programs

Docket No.: 040-09075

Enclosure: Powertech Figure 1.4-1

cc:
Stan Michals
South Dakota Game Fish and Parks
523 East Capitol Avenue
Pierre, SD 57501

March 19, 2010

Joseph Brings Plenty, Chairman
Cheyenne River Sioux Tribe
P.O. Box 590
Eagle Butte, SD 57625-0590

SUBJECT: INVITATION FOR FORMAL CONSULTATION UNDER NATIONAL HISTORIC
            PRESERVATION ACT SECTION 106 AS WELL AS REQUEST FOR
            INFORMATION REGARDING TRIBAL HISTORIC AND CULTURAL
            RESOURCES POTENTIALLY AFFECTED BY THE POWERTECH INC.
            PROPOSED DEWEY-BURDOCK IN-SITU RECOVERY FACILITY NEAR
            EDGEMONT SOUTH DAKOTA

Dear Chairman Plenty:

The U.S. Nuclear Regulatory Commission (NRC) has received an application from Powertech
Inc. (Powertech) for a new radioactive source materials license to develop and operate the
Dewey-Burdock Project located near Edgemont, South Dakota in Fall River and Custer
Counties. The facility, if licensed, would use an *in-situ* recovery methodology to extract uranium
at the Dewey-Burdock site. The proposed project area consists of approximately 10,580 acres
(4,282 ha) located on both sides of Dewey Road (County Road 6463) and portions of Sections
1-5, 10-12, 14, and 15, Township 7 South, Range 1 East and Sections 20, 21, 27, 28, 29, and
30-35, Township 6 South, Range 1 East, Black Hill Meridian. A map showing the proposed
project boundary is enclosed (Powertech Figure 1.4-1).

The South Dakota State Historic Preservation Officer identified the Cheyenne River Sioux Tribe
as potentially attaching religious and cultural significance to historic properties in the project
area. By this letter, the NRC invites the Cheyenne River Sioux Tribe to participate as a
consulting party in the National Historic Preservation Act Section 106 process. If the Tribe
would like to participate as a consulting party, please respond by writing to Mr. Kevin Hsueh.

As established in Title 10 Code of Federal Regulations Part 51 (10 CFR 51), the NRC regulation
that implements the National Environmental Policy Act of 1969, as amended, the NRC is
preparing a Supplemental Environmental Impact Statement (SEIS) for the proposed action.
As part of the environmental review, the SEIS will include an analysis of potential impacts to
historic and cultural properties and is therefore requesting input from the Cheyenne River Sioux
Tribe to facilitate the identification of tribal historic sites or cultural resources that may be
affected by the proposed action. Specifically, the NRC is interested in learning of any areas on
the Dewey-Burdock site that you believe have traditional religious or cultural significance. After
a careful review and assessment of all the comments received, the NRC will determine what
additional actions are necessary to comply with 10 CFR 51 and 36 CFR 800, the implementing
regulation for Section 106 of the National Historic Preservation Act.

J. B. Plenty   2

The Powertech Dewey-Burdock Project license application is publicly available in the NRC Public Document Room located at One White Flint North, 11555 Rockville Pike, Rockville, Maryland 20852, or from the NRC's Agency Wide Documents Access and Management System (ADAMS). The ADAMS Public Electronic Reading Room is accessible at http://www.nrc.gov/reading-rm/adams.html. The accession numbers for the Powertech application including the Environmental report is ML092870160.

Please submit your request to be a consulting party as well as any information that you may have regarding this environmental review within 30 days of the receipt of this letter to NRC, Attention: Mr. Kevin Hsueh, Mail Stop T8F05, Washington, DC 20555. If you have any questions or comments, or need any additional information, please contact Ms. Haimanot Yilma of my staff by telephone at 301-415-8029, or email at Haimanot.Yilma@nrc.gov.

Sincerely,

/RA/

Kevin Hsueh, Branch Chief
Environmental Review Branch-B
Environmental Protection and Performance
 Assessment Directorate
Division of Waste Management
 and Environmental Protection
Office of Federal and State Materials
 and Environmental Management Programs

Enclosure:
Figure 1.4-1

cc w/enclosure:
D. Dupris

United States Department of the Interior

FISH AND WILDLIFE SERVICE
Ecological Services
420 South Garfield Avenue, Suite 400
Pierre, South Dakota 57501-5408

March 29, 2010

Mr. Kevin Hsueh
Nuclear Regulatory Commission
Mail Stop T8F05
Washington, DC   20555

> Re   Powertech Dewey-Burdock Project,
> Docket 040-09075, Custer and Fall River
> County, South Dakota

Dear Mr. Hsueh:

This letter is to provide environmental comments on your March 15, 2010, letter regarding the above referenced project which proposes granting a radioactive source materials license to develop and operate the Dewey-Burdock *in-situ* recovery facility near Edgemont, South Dakota. The Powertech Dewey-Burdock Project is proposed within portions of Sections 1-5, 10-12, 14 and 15, Township 7 South, Range 1 East, Fall River County, South Dakota, and Sections 20, 21, and 27-35, Township 6 South, Range 1 East, Custer County, South Dakota.

These comments have been prepared in accordance with provisions of the Fish and Wildlife Coordination Act (16 U.S.C. 661 et seq.) and the Endangered Species Act of 1973 (16 U.S.C. 1531 et seq.), as amended. This constitutes the report of the Department of the Interior on the proposed project and is to be used in your determination of 404 (b)(1) guidelines (40 CFR 230) and in your public interest review (33 CFR 320.4) as they relate to the protection of fish and wildlife resources.

The National Wetlands Inventory indicates that numerous wetlands exist within the proposed permit boundary area. Two maps are enclosed showing wetlands found within the two initial mine units. The locations of other wetlands within the proposed permit boundary can be accessed at http://www.fws.gov/wetlands/Data/Mapper.html.

The U.S. Fish and Wildlife Service (Service), in accordance with the National Environmental Policy Act of 1969 (42 U.S.C. 4321-4347) and other environmental laws and rules, recommends complete avoidance of these areas, if possible. If this is not possible, attempts should be made to minimize adverse impacts. Finally, if adverse impacts are unavoidable, then measures should be undertaken to replace the impacted areas.

2

The Fish and Wildlife Coordination Act and Executive Order 11990 (Protection of Wetlands) encourages the protection and conservation of wetlands. In reviewing projects that may impact wetlands, the Service encourages: 1) avoidance of wetlands, if possible, 2) minimization of impacts to wetlands if they cannot be avoided, and 3) replacement of wetland values that may be impacted by a project.

In accordance with section 7(c) of the Endangered Species Act, as amended, 16 U.S.C. 1531 et seq., we have determined that the following federally listed species may occur (this list is considered valid for 90 days) within Custer County:

Species	Status	Expected Occurrence
Whooping crane (Grus americana)	Endangered	Migration.
Black-footed ferret (Mustela nigripes)	Endangered	None.

The whooping crane generally migrates through the eastern portion of Custer County, and the black-footed ferret is currently only found in the Wind Cave National Park. We have no information to indicate that these species are located within the project boundaries.

Whooping cranes migrate through South Dakota on their way to northern breeding grounds and southern wintering areas. They occupy numerous habitats such as cropland and pastures; wet meadows; shallow marshes; shallow portions of rivers, lakes, reservoirs, and stock ponds; and both freshwater and alkaline basins for feeding and loafing. Overnight roosting sites frequently require shallow water in which they stand and rest. Additionally, should mining activities occur during spring or fall migration, the potential for disturbances to whooping cranes exists. Disturbance (flushing the birds) stresses them at critical times of the year. We recommend that you remain vigilant for these birds. There is little that can be done to reduce disturbance besides ceasing activities at sites where the birds have been observed. The birds normally do not stay in any one area for long during migration. Any whooping crane sightings should be reported to this office.

Fall River County does not have any federally listed species, but the greater sage-grouse (Centrocercus urophasianus) is a candidate species that historically occurred in the area and has a potential to be present within the proposed area of review.

The Service determined that the species is warranted for listing pursuant to the Endangered Species Act; however, efforts in that regard will be precluded in deference to higher priority species. This finding means that the greater sage-grouse is now a candidate species for future reclassification as a threatened or endangered species under the Endangered Species Act. The 12-Month Findings for Petitions to List the Greater Sage-Grouse (Centrocercus urophasianus) as Threatened or Endangered was published in the Federal Register 75: 13909-13958 on March 23, 2010.

3

Candidate species have no legal protection under the Endangered Species Act at the present time; however, some agencies may consider their status in their management efforts.

If the Nuclear Regulatory Agency or their designated representative determines that the project "may adversely affect" listed species in South Dakota, it should request formal consultation from this office. If a "may affect - not likely to adversely affect" determination is made for this project, it should be submitted to this office for concurrence. If a "no effect" determination is made, further consultation may not be necessary. However, a copy of the determination should be sent to this office. For more information regarding Federal action agency responsibilities as related to section 7 of the Endangered Species Act, please refer to the Service's Endangered Species Act Consultation Handbook, available online at http://endangered.fws.gov/consultations/index.html.

If changes are made in the project plans or operating criteria, or if additional information becomes available, the Service should be informed so that the above determinations can be reconsidered.

The Service appreciates the opportunity to provide comments. If you have any questions regarding these comments, please contact Terry Quesinberry of this office at (605) 224-8693, Extension 237.

Sincerely,

Scott Larson
Acting Field Supervisor
South Dakota Field Office

Enclosures

cc: Corps of Engineers/Regulatory; Pierre, SD

September 8, 2010

Mrs. Theresa Two Bulls
Oglala Sioux Tribe
P.O. Box 2070
Pine Ridge, SD 57770-2070

SUBJECT: REQUEST FOR UPDATED TRIBAL COUNCIL MEMBERS FOR THE OGLALA
            SIOUX TRIBE

Dear President Two Bulls:

As established in Title 10 Code of Federal Regulations Part 51 (10 CFR 51), the U.S. Nuclear
Commission (NRC) regulation implementing the National Environmental Policy Act of 1969, as
amended, the NRC is preparing a Supplemental Environmental Impact Statement (SEIS) for the
proposed action. As part of the environmental review, the SEIS will include an analysis of
potential impacts of the proposed action to historic and cultural properties.

On March 19, 2010, the NRC sent a letter to your office inviting the Oglala Sioux to participate
as a consulting party and requested information regarding tribal historic and cultural resources
potentially affected by the proposed Dewey-Burdock ISR facility. This letter was also forwarded
to the Cultural Resources Director, Michael Catches Enemy. The March 19th letter is enclosed,
for your convenience.

The NRC Staff remains committed to receiving information from the Oglala Sioux concerning
cultural and historical resources at the proposed Dewey-Burdock ISR facility. During the recent
hearing process, the NRC staff was advised the Tribal Historic Preservation Officer and Cultural
Resources Director at the Oglala Sioux tribe had changed. In an effort to ensure continued
open and prompt communication between the Oglala Sioux tribe and the NRC, the NRC kindly
requests the names and contact information for the new Tribal Historic Preservation Officer and
the Cultural Resources Director.

T. Two Bulls   2

The NRC staff requests that you submit information regarding the new tribal officers of the Oglala Sioux tribe to the following address: Attention: Mr. Kevin Hsueh, Mail Stop T8F05, Washington, D.C. 20555. If you have any questions or comments, or need any additional information, please contact the environmental project manager, Ms. Haimanot Yilma by telephone at 301-415-8029, or email at Haimanot.Yilma@nrc.gov.

Sincerely,

/RA/

Kevin Hsueh, Branch Chief
Environmental Review Branch - B
Environmental Protection and Performance
 Assessment Directorate
Division of Waste Management
 and Environmental Protection
Office of Federal and State Materials
 and Environmental Management Programs

cc: Counsel for the Oglala Sioux Tribe   Counsel for the Oglala Sioux Tribe
Western Mining Action Project   Gonzales Law Firm
P. O. Box 349   522 7th Street, Suite 202
Lyons, CO 80540   Rapid City, SD 57701
Jeffrey C. Parsons   Grace Dugan Esq

Kade

UNITED STATES
NUCLEAR REGULATORY COMMISSION
WASHINGTON, D.C. 20555-0001

March 19, 2010

It is the determination of the Turtle Mountain Tribal Historic Preservation Office that
this project will have no effect on historic properties of importance to the Turtle
Mountain Band of Chippewa Indians. A determination of No Historic Properties
Affected is granted for the project to proceed.

4-7-1c

Twila Martin-Kekabah, Chairperson
Turtle Mountain Band of Chippewa
P.O. Box 900
Belcourt, ND 58316

SUBJECT: INVITATION FOR FORMAL CONSULTATION UNDER NATIONAL HISTORIC
PRESERVATION ACT SECTION 106 AS WELL AS REQUEST FOR
INFORMATION REGARDING TRIBAL HISTORIC AND CULTURAL
RESOURCES POTENTIALLY AFFECTED BY THE POWERTECH INC.
PROPOSED DEWEY-BURDOCK IN-SITU RECOVERY FACILITY NEAR
EDGEMONT SOUTH DAKOTA

Dear Chairperson Martin-Kekabah:

The U.S. Nuclear Regulatory Commission (NRC) has received an application from Powertech
Inc. (Powertech) for a new radioactive source materials license to develop and operate the
Dewey-Burdock Project located near Edgemont, South Dakota in Fall River and Custer
Counties. The facility, if licensed, would use an *in-situ* recovery methodology to extract uranium
at the Dewey-Burdock site. The proposed project area consists of approximately 10,580 acres
(4,282 ha) located on both sides of Dewey Road (County Road 6463) and portions of Sections
1-5, 10-12, 14, and 15, Township 7 South, Range 1 East and Sections 20, 21, 27, 28, 29, and
30-35, Township 6 South, Range 1 East, Black Hill Meridian. A map showing the proposed
project boundary is enclosed (Powertech Figure 1.4-1).

The South Dakota State Historic Preservation Officer identified the Turtle Mountain Band of
Chippewa as potentially attaching religious and cultural significance to historic properties in the
project area. By this letter, the NRC invites the Turtle Mountain Band of Chippewa to participate
as a consulting party in the National Historic Preservation Act Section 106 process. If the Band
would like to participate as a consulting party, please respond by writing to Mr. Kevin Hsueh.

As established in Title 10 Code of Federal Regulations Part 51 (10 CFR 51), the NRC regulation
that implements the National Environmental Policy Act of 1969, as amended, the NRC is
preparing a Supplemental Environmental Impact Statement (SEIS) for the proposed action.
As part of the environmental review, the SEIS will include an analysis of potential impacts to
historic and cultural properties and is therefore requesting input from the Turtle Mountain Band
of Chippewa to facilitate the identification of tribal historic sites or cultural resources that may be
affected by the proposed action. Specifically, the NRC is interested in learning of any areas on
the Dewey-Burdock site that you believe have traditional religious or cultural significance. After
a careful review and assessment of all the comments received, the NRC will determine what
additional actions are necessary to comply with 10 CFR 51 and 36 CFR 800, the implementing
regulation for Section 106 of the National Historic Preservation Act.

September 10, 2010

Ron His Horse Is Thunder, Chairman
Standing Rock Sioux Tribe
P.O Box D
Ft. Yates, ND 58538-0522

SUBJECT: INVITATION FOR FORMAL CONSULTATION UNDER THE SECTION 106 OF
             THE NATIONAL HISTORIC PRESERVATION ACT

Dear Chairman Thunder:

As established in Title 10 *Code of Federal Regulations* Part 51 (10 CFR 51), the U.S. Nuclear
Regulatory Commission (NRC) regulations that implement the National Environmental Policy
Act (NEPA) of 1969, as amended, the NRC is preparing a Supplemental Environmental Impact
Statement (SEIS) for the proposed Powertech Inc. Dewey-Burdock In-Situ Recovery (ISR)
Facility near Edgemont, South Dakota. As part of the environmental review, the SEIS will
include an analysis of potential impacts of the proposed action to historic and cultural properties.

On March 19, 2010, the NRC sent a letter to your office inviting the Standing Rock Sioux Tribe
to participate as a consulting party and requested information regarding tribal historic and
cultural resources potentially affected by the proposed Dewey-Burdock ISR facility. A copy of
the March 19th letter is enclosed, for your convenience.

To date, the NRC has not received any response from your office regarding the Tribe's interest
in becoming a consulting party for the proposed Dewey-Burdock ISR facility near Edgemont
South Dakota.

The NRC again extends an invitation to the Standing Rock Sioux Tribe to participate as a
consulting party for the proposed Dewey Burdock ISR facility. Specifically, the NRC is
interested in learning of any areas on the proposed Dewey-Burdock site that you believe have
traditional religious or cultural significance and whether there are specialized concerns or
information known to the Tribe that should be considered by the staff during the development of
the SEIS.

The NRC staff understands that the Tribe may raise issues in consultation that should be kept
confidential and nonpublic; the staff is committed to maintaining confidentiality of said
information.

After a careful review and assessment of all information and comments received, the NRC will
determine what additional actions are necessary to comply with 10 CFR Part 51 and 36 CFR
800, the implementing regulations for Section 106 of the National Historic Preservation Act.

R. Thunder   2

If the Tribe would like to participate as a consulting party pursuant to Section 106, the Tribe should express its interest in participating and identify areas of concern, within 60 days of receipt of this letter, to ensure that the parties will have the opportunity to engage in meaningful and productive consultation. The Tribe should forward its response to the following address: Mr. Kevin Hsueh, Mail Stop T-8F05, Washington, DC. 20555.

If you have any questions or comments, or need any additional information, please contact the environmental Project Manager, Ms. Haimanot Yilma by telephone at 301-415-8029, or email at Haimanot.Yilma@nrc.gov.

Sincerely,

/RA/

Kevin Hsueh, Branch Chief
Environmental Review Branch B
Environmental Protection and Performance
  Assessment Directorate
Division of Waste Management
  and Environmental Protection
Office of Federal and State Materials
  and Environmental Management Programs

Docket No.: 040-09075

Enclosure: Letter of March 19, 2010

cc: w/enclosure
Waste' Win Young, THPO
Standing Rock Sioux Tribe
P.O Box D
Fort Yates, ND 58538

Three Affiliated Tribes
MANDAN * HIDATSA * ARIKARA

TRIBAL HISTORIC PRESERVATION
Mandan Hidatsa Arikara
Perry 'No Tears' Brady, Director.
404 Frontage Road,
New Town, North Dakota 58763
Ph/701-862-2474 fax/701-862-2490
pbrady@mhanation.com

September 20, 2010

Mr. Kevin Hsueh,
Mail Stop T-8F05,
Washington, DC 20555

Subject: Invitation for formal consultation under the section 106 of the National Historic
Preservation ACT

Dear Mr. Hsueh

After review of the documentation provided by your Office, the Mandan Hidatsa Arikara
Nations. Tribal Historic Preservation Office determines there will be 'No Adverse Affect/No
Historic Properties Affected' in regard to any pre and post-historic relics, artifacts or sacred
In addition, cultural resources in the proposed Project area.

We respectfully request to be notified should any cultural/tribal issue or others arise as the
Project progresses.

Sincerely,

Perry "No Tears" Brady,
Tribal Historic Preservation Officer
Mandan Hidatsa Arikara Nations.

DeweyBurdPubEm Resource

From: Dianne Desrosiers [DianneD@SWO-NSN.GOV]
Sent: Friday, October 01, 2010 11:40 AM
To: Yilma, Haimanot
Subject: Consultation

Good morning,

My name is Dianne Desrosiers and I am the Tribal Historic Preservation Officer for the Sisseton Wahpeton Oyate in NE South Dakota. I am in receipt of your correspondence dated September 10, 2010, invitation for consultation.

At this time the Sisseton Wahpeton Oyate would like to participate in consultation for Section 106 of the NHPA, for the Dewey Burdock ISR facility. We will be sending our request to consult to Mr. Kevin Hsueh.

Thank you for your attention in this matter.

1

Tribal Historic Preservation Office

P.O. Box 907
205 Oak St. East, Suite 121
Sisseton, SD 57262
(605) 698-3584 phone
(605) 698-4283 fax

November 2, 2010

Kevin Hsueh, Branch Chief
Environmental Review Branch B
United States Nuclear Regulatory Commission
Mail Stop T-8F05
Washington, DC 20555

Re: *Invitation for Formal Consultation Under the National Historic Preservation Act;*
Proposed Dewey Burdock In Situ Recovery Facility near Edgemont South Dakota

Dear Mr. Hsueh,

We are in receipt of your correspondence dated September 20, 2010. *Invitation for Formal Consultation Under Section 106 of the National Historic Preservation Act.*

Thank you for inviting the Sisseton Wahpeton Oyate (SWO) to participate as a consulting party as the U.S. Nuclear Regulatory Commission (NRC) works to satisfy its statutory obligations under the National Historic Preservation Act (NHPA) of 1966 (as amended), and National Environmental Policy Act (NEPA) to review impacts to cultural and historic resources potentially impacted by the proposed Powertech, Inc. Dewey-Burdock In-Situ Leach Uranium Mine. As you are aware, the proposed mine is located within the traditional and treaty lands of the Great Sioux Nation, which includes the Dakota bands. The SWO is interested in working with the NRC to identify and protect the cultural and historic resources threatened by the project.

As you may be aware the project is in a high probability and highly sensitive area. The effects to the land and resources include not only site-specific physical impacts, but also broader landscape-level impacts along with more intangible impacts to the integrity of the area from cultural, historical, spiritual, and religious perspectives.

1

The area of potential effect is located within the traditional and treaty lands of the Great Sioux Nation and other nations which consider this area, traditional homelands. We are interested in working with the NRC to identify and protect the cultural and historic resources threatened by the project.

At this time the Sisseton Wahpeton Oyate is requesting formal group consultation with regard to the above mentioned project. I would encourage you to make every effort to contact tribes that place religious and cultural significance to properties within the Black Hills (APE) which they deem significant to their existence. In an earlier conversation with a representative from your office (Haimanot Yilma), I suggested contacting the Oglala Sioux Tribe to afford them the opportunity to host such a meeting at the Prairie Winds Casino (which is the tribal nation in closest proximity to the site) and invite tribes to this group consultation. This would offer tribes an opportunity to discuss and share information to better preserve areas of religious and cultural significance.

We look forward to meeting with you in the near future. Together, we will protect and preserve our irreplaceable cultural resources. If you have any questions, please contact our office.

Thank you for your attention in this matter.

Dianne Desrosiers
Dianne Desrosiers
Tribal Historic Preservation Officer

Cc: Rosebud Sioux Tribe THPO
       Oglala Sioux Tribe THPO
       Cheyenne River Sioux Tribe THPO
       Santee Sioux Tribe THPO
       Yankton Sioux Tribe THPO
       Northern Cheyenne, THPO
       Standing Rock Sioux Tribe THPO
       Crow Tribe THPO

2

Protecting the Land, Cultural,
Heritage and Tradition for
the Future Generation

Tribal Historic Preservation Office

P.O. Box 809
Rosebud, South Dakota
Telephone: (605) 747-4255
Fax: (605) 747-4211
Email: rsthpo@yahoo.com

Russell Eagle Bear
Officer

Kathy Arcoren
Administrative Assistant

November 7, 2010

Kevin Hsueh, Branch Chief
Environmental Review Branch B
United States Nuclear Regulatory Commission
Mail Stop T-8F05
Washington, DC 20555

Re: *Invitation for Formal Consultation Under the National Historic Preservation Act:*
Proposed Dewey Burdock In Situ Recovery Facility near Edgemont South Dakota

Dear Mr. Hsueh,

We are in receipt of your correspondence dated September 20, 2010. *Invitation for Formal*
Consultation Under Section 106 of the National Historic Preservation Act.

Thank you for inviting the Rosebud Sioux Tribe (RST) to participate as a consulting party with
the U.S. Nuclear Regulatory Commission (NRC). We realize that the NRC is attempting to work
at satisfying its statutory obligations under the National Historic Preservation Act (NHPA) of
1966 (as amended), and National Environmental Policy Act (NEPA) to review impacts to
cultural and historic resources potentially impacted by the proposed Powertech, Inc. Dewey-
Burdock In-Situ Leach Uranium Mine. As you are aware, the proposed mine is located within
the traditional homelands and treaty set aside lands of the Great Sioux Nation. The RST is
interested in working with the NRC to identify and protect the cultural and historic resources
threatened by the proposed project.

As you are undoubtedly aware, the proposed project is in a highly probable and sensitive area
regarding cultural resources. The effects to the land and resources include not only site-specific
physical impacts, but also broader landscape-level impacts including intangible and tangible
impacts to the integrity of the area from cultural, historical, spiritual, and religious perspectives.

At this time the RST is requesting formal group consultation with regard to the above mentioned
project. We encourage you to make every effort to contact tribes that place religious and cultural
significance to properties within the Black Hills (APE) that they deem significant to their
continuity and existence. We suggest contacting the Oglala Sioux Tribe (OST) to afford them
the opportunity to host a meeting at the Prairie Winds Casino and invite tribes to this group

consultation. As a caveat to this suggestion, understand that the OST is the tribe nearest in physical proximity to the APE and what effects' one of our allies, effect's us all. This action would offer tribes an opportunity to discuss and share information to better preserve areas of religious and cultural significance.

We look forward to meeting with you in the near future. Together, we can and will protect and preserve our irreplaceable cultural resources. If you have any questions, please contact our office as soon as possible.

Thank you for your attention in this matter.

Sincerely,

Kathy Arneu

6" Mr. Russell Eagle Bear
Tribal Historic Preservation Officer
Rosebud Sioux Tribe
PO Box 809
Rosebud, South Dakota 57570
Ph.- (605) 747-4255
Email: rstthpo@yahoo.com

DeweyBurdPubEm Resource

From:	Yilma, Haimanot
Sent:	Friday, November 12, 2010 4:32 PM
To:	clairsgreen@yahoo.com
Cc:	DeweyBurdHrgFile Resource
Attachments:	Lower Brule Sioux.pdf

Ms. Green,

Per your request, attached please find the consultation letter sent to the Lower Brule Sioux Tribe on September 2010.

Thanks

Haimanot Yilma
Project Manager
FSME/DWMEP/EPPAD/ERB
U.S Nuclear Regulatory Commission
Phone: 301-415-8029
email: haimanot.yilma@nrc.gov
Mail Stop : T8H09

1

UNITED STATES
NUCLEAR REGULATORY COMMISSION
WASHINGTON, D.C. 20555-0001

September 10, 2010

Michael Jandreau, Chairman
Lower Brule Sioux Tribe
P O Box 187
Lower Brule, SD 57548-0187

SUBJECT: INVITATION FOR FORMAL CONSULTATION UNDER THE SECTION 106 OF
THE NATIONAL HISTORIC PRESERVATION ACT

Dear Chairman Jandreau:

As established in Title 10 *Code of Federal Regulations* Part 51 (10 CFR 51), the U.S. Nuclear
Regulatory Commission (NRC) regulations that implement the National Environmental Policy
Act (NEPA) of 1969, as amended, the NRC is preparing a Supplemental Environmental Impact
Statement (SEIS) for the proposed Powertech Inc. Dewey-Burdock In-Situ Recovery (ISR)
Facility near Edgemont, South Dakota. As part of the environmental review, the SEIS will
include an analysis of potential impacts of the proposed action to historic and cultural properties.

On March 19, 2010, the NRC sent a letter to your office inviting the Lower Brule Sioux Tribe to
participate as a consulting party and requested information regarding tribal historic and cultural
resources potentially affected by the proposed Dewey-Burdock ISR facility. A copy of the
March 19th letter is enclosed, for your convenience.

To date, the NRC has not received any response from your office regarding the Tribe's interest
in becoming a consulting party for the proposed Dewey-Burdock ISR facility near Edgemont
South Dakota.

The NRC again extends an invitation to the Lower Brule Sioux Tribe to participate as a
consulting party for the proposed Dewey Burdock ISR facility. Specifically, the NRC is
interested in learning of any areas on the proposed Dewey-Burdock site that you believe have
traditional religious or cultural significance and whether there are specialized concerns or
information known to the Tribe that should be considered by the staff during the development of
the SEIS.

The NRC staff understands that the Tribe may raise issues in consultation that should be kept
confidential and nonpublic; the staff is committed to maintaining confidentiality of said
information.

After a careful review and assessment of all information and comments received, the NRC will
determine what additional actions are necessary to comply with 10 CFR Part 51 and 36 CFR
800, the implementing regulations for Section 106 of the National Historic Preservation Act.

M. Jandreau   2

If the Tribe would like to participate as a consulting party pursuant to Section 106, the Tribe should express its interest in participating and identify areas of concern, within 60 days of receipt of this letter, to ensure that the parties will have the opportunity to engage in meaningful and productive consultation. The Tribe should forward its response to the following address: Mr. Kevin Hsueh, Mail Stop T-8F05, Washington, DC. 20555.

If you have any questions or comments, or need any additional information, please contact the environmental Project Manager, Ms. Haimanot Yilma by telephone at 301-415-8029, or email at Haimanot.Yilma@nrc.gov.

Sincerely,

Kevin Hsueh, Branch Chief
Environmental Review Branch B
Environmental Protection and Performance
 Assessment Directorate
Division of Waste Management
 and Environmental Protection
Office of Federal and State Materials
 and Environmental Management Programs

Docket No.: 040-09075

Enclosure: Letter of March 19, 2010

cc: w/enclosure
Clair Green
Cultural Resources
Lower Brule Sioux Tribe
P.O Box 187
Lower Brule, SD 57548-0187

LOWER BRULE SIOUX TRIBE ————————————

November 15, 2010

Kevin Hsueh, Branch Chief
Environmental Review, Branch B
Nuclear Regulatory Commission
Mail Stop T-8F05
Washington, DC. 20555

Dear Mr. Hsueh:

The Lower Brule Sioux Tribe requests to participate as a consulting party in both the NHPA
Section 106 consultation as well as consultation on the Supplemental EIS for the Dewey-Burdock
ISR facility near Edgemont South Dakota. Ms Clair Green, Lower Brule Cultural
Resource/Public Information Office (605) 473-8037 is our contact for this project.

Thank you for your consideration.

Sincerely,

Michael B. Jandreau
Chairman

187 Oyate Circle • Lower Brule, SD 57548 • Phone 605-473-5561 • Fax 605-473-5554

DeweyBurdPubEm Resource

From:	Yilma, Haimanot
Sent:	Monday, November 22, 2010 5:15 PM
To:	gravattiana@yahoo.com
Cc:	Hsueh, Kevin
Subject:	NRC's Consultation letter for proposed Dewey-Burdock ISR facility near Edgemont, SD
Attachments:	Yankton Sioux Tribe.pdf

Ms. Gravatt,

Thank you for taking the time to speak with me today. Per our conversation just now and your request, I have attached the consultation letter sent on Sept 10, 2010. Please address your response to the attached letter to my Branch Chief Kevin Hsueh at the following address:

Mr. Kevin Hsueh,
Mail Stop T-8HF05,
Washington, DC 20555.

You can also find this information inside the attached letter.

Sincerely,

Haimanot Yilma
Project Manager
FSME/DWMEP/EPPAD/ERB
U.S Nuclear Regulatory Commission
Phone: 301-415-8029
email: haimanot.yilma@nrc.gov
Mail Stop : T8H09

From: Yilma, Haimanot
Sent: Monday, November 22, 2010 5:01 PM
To: Yilma, Haimanot
Subject:

1

UNITED STATES
NUCLEAR REGULATORY COMMISSION
WASHINGTON, D.C. 20555-0001

September 10, 2010

Robert Cournoyer, Chairman
Yankton Sioux Tribe
P.O Box 248
Marty, SD 57361-0248

SUBJECT: INVITATION FOR FORMAL CONSULTATION UNDER THE SECTION 106 OF
            THE NATIONAL HISTORIC PRESERVATION ACT

Dear Chairman Cournoyer:

As established in Title 10 *Code of Federal Regulations* Part 51 (10 CFR 51), the U.S. Nuclear
Regulatory Commission (NRC) regulations that implement the National Environmental Policy
Act (NEPA) of 1969, as amended, the NRC is preparing a Supplemental Environmental Impact
Statement (SEIS) for the proposed Powertech Inc. Dewey-Burdock In-Situ Recovery (ISR)
Facility near Edgemont, South Dakota. As part of the environmental review, the SEIS will
include an analysis of potential impacts of the proposed action to historic and cultural properties.

On March 19, 2010, the NRC sent a letter to your office inviting the Yankton Sioux Tribe
to participate as a consulting party and requested information regarding tribal historic and
cultural resources potentially affected by the proposed Dewey-Burdock ISR facility. A copy of
the March 19th letter is enclosed, for your convenience.

To date, the NRC has not received any response from your office regarding the Tribe's interest
in becoming a consulting party for the proposed Dewey-Burdock ISR facility near Edgemont
South Dakota.

The NRC again extends an invitation to the Yankton Sioux Tribe to participate as a consulting
party for the proposed Dewey-Burdock ISR facility. Specifically, the NRC is interested in
learning of any areas on the proposed Dewey-Burdock site that you believe have traditional
religious or cultural significance and whether there are specialized concerns or information
known to the Tribe that should be considered by the staff during the development of the SEIS.

The NRC staff understands that the Tribe may raise issues in consultation that should be kept
confidential and nonpublic; the staff is committed to maintaining confidentiality of said
information.

After a careful review and assessment of all information and comments received, the NRC will
determine what additional actions are necessary to comply with 10 CFR Part 51 and 36 CFR
800, the implementing regulations for Section 106 of the National Historic Preservation Act.

R. Cournoyer   2

If the Tribe would like to participate as a consulting party pursuant to Section 106, the Tribe should express its interest in participating and identify areas of concern, within 60 days of receipt of this letter, to ensure that the parties will have the opportunity to engage in meaningful and productive consultation. The Tribe should forward its response to the following address: Mr. Kevin Hsueh, Mail Stop T-8F05, Washington, DC. 20555.

If you have any questions or comments, or need any additional information, please contact the environmental Project Manager, Ms. Haimanot Yilma by telephone at 301-415-8029, or email at Haimanot.Yilma@nrc.gov.

                                    Sincerely,

                                    Kevin Hsueh, Branch Chief
                                    Environmental Review Branch B
                                    Environmental Protection and Performance
                                      Assessment Directorate
                                    Division of Waste Management
                                      and Environmental Protection
                                    Office of Federal and State Materials
                                      and Environmental Management Programs

Docket No.: 040-09075

Enclosure: Letter of March 19, 2010

Box 248
Marty, SD 57361

(605) 384-3804 / 384-3641
FAX (605) 384-5687

December 3, 2010

Branch Chief Kevin Hsueh
U.S. Nuclear Regulatory Commission
Mail Stop T-8HF05
Washington, DC 20555

Mr. Hsueh,

The Yankton Sioux Tribe is requesting a face to face consultation on all past and current projects at your earliest convenience. The Yankton Sioux Tribe will need to survey projects to protect Traditional Cultural Properties. We want to know what your position is on this. An archeological review is not the same as a TCP survey it is different in knowledge and methodology. We know we have properties of such within your project areas. There is nothing from the Yankton Sioux tribe that states no response means concurrence within 30 days. Please reach me at 1-605-384-3641 or email me at gravattlana@yahoo.com.

Respectfully,

Lana M. Gravatt
Yankton Sioux Tribal Historic Preservation Officer

DeweyBurdPubEm Resource

From:	Adrienne Swallow [aswallow@standingrock.org]
Sent:	Wednesday, December 08, 2010 3:46 PM
To:	Yilma, Haimanot
Cc:	Hsueh, Kevin
Subject:	Dewey-Burdock Project

Dear Mr. Hsueh,

The Standing Rock Tribe is in receipt of you letter dated September 10, 201 regarding our participation as a consulting party under Section 106 of the National Historic Preservation Act for the proposed Dewey Burdock In-Situ Recovery (ISR) facility near Edgemont, SD.

Please note that Ron His Horse is Thunder is no longer our Tribal Chairman and direct all future correspondence to our current Chairman, Charles W. Murphy.

We are not interested in becoming a consulting party for the proposed Dewey-Burdock ISR. In fact, we are opposed to the project because of its proximity to the Black Hills. The Black Hills are considered sacred by the Lakota and Dakota people and we are very concerned that there could be accidental environmental contamination of the area during the operation of in-situ recovery.

We are particularly concerned about contamination of groundwater. What steps will be taken to ensure that groundwater will not be contaminated before, during and after the mine has been completed? How will groundwater be restored to its original pre-mining condition? How will large volumes of waste water be disposed of? How will Native American cultural resources be protected? We hope there will be a satisfactory response to these concerns prior to construction.

We ask that prior to any ground disturbance, a Class III survey be conducted by a Tribal member.

Please keep us informed of all activities regarding the Dewey-Burdock In-Situ Recovery facility.

Sincerely,
Adrienne Swallow
Environmental Protection Specialist
Standing Rock Sioux Tribe
PO Box D
Fort Yates, ND 58538
701-854-8582
cell: 701-226-0291
fax:701-854-3488
aswallow@standingrock.org

1

December 15, 2010

Mr. John M. Fowler, Executive Director
Advisory Council on Historic Preservation
Office of Federal Agency Programs
1100 Pennsylvania Ave, NW, Suite 803
Washington, DC 20004

SUBJECT: POWERTECH INC. PROPOSED DEWEY-BURDOCK IN-SITU RECOVERY
            FACILITY NEAR EDGEMONT, SOUTH DAKOTA (DOCKET 040-09075)

Dear Mr. Fowler:

The U.S. Nuclear Regulatory Commission (NRC) has received an application from Powertech
Inc. (Powertech) for a new radioactive source materials license to develop and operate the
Dewey-Burdock Project located near Edgemont, South Dakota in Fall River and Custer
Counties. The facility, if licensed, would use an *in-situ* recovery (ISR) methodology to extract
uranium at the Dewey-Burdock site. The proposed project boundary consists of approximately
10,580 acres (4,282 ha) located on both sides of Dewey Road (County Road 6463) and portions
of Sections 1-5, 10-12, 14, and 15, Township 7 South, Range 1 East and Sections 20, 21, 27,
28, 29, and 30-35, Township 6 South, Range 1 East, Black Hill Meridian. A map showing the
proposed project boundary is enclosed (Powertech Figure 1.4-1).

As established in Title 10 *Code of Federal Regulations* Part 51 (10 CFR 51), the NRC regulation
that implements the National Environmental Policy Act of 1969 (NEPA), as amended, the NRC
is preparing a Supplemental Environmental Impact Statement (SEIS) for the proposed action.
The SEIS will address the impacts associated with the construction, operation, and
decommissioning of the proposed facility. As outlined in 36 CFR 800, to comply with Section
106 of the National Historic Preservation Act of 1966 through the requirements of the NEPA, the
SEIS will include analyses of potential impacts to historic and cultural resources.

To enhance the scope and quality of our review and facilitate the identification of tribal historic
sites and/or cultural resources, specifically, sites that may have traditional religious or cultural
significance to Native American Tribes that may be interested in and/or affected by the
proposed action, the NRC sent consultation letters to 17 tribes, including the Oglala Sioux, on
March 19, 2010 and September 10, 2010. The NRC has also made additional contacts with
tribal officials offering consultation and seeking information by other means such as telephone
calls and emails.

To date, Turtle Mountain Band of Chippewa and Three Affiliated Tribes (Mandan, Hidatsa &
Arikara Nation) responded in writing to the consultation letters and stated they anticipate no
adverse effect on cultural resources by the proposed action. Fort Peck Assiniboine & Sioux;
Sisseton-Wahpeton Oyate, Rosebud Sioux Tribe, and Lower Brule Sioux responded requesting
formal consultation. Eastern Shoshone tribe informally indicated that they are interested in
formal consultation. The NRC has not yet received responses from the Oglala Sioux; Cheyenne
River Sioux; Crow Creek Sioux; Flandreau-Santee Sioux; Standing Rock Sioux; Yankton Sioux;
Spirit Lake Tribe; Lower Sioux Indian Community; Northern Cheyenne; and Northern Arapaho
tribes.

J.M. Fowler   2

The NRC plans on issuing the draft SEIS in summer 2011; when the draft becomes available, a
copy will be sent to your office for your review and comment. The NRC will also notify the
17 tribes when the draft SEIS is available and request their comments.

The Powertech Dewey-Burdock Project license application is publicly available in the NRC
Public Document Room (PDR) located at One White Flint North, 11555 Rockville Pike,
Rockville, Maryland 20852, or from the NRC's Agency Wide Documents Access and
Management System (ADAMS). The ADAMS Public Electronic Reading Room is accessible at
http://www.nrc.gov/reading-rm/adams.html. The accession numbers for the Powertech
application including the Environmental Report is ML092870160.

If you have any questions or comments, or need any additional information, please contact
Ms. Haimanot Yilma of my staff by telephone at 301-415-8029 or email at
haimanot.yilma@nrc.gov

                              Sincerely,

                              /RA/

                              Kevin Hsueh, Branch Chief
                              Environmental Review Branch B
                              Environmental Protection
                                and Performance Assessment Directorate
                              Division of Waste Management
                                and Environmental Protection
                              Office of Federal and State Materials
                                and Environmental Management Programs

Docket No.: 040-09075

Enclosure: Map

CC: Marian Atkins, Field Office Manager
     South Dakota Field Office - BLM
     310 Roundup Street
     Belle Fourche, SD 57717-1698

     Gregory R. Fesko P.G.
     Coal Program Coordinator
     Branch of Solid Minerals - BLM
     Montana State Office
     5001 Southgate Drive
     Billings, MT 59001

Oglala Sioux Tribe
Tribal Historic Preservation Office
P.O. Box 320, W. Hwy 18
Pine Ridge, SD 57770
Phone: (605) 867-5969
Fax: (605) 867-2818
ustnrrathpo@gwtc.net

TRIBAL HISTORIC PRESERVATION ADVISORY COUNCIL:
Mr. Tom Bad Heart Bull - Oglala District
Mr. Francis "Chubbs" Thunder Hawk - Porcupine District
Mr. Garvard Good Plume, Jr. - Wakpamni District

STAFF:
Tribal Historic Preservation Officer – Mr. Wilmer Mesteth
Project Review Officer - Ms. Roberta Joyce Whiting
Natural Resources Director – Mr. Michael Catches Enemy

January 31, 2011

Kevin Hsueh, Branch Chief
Environmental Review Branch B
Environmental Protection and Performance
  Assessment Directorate
Division of Waste Management and
  Environmental Protection
Office of Federal and State Materials and
  Environmental Management Programs
United States Nuclear Regulatory Commission
Mail Stop T-8F05
Washington, DC 20555

Re: Invitation for Formal Consultation under the National Historic Preservation Act; Request for Information under the National Environmental Policy Act; Proposed Powertech Inc. Dewey-Burdock In-Situ Leach Uranium Mine (NRC Docket No. 040-09075)

Dear Mr. Hsueh:

Thank you for your letters dated September 8, 2010, and September 10, 2010, inviting the Oglala Sioux Tribe to participate as a consulting party as the U.S. Nuclear Regulatory Commission (NRC) works to satisfy its statutory obligations under the National Historic Preservation Act (NHPA) and National Environmental Policy Act (NEPA) to review impacts to cultural and historic resources potentially impacted by the proposed Powertech, Inc. Dewey-Burdock In-Situ Leach Uranium Mine. As you are aware, the proposed mine is located within the traditional and treaty lands of the Great Sioux Nation, which includes the Oglala Sioux Tribe. The Tribe is committed to working with the NRC to identify and protect the cultural and historic resources threatened by the project.

Currently, the Oglala Sioux Tribal Preservation Historic Office is directed by Mr. Wilmer Mesteth. The Tribal Historic Preservation Office looks forward to any support the NRC Staff can provide in facilitating this review, including providing the Tribe an ongoing opportunity to review and comment on the agency's review as it is developed. Please note that the responsibilities and resources of other federal agencies to protect the cultural and historical resources of the Oglala Sioux Tribe which are located on and near the adjacent Black Hills National Forest are also implicated by the location of this project, including the U.S. Forest Service and the Bureau of Land Management

Page 1 of 3

From information obtained through the application submitted to the NRC by Powertech, Inc., the proposed Dewey-Burdock In-Situ Leach Uranium Mine project represents a substantial potential threat to the preservation of cultural and historic resources of the Oglala Sioux Tribe. These impacts include not only site-specific physical impacts, but also broader landscape-level impacts along with more intangible impacts to the integrity of the area from cultural, historical, spiritual, and religious perspectives.

Importantly, the impact from the proposed mine extends not just from the disturbance associated with the Dewey-Burdock site, which could be substantial in itself, but also the impacts associated with the foreseeable use of the Dewey-Burdock site as a regional uranium processing center for potential mining operations across the region. These broader effects are further compounded by the substantial impacts associated with the large open pit uranium mines at the project area that have been egregiously left entirely unreclaimed since the last uranium boom. It is critical for any credible cultural and historic resource impact analysis to consider the entirety of these past and reasonably foreseeable activities.

A review of the application materials submitted by Powertech to the NRC reveals an incomplete analysis. Nowhere does the application recognize the cumulative impacts associated with regional uranium development or past uranium development. Indeed, the applicant's materials represent that the cultural and historic impacts associated with the entire proposal, even when combined with reasonably foreseeable future and past actions, are "none." This conclusion is unsupportable, and appears to have resulted in part from the incomplete methodology employed in reaching it. It appears that the review prepared by Powertech failed to include any direct input from any tribal sources, whether written or oral.

It should be noted that a primary source of credible information in this case are oral histories and ethnographic information of those knowledgeable about the impacted area, whether through personal, family, or ancestral connections. Instead, the application cites only to a handful of studies, most prepared for other projects within some undefined geographic proximity to the proposed mine site. Incorporation of all credible and relevant written and oral sources is necessary, with appropriate measures taken to respect the integrity and confidentiality of such information. In this case, the application fails to assess even the detailed information contained in sworn oral testimony during hearings at the early stages of the State of South Dakota permitting process. These same gaps in information bring into question the reliability and completeness of the application, including the site visit analysis conducted by Powertech. Indeed, the site-visit analysis itself was conducted without any tribal participation and identifies a significant number of archaeological, historical, and traditional cultural resources within the project area that have not yet been evaluated at all.

Page 2 of 3

Critically, information on the historic and cultural significance of the proposed project area is not limited to that held by members of the Oglala Sioux Tribe. Rather, this area is within an area within which other Sioux tribes, in addition to the Cheyenne, Arapahoe, Crow, and Arikara Tribes, among others, also possess intimate cultural knowledge. As such, any credible impact review must assess the historic and cultural impacts associated with these other cultures. For example, the Oglala Sioux Tribal Historic Preservation Office is working toward establishing such a study, with culturally appropriate protocols to protect the information acquired and to incorporate necessary protections where information about such persons is involved in the collection of oral histories and ethnographies.

We look forward to working with the NRC on these important issues. To the fullest extent possible, we ask that the agency share any information that it has collected, and work with the Tribe to identify additional sources of information to include in its analysis. This request includes making available to the Tribe physical copies of all of the cited references in Powertech's application materials, as well as any additional resources NRC Staff may have available. Further, the Tribe requests NRC's help in facilitating a timely site visit and review so as to provide the Tribe an opportunity to conduct a full review of the cultural and historic resources at stake. Lastly, the Oglala Sioux Tribe requests that the NRC sponsor and conduct a regional meeting of Tribal Historic Preservation Officers from all affected tribes in order to encourage effective communication to the NRC.

Thank you for the invitation to conduct a thorough consultation process on this important matter.

Sincerely,

Michael Catches Enemy
Natural Resources Director

Sincerely,

Wilmer Mesteth
Tribal Historic Preservation Officer

Cc: Honorable President John Yellowbird Steele, Oglala Sioux Tribe
       Oglala Sioux Tribal Land & Natural Resources Committee
       SSR Law
       File

March 4, 2011

Cedric Black Eagle, Chairman
Crow Tribe of Montana
Baacheeitche Avenue
P.O. Box 159
Crow Agency, MT 59022

SUBJECT: INVITATION FOR FORMAL CONSULTATION UNDER SECTION 106 OF THE
            NATIONAL HISTORIC PRESERVATION ACT

Chairman Black Eagle:

The U.S. Nuclear Regulatory Commission (NRC) has received an application from Powertech
Inc. (Powertech) for a new radioactive source materials license to develop and operate the
Dewey-Burdock Project located near Edgemont, South Dakota in Fall River and Custer
Counties. The facility, if licensed, would use an *in-situ* recovery methodology to extract uranium
at the Dewey-Burdock site. The proposed project area consists of approximately 10,580 acres
(4,282 ha) located on both sides of Dewey Road (County Road 6463) and portions of Sections
1-5, 10-12, 14, and 15, Township 7 South, Range 1 East and Sections 20, 21, 27, 28, 29, and
30-35, Township 6 South, Range 1 East, Black Hill Meridian. A map showing the proposed
project boundary is enclosed (Powertech Figure 1.4-1).

The South Dakota State Historic Preservation Officer identified the Crow Tribe of Montana as
potentially attaching religious and cultural significance to historic properties in the project area.
By this letter, the NRC invites the Crow Tribe to participate as a consulting party in the National
Historic Preservation Act Section 106 process. If the Tribe would like to participate as a
consulting party, please respond to this letter.

In regards to the proposed project, the NRC is also engaged in an environmental review and is
preparing a Supplemental Environmental Impact Statement (SEIS) pursuant to the National
Environmental Policy Act. As part of this review, the SEIS will include an analysis of potential
impacts to historic and cultural properties and is therefore requesting input from the Crow Tribe
to facilitate the identification of tribal historic sites or cultural resources that may be affected by
the proposed action. Specifically, the NRC is interested in learning of any areas on the Dewey-
Burdock site that you believe have traditional religious or cultural significance.

The NRC staff understands that the Tribe may raise issues in consultation that should be kept
confidential and nonpublic; the staff is committed to maintaining confidentiality of said
information.

After a careful review and assessment of all information and comments received, the NRC will
determine what additional actions are necessary to comply with 10 CFR Part 51 and 36 CFR
800, the implementing regulations for Section 106 of the National Historic Preservation Act. If
the Tribe would like to participate as a consulting party pursuant to Section 106, the Tribe
should express its interest in participating and identify areas of concern, within 60 days of
receipt of this letter, to ensure that the parties will have the opportunity to engage in meaningful
and productive consultation. The Tribe should forward its response to the following address:
Mr. Larry Camper, Mail Stop T-8F05, Washington, DC 20555.

C. B. Eagle   2

The Powertech Dewey-Burdock Project license application is publicly available in the NRC
Public Document Room located at One White Flint North, 11555 Rockville Pike, Rockville,
Maryland 20852, or from the NRC's Agency Wide Documents Access and Management
System (ADAMS). The ADAMS Public Electronic Reading Room is accessible at
http://www.nrc.gov/reading-rm/adams.html. The accession numbers for the Powertech
application including the Environmental report is ML092870160.

If you have any questions or comments, or need any additional information, please contact the
Environmental Project Manager, Ms. Haimanot Yilma by telephone at 301-415-8029, or email at
Haimanot.Yilma@nrc.gov.

                                    Sincerely,

                                    /RA/ by K. McConnell for

                                    Larry W. Camper, Director
                                    Division of Waste Management
                                     and Environmental Protection
                                    Office of Federal and State Materials
                                     and Environmental Management Programs

Docket No.: 040-09075

Enclosure:
Figure 1.4-1

cc with enclosure:

Dale Old Horn, THPO,
Crow Tribe of Montana   Marian Atkins
Baacheeitche Avenue   Field Office Manager - BLM
P.O. Box 159   South Dakota Field Office
Crow Agency, MT 59022   310 Roundup Street
                                         Belle Fourche, SD 57717-1698

Hubert Two Leggings   Gregory R. Fesko P.G.
Cultural Resource Officer   Coal Program Coordinator
Crow Tribe of Montana   Branch of Solid Minerals - BLM
Baacheeitche Avenue   Montana State Office
P.O. Box 159   5001 Southgate Drive
Crow Agency, MT 59022   Billings, MT 59001

From:	Yilma, Haimanot
Sent:	Thursday, March 10, 2011 7:51 AM
To:	Miller, Debra; Rajapakse, Champa
Cc:	Hsueh, Kevin
Subject:	FW: Dewey-Burdock Project

Deb and / or Champa,

Can you please put this email in ADAMS and give me the ML number for my records. I am working from home today.

Kevin,

Just to let you know, Crow tribe of MT is interested in becoming a consulting party for Dewey. I will update the status report.

Thanks
Haimanot

From: Yilma, Haimanot
Sent: Thursday, March 10, 2011 7:46 AM
To: 'Hubert Two Leggins'
Cc: Yilma, Haimanot
Subject: RE: Dewey-Burdock Project

Mr. Two Leggins

Thank you for responding to our invitation letter. This email is sufficient for us to know your interest in becoming a consulting party under the section 106 consultation process. I will share your interest with my management.

Regards,

Haimanot Yilma
Project Manager
301-415-8029

From: Hubert Two Leggins [mailto:hubertt@crownations.net]
Sent: Wednesday, March 09, 2011 7:22 PM
To: Yilma, Haimanot
Subject: Dewey-Burdock Project

Hello Ms. Haimanot Yilma,

I accept the invitation for the formal consultation under section 106 of NHPA. The Crow Tribe has religious and cultural significance to the project area and wants to be a consulting party. I don't know if this is going to

Work for Mr. Larry Camper or if I need to send a separate letter to him please let me know.

Thank You

Hubert B. Two Leggins
Crow Tribal Cultural Resource Director/Renewable Resource Supervisor
P. O. Box 159
Crow Agency, Mt. 59022
(406) 638-3793 work
(406) 678-1677 cell

hubertt@crownations.net

May 12, 2011

Ms. Lana Gravatt
Yankton Sioux Tribe
P.O. Box 248
Marty, SD 57631-0248

SUBJECT: INVITATION FOR INFORMAL INFORMATION-GATHERING MEETING
            PERTAINING TO THE DEWEY-BURDOCK, CROW BUTTE NORTH TREND,
            AND CROW BUTTE LICENSE RENEWAL, *IN-SITU* URANIUM RECOVERY
            PROJECTS

Dear Ms. Gravatt:

The U.S. Nuclear Regulatory Commission (NRC) staff would like to extend an invitation to the
Yankton Sioux Tribe officials (Tribal Historic Preservation Officers and/or Cultural Resources
Officers) to assist the NRC in the identification of tribal historic sites, traditional cultural
properties, and cultural resources that may be affected by the actions proposed by Cameco
Resources Inc. and Powertech Inc. We are extending this invitation because you indicated that
you would like to be a consulting party. The NRC will hold an informal information-gathering
meeting on June 7, 8, and 9, 2011, at the Prairie Wind Casino and Hotel on the Pine Ridge
Reservation in South Dakota.

The information-gathering meeting will include a staff-to-staff session on June 8, 2011, at the
Prairie Wind Casino and Hotel on the Pine Ridge Reservation and two days of site visits (June 7
and 9, 2011) to the proposed facilities. The NRC staff has coordinated with Cameco Resources
Inc. and Powertech Inc. to arrange for visits to the proposed facilities. The itinerary for the site
visit is included in Enclosure 4. Although attendance at the site visits is optional, Tribal
representatives are encouraged to participate because the visits will provide an opportunity to
tour an existing facility and to view the proposed project areas.

The NRC is in the process of conducting environmental reviews for a number of license
applications involving in-situ uranium recovery facilities. These applications include Cameco
Resources Inc.'s applications for Crow Butte North Trend and Crow Butte License Renewal, as
well as Powertech (USA) Inc.'s application for Dewey-Burdock. The NRC is undertaking these
reviews as part of our responsibilities under the National Environmental Policy Act (NEPA). Our
reviews will culminate in the issuance of an Environmental Assessment, Environmental Impact
Statement, or a Supplemental Environmental Impact Statement. These documents will include
analyses of potential impacts to historic and cultural properties. In accordance with the National
Historic Preservation Act regulation 36 CFR 800.8(c), we are coordinating our Section 106
review with our NEPA assessment. As part of our reviews, the NRC requests the input of
Tribes concerned with the effects the proposed actions may have on historic and cultural
properties, so their views may be considered in the decision-making process.

The NRC staff is interested in identifying areas within the Crow Butte sites and/or Dewey-
Burdock site that have traditional religious or cultural significance to the Yankton Sioux Tribe of
South Dakota, so these may be considered in our environmental reviews. Maps identifying the
specific locations of each proposed project are enclosed for your reference (Enclosure 1).

L. Gravatt   2

Digital copies of the publicly available archaeological surveys prepared for the Crow Butte Resources and Dewey-Burdock projects are enclosed. These reports are contained in 3 -diskettes (one for Crow Butte and two for Dewey-Burdock- Enclosure 3).

Please provide the NRC staff with copies of, or references to, documentary or published materials you like the staff to review.

NRC staff, in conjunction with the Oglala Sioux Tribe has prepared a tentative agenda for this meeting. The draft agenda is enclosed in this letter for your review (Enclosure 2).

Please identify any additional topics you would like to discuss and let us know by May 20, 2011, how many of your tribal representatives will be attending this meeting. We request your reply be addressed to Mr. Larry Camper, Mail Stop T-8F05, Washington, D.C. 20555. In order to attend the site visit, NRC must have your confirmation by this date to ensure adequate transportation is available for all participants. If possible, provide us with the name and job descriptions of tribal representatives who plan to participate. Ms. Haimanot Yilma and Mr. Nathan Goodman will follow up with you via a phone call to finalize meeting logistics.

The applicants will be providing roundtrip transportation between the Prairie Wind Casino and Hotel on the Pine Ridge Reservation and the Crow Butte and Dewey-Burdock sites. However, other travel costs associated with the June 7–9, 2011, meeting will not be covered.

If you have any questions regarding this meeting, please contact my staff members, Ms. Yilma (via email at Haimanot.Yilma@nrc.gov or via phone at 301-415-8029) or Mr. Goodman (via email at Nathan.Goodman@nrc.gov or via phone at 301-415-2703).

                              Sincerely,

                              /RA/ APersinko for LCamper

                              Larry W. Camper, Director
                              Division of Waste Management
                               and Environmental Protection
                              Office of Federal and State Materials
                               and Environmental Management Programs

Docket No.: 040-09075
Docket No.: 040-08943

Enclosures:
1. Map of Crow Butte and Dewey-Burdock
    Proposed Project Areas/Boundary
2. Draft Agenda
3. Archeological Surveys of the Proposed
    Crow Butte & Dewey-Burdock Projects
4. Itinerary for Site Visits

cc: Chairman Robert Cournoyer

August 12, 2011

Mr. Richard Blubaugh
VP-HS&E Resources
Powertech (USA) Inc.
5575 DTC Parkway
Suite 140
Greenwood Village, CO 80111

SUBJECT: INFORMATION REQUIRED BY THE U.S NUCLEAR REGULATORY
COMMISSION STAFF TO SATISFY ITS OBLIGATIONS UNDER SECTION 106
OF THE NATIONAL HISTORIC PRESERVATION ACT AND NATIONAL
ENVIRONMENTAL POLICY ACT TO COMPLETE ITS REVIEW OF THE
IMPACTS TO THE CULTURAL RESOURCES FROM THE PROPOSED
DEWEY-BURDOCK PROJECT

Dear Mr. Blubaugh:

The U.S. Nuclear Regulatory Commission (NRC) has received an application from Powertech
(USA) Inc. for a new source material license to permit Powertech to operate the proposed
Dewey-Burdock *In-Situ* recovery (ISR) facility. As part of our responsibilities under the National
Environmental Policy Act (NEPA), the NRC is conducting an environmental review of
Powertech's application. In addition to our NEPA review, the NRC must comply with the
National Historic Preservation Act (NHPA). Under Section 106 of the NHPA and its
implementing regulations (36 CFR. Part 800), the NRC must take into account the effects that
issuing a license to Powertech would have on historic properties and afford the Advisory Council
on Historic Preservation (ACHP) a reasonable opportunity to comment on the NRC's findings.

In order to comply with NEPA and Section 106 of the NHPA, the NRC must make reasonable
and good faith efforts to identify historic properties within the area of potential effects for the
proposed Dewey-Burdock ISR facility. Historic properties include properties of traditional
religious and cultural importance to one or more Indian Tribes. Based on information gathered
during a June 2011 meeting between the NRC and representatives from six Indian Tribes, and
based on consultation with the South Dakota State Historic Preservation Officer (SHPO) and the
ACHP, the NRC staff has determined that it requires additional information on Traditional
Cultural Properties (TCPs) in the area of potential effect for the proposed Dewey-Burdock
facility. The NRC staff needs this information not only to fulfill our obligations under Section 106
of the NHPA, but also to fulfill our obligation under NEPA that we assess potential impacts to
cultural resources. This information on TCPs would be in addition to the extensive
archeological surveys that Powertech has already submitted in support of its license application.

Although the NRC believes that a traditional cultural property survey of the area of potential
effect is an effective method to identify these properties; information on TCPs can also be
obtained in a variety of ways. For example, site visits by tribal representatives could be used to

R. Blubaugh   2

identify TCPs, or an applicant could hire an archaeologist with experience identifying and evaluating potential TCPs. Alternatively, an applicant could use a combination of these or other methods.

The NRC requests that by August 31, 2011, Powertech submit a written plan for acquiring information on TCPs. Upon receipt of this TCP identification plan, the NRC will determine whether the actions outlined in the plan will provide information sufficient for the NRC to meet any applicable requirements under NEPA and the NHPA.

Attached to this letter for your possible use is a list of Indian Tribes that have expressed interest in historic properties in the area of potential effect for the proposed Dewey-Burdock facility.

Please submit your TCP identification plan to NRC, Attention: Mr. Kevin Hsueh, Mail Stop T8F05, Washington, DC 20555. If you have any questions or comments, or need any additional information, please contact Ms. Haimanot Yilma of my staff by telephone at 301-415-8029, or by email at Haimanot.Yilma@NRC.gov.

Sincerely,

/RA/

Kevin Hsueh, Branch Chief
Environmental Review Branch-B
Environmental Protection and Performance
  Assessment Directorate
Division of Waste Management
  and Environmental Protection
Office of Federal and State Materials
  and Environmental Management Programs

Docket No. 040-09075

cc: See Attached List

RICHARD E. BLUBAUGH
Vice President – Health Safety
& Environmental Resources

POWERTECH (USA) INC.

August 31, 2011

Kevin Hsueh, Branch Chief
Environmental Review Branch B
Environmental Protection and Performance Assessment Directorate
Division of Waste Management and Environmental Protection
Office of Federal and State Materials and Environmental Management Programs
U.S. Nuclear Regulatory Commission
Mail Stop T8-F05
Washington, DC 20555-0001

Re: Powertech (USA) Inc.'s Response to NRC Request for National Historic
Preservation Act Section 106 Information; Docket No. 040-09075

Dear Mr. Hsueh:

Attached please find a proposal, developed by SRI Foundation, to support Cameco
Resources and Powertech (USA) Inc. efforts to collect National Historic Preservation Act
Section 105 information required for the NRC evaluation of proposed Powertech (USA)
Inc. operations at the proposed Dewey-Burdock Project.

Should you have any questions, please do not hesitate to contact the undersigned at (303)
790-7528.

Respectfully yours,

Richard E. Blubaugh

Enclosure
cc: R.F. Clement, President and CEO

        Thompson &Pugsley, PLLC
        1225 19th Street, NW
        Suite 300
        Washington, DC 20036

5575 DTC Parkway, Suite 140   Telephone: 303-790-7528   Website: www.powertechuranium.com
Greenwood Village, CO 80111 USA   Facsimile: 303-790-3885   Email: info@powertechuranium.com

Draft Plan August 30, 2011

Proposal from Cameco Resources and Powertech Inc.
to the U.S. Nuclear Regulatory Commission

*A plan for assisting NRC, as the Federal lead agency for Section 106, by gathering
information about properties of religious and cultural significance to Federally-
recognized Indian tribes that may be affected by their proposed undertakings*

Cameco Resources and Powertech Inc. (hereafter "the companies") propose to carry out a
phased program of information gathering with Indian tribes, as described below, in order
to identify places of religious and cultural significance to those tribes that may be
affected by the proposed Three Crow, Crow Butte, North Trend, Marsland (Cameco), and
Dewey-Burdock (Powertech) projects. These efforts will be carried out in response to
NRC's letter of August 5, 2011, to John Schmuck at Cameco and of August 12, 2011 to
Richard Blubaugh of Powertech, requesting additional information on historic properties
in support of compliance with Section 106 of the National Historic Preservation Act of
1966 and its implementing regulation, 36 CFR part 800. The companies have secured the
services of the SRI Foundation (SRIF) of Rio Rancho, NM to assist them in this effort.
Unless otherwise indicated, all tasks below will be carried out by SRIF under the
direction of the companies.

Phase 1
- Prepare a written plan detailing how the companies propose to proceed and a final list
  of tribes to be contacted. The companies will submit the plan and list of tribes for
  review by NRC (and by BLM and SHPOs if they wish to review).
- Define a general study area, encompassing all five proposed undertakings, and two
  proposed expanded areas of potential effects (APEs; 36 CFR §800.4(a)), one
  including all areas from which the Dewey-Burdock project will be visible and one
  encompassing all areas from which any of the four Cameco project areas will be
  visible. These expanded APEs will take into account the potential sensitivity of places
  of religious and cultural significance to indirect effects.
- Touch base with NRC review and cultural resource staff; Wyoming, South Dakota,
  and Nebraska State Historic Preservation Officers (SHPOs); and Bureau of Land
  Management (BLM) South Dakota Field Office personnel to introduce the
  information-gathering effort and SRIF staff, and secure agency input on how this
  effort should proceed. Among the issues to raise: the list of tribes to be contacted,
  proposed expanded APEs, Advisory Council on Historic Preservation (ACHP)
  involvement, SHPO preferences for reviewing documents, BLM's preference for
  level of involvement, and available information about previously conducted
  ethnographic research.

1

Draft Plan August 30, 2011

- Review NRC documentation concerning previous consultation with tribes and any NRC policies or protocols for tribal consultation.
- Touch base with company attorneys to identify any sensitive issues relative to current administrative appeals or future litigation.
- Revise the plan and the list of tribes to reflect NRC (and SHPO and BLM, if participating) comments.

Phase 2
- Assist the companies in developing joint or separate RFPs to secure the services of appropriate ethnographers to complete identification of properties of religious and cultural significance (36 CFR §800.3(c)(2)(B)(ii)) within the two APEs. Assistance may include identifying potential ethnographic consultants, developing scopes of work, and reviewing and commenting on proposals received.
- Develop a brief overview of Native American use and practices in the study area encompassing the companies' project areas to serve as a context; this overview will include information on types of traditional cultural properties typically encountered in this region.
- Develop a script for initial tribal contacts concerning the information-gathering project, and identify supporting materials to be included. The supporting materials may include maps of the projects and areas to be studied (the APEs), photographs of what developed in situ uranium recovery projects look like, an animation of how the in situ recovery process works, etc. Parameters: Information provided should be brief, clear, and nontechnical. Tribes will be provided with a choice among several possible levels of future participation ranging from "not interested in being consulted about these projects" through "would like to be informed about results of efforts to identify archaeological sites and traditional cultural places" through "wish to participate in field visits and ethnographic interviews." Tribes will be encouraged to offer any comments on the proposed areas to be studied.
- Submit script and materials for review by the companies
- Companies submit script and accompanying materials for review by NRC (and BLM if they wish to participate)
- Revise script and materials per NRC (and BLM) comments

Phase 3
- Make initial contacts with all tribes on the final list. Although NRC has already initiated consultation about these undertakings by letter, it would be most effective if the initial contact about this information-gathering effort were to come from NRC based on draft letters and materials provided by the companies and developed by SRIF. Alternatively, the information sent to the tribes could include copies of the

2

Draft Plan August 30, 2011

letters from NRC to the companies requesting that the companies gather additional information on traditional cultural properties.

- Follow up with the tribes as needed to secure a response and decision from as many tribes on the list as possible.
- Maintain a detailed record of tribal contacts and responses.
- Provide information to the companies and NRC about tribes wishing to participate in field visits and ethnographic interviews, tribes wishing to participate at lesser levels of consultation, and tribes not wishing to participate.
- Adjust boundaries of the two expanded APEs in response to tribal comments if needed, and coordinate this with SHPOs and NRC.
- Prepare draft letters for NRC's use (if they so wish) to formally invite the interested tribes to be consulting parties for the appropriate Section 106 undertaking or undertakings.

Phase 4

- Provide assistance to the companies in managing the ethnographic contracts: monitor schedules, recordkeeping, and results; assist contractors with any problems; review reports; ensure that contractors are gathering the needed information about identification, eligibility, effects, and potential measures to resolve any adverse effects.
- Provide monthly updates on the progress of the project to the companies for submission to NRC (and BLM if they wish to receive these reports).
- Communicate with those tribes who asked to be kept informed as the projects proceed.

Phase 5

- Assemble information from contractors and prepare eligibility recommendations (36 CFR §800.4(c)) for traditional cultural properties in the Cameco and Powertech APEs.
- The companies then submit these eligibility recommendations to NRC for consultation with SHPOs and tribes (and BLM if any properties are on BLM-managed lands).
- Assemble information from contractors, apply the criteria of adverse effect (36 CFR §800.5(a)), and prepare recommendations concerning the effects of in situ recovery development activities on eligible or listed historic properties within the APEs.
- The companies then submit these effect recommendations to NRC for consultation with SHPOs and tribes (and BLM if any properties are on BLM-managed lands).

3

Draft Plan August 30, 2011

Optional Phase 6

- If any adverse effects are identified during Phase 5, assist the company or companies and the NRC to complete consultations with BLM, the SHPO(s) and tribal consulting parties (and ACHP, if they choose to participate) to identify measures to resolve the adverse effects (36 CFR §800.6).

- Prepare a draft Section 106 agreement document (or documents, if multiple Section 106 undertakings are found to have adverse effects (36 CRF §800.6(c) or §800.14(b)(3)) and submit to NRC.

4

October 20, 2011

Mr. James Laysbad, THPO
Oglala Sioux Tribe
P.O. Box 320
Pine Ridge, SD 57770

SUBJECT: TRANSCRIPT OF INFORMAL INFORMATION-GATHERING MEETING,
             PERTAINING TO THE DEWEY-BURDOCK, CROW BUTTE NORTH TREND,
             AND CROW BUTTE LICENSE RENEWAL *IN-SITU* URANIUM RECOVERY
             PROJECTS, HELD AT PRAIRIE WIND CASINO AND HOTEL ON JUNE 8, 2011;
             INFORMATION PERTAINING TO TRADITIONAL CULTURAL PROPERTIES;
             AND ATTACHED UNREDACTED PORTIONS OF ARCHEOLOGICAL SURVEYS

Dear Mr. Laysbad:

The U.S. Nuclear Regulatory Commission (NRC) extends its thanks to all the Tribal
representatives who participated in the Information-Gathering Meeting and associated site visits.
We would especially like to thank the Oglala Sioux Tribe for allowing the NRC to hold the meeting
on the Pine Ridge Reservation and for hosting meetings. We would also like to thank Mr. Michael
Catches Enemy, Oglala Sioux Natural Resources Officer, for serving as the Co-facilitator for the
meeting. These efforts culminated in a productive meeting where all parties were able to share
their respective viewpoints, comments, and concerns about the uranium recovery process and the
facets of the NRC's technical and environmental review process for the *in-situ* recovery facilities.

Enclosed is a paper of the Official Transcript of the Information-Gathering Meeting. If you would
like an electronic copy, please contact Ms. Haimanot Yilma or Mr. Nathan Goodman. Please note
corrections were not made to the transcript to maintain its originality. If you believe your
statements were misrepresented in the transcript, please advise the NRC staff and your
comments will be added to the record.

Additionally, at the June 8, 2011 meeting, several Tribal Historic Preservation Officers (THPOs)
requested unredacted versions of the archeological surveys submitted as part of the license
application. In response to those requests and to facilitate government-to-government
consultation, the NRC is enclosing a paper copy of the unredacted portions of the archeological
surveys for the Dewey-Burdock, Crow Butte North Trend, and Crow Butte License Renewal
projects for your review. A map of all archeological sites on the proposed Dewey-Burdock project
area is also included.

As a reminder, this information is sensitive, and the NRC requests the proper storage of these
documents, consistent with applicable laws and regulations pertaining to sensitive information.

At the June 8, 2011 meeting, a number of the tribal representatives requested more information
on traditional cultural properties (TCPs) be collected. Based on those requests and in accordance

J. Laysbad   2

with the requirement of 36 CFR 800.4 (a) the NRC staff has asked the applicants to provide information on how the proposed projects may affect TCPs. In order to address the NRC's request, the applicants may contact your office directly to seek your assistance on developing information on TCPs, in the near future. Once the NRC receives additional information regarding TCPs from the applicants, the NRC plans to distribute the information to all interested THPOs and State Historic Preservation Officers for review and comments.

If you have any questions regarding this letter, please contact my staff members, Ms. Yilma (via email at Haimanot.Yilma@nrc.gov or via phone at 301-415-8029) or Mr. Goodman (via email at Nathan.Goodman@nrc.gov or via phone at 301-415-2703).

Sincerely,

/RA/

Kevin Hsueh, Branch Chief
Environmental Review Branch-B
Environmental Protection and Performance
  Assessment Directorate
Division of Waste Management
  and Environmental Protection
Office of Federal and State Materials
  and Environmental Management Programs

Docket No. 40-8943, 40-9075

Enclosures:
1.   Official Transcript
2.   Unredacted Portions of the Archeological Survey for Dewey-Burdock Project
3.   Unredacted Portions of the Archeological Survey for Crow Butte North Trend Project
4.   Unredacted Portions of the Archeological Survey for License Renewal Project
5.   Archeological Sites on the Proposed Dewey-Burdock Project Area

cc:
Mr. Richard E. Blubaugh
Vice President of Environmental Health
  and Safety Resources
Powertech (USA), Inc.
5575 DTC Parkway, Suite 140
Greenwood Village, CO 80111

Jill Dolberg
Nebraska State Historical Society
P.O. Box 82554
1500 R Street
Lincoln, NE 68501

Paige Olson
Review and Compliance Coordinator
South Dakota State Historic Society
900 Governors Drive
Pierre, SD 57501

October 28, 2011

Mr. James Laysbad, THPO
Oglala Sioux Tribe
P.O. Box 320
Pine Ridge, SD 57770

SUBJECT: INFORMATION RELATED TO TRADITIONAL CULTURAL PROPERTIES;
PERTAINING TO THE DEWEY-BURDOCK, CROW BUTTE NORTH TREND,
AND CROW BUTTE LICENSE RENEWAL *IN-SITU* URANIUM RECOVERY
PROJECTS

Dear Mr. Laysbad:

The U.S. Nuclear Regulatory Commission (NRC) staff wants to update your office on NRC's
ongoing consultation activities, per Section 106 of the National Historic Preservation Act
(NHPA), for the proposed Dewey-Burdock and Crow-Butte projects. In response to requests for
a survey of traditional cultural properties (TCPs) raised by many Tribal representatives at the
June 8, 2011, Information Gathering Meeting, the NRC staff has determined that further work is
needed to identify properties of religious and cultural significance to the Tribes. The staff has
asked the applicants to undertake studies and surveys to provide the NRC with this information,
as is permissible under 36 CFR § 800.2(c)(4).

Powertech (USA), Inc. and Cameco Resources, the respective applicants, have engaged the
services of SRI Foundation (SRIF) of Rio Rancho, New Mexico to collect information concerning
TCPs that may be located in the proposed Dewey-Burdock and Crow Butte project areas.
Dr. Lynne Sebastian will direct these investigations for SRIF; Dr. Martha Graham will contact all
consulting Tribes in the near future to develop a plan for gathering information. Attached to this
letter are brief biographies of the SRIF lead researchers and a link to the SRIF website.

The NRC remains the lead in carrying out Tribal consultation efforts for both projects, pursuant
to its obligation under the regulations of 36 CFR Part 800. Although the NRC has authorized
the applicants and, thereby, SRIF, acting on the behalf of the applicants, to contact Tribes to
obtain needed information, the NRC, nonetheless, remains legally responsible for all findings
and determinations and for maintaining government-to-government relationships with the
involved Indian Tribes. In keeping with that, the NRC staff will continue to be involved in
consultation activities and will coordinate with SRIF and the Tribes, as necessary, to facilitate
SRIF's informational gathering efforts. With that said, the NRC staff invites your office and
Tribal leadership to work with SRIF as they reach out to you regarding the Oglala Sioux Tribe.
Specifically:

- SRIF will contact all Tribes to determine: (1) if Tribes are interested in participating in
field visits and ethnographic interviews, or (2) if the Tribes wish to conduct their own
research, with facilitation provided by SRIF. If your office requests the services of an
ethnographer or ethno-historian, SRIF will arrange for those services.

J. Laysbad   2

- SRIF will coordinate with the NRC, the United States Bureau of Land Management (BLM) (for the proposed Dewey Burdock project), State Historic Preservation Officers (SHPOs), and interested Tribes to adjust boundaries of the area of potential effects (APE) if discussions with Tribes indicate that historic properties outside project boundaries may be affected.

- SRIF will assemble information gathered from interested Tribes, ethnographers, and ethno-historians, prepare preliminary eligibility recommendations for TCPs identified in the proposed Crowe Butte and Dewey-Burdock APEs, and submit the information to the NRC for consideration.

- NRC will review all information and comments provided by consulting parties and will apply the criteria of adverse effect found in 36 CFR § 800.5(a). Based on its review of the information, the NRC will prepare determinations concerning the effects of *in-situ* recovery development activities on eligible or listed historic properties within the APEs.

- The NRC will consult with SHPOs and Tribes, and BLM (when properties are located on BLM-managed lands) on the effect determinations before finalizing its position on such.

- If the NRC determines there are adverse effects to historic properties, the NRC will consult further with interested parties to develop methods to resolve adverse effects.

The NRC staff understands that information provided to SRIF and/or the NRC staff may be sensitive in nature and, as such, you may want the NRC and SRIF to treat provided information as confidential. Both the NRC and SRIF will protect any information identified as confidential, in accordance with 36 CFR § 800.11(c).

If you have any questions or have comments regarding this letter, please contact the following members of my staff: Ms. Haimanot Yilma (via email at Haimanot.Yilma@nrc.gov. or via phone at 301-415-8029), or Mr. Nathan Goodman (via email at Nathan.Goodman@nrc.gov. or via phone at 301-415-2703).

Sincerely,

/RA/

Kevin Hsueh, Branch Chief
Environmental Review Branch-B
Environmental Protection and Performance
  Assessment Directorate
Division of Waste Management
  and Environmental Protection
Office of Federal and State Materials
  and Environmental Management Programs

Docket Nos. 40-8943, 40-9075

Enclosure: SRIF Biographies

cc: See Next Page

cc: With Enclosures:

Mr. John Schmuck
Cameco Resources
202 Carey Avenue, Suite 600
Cheyenne, WY 82001

Ms. Jill Dolberg
Nebraska State Historical Society
P.O. Box 82554
1500 R Street
Lincoln, NE 68501

Mr. Richard E. Blubaugh
Vice President of Environmental Health
 and Safety Resources
Powertech (USA), Inc.
5575 DTC Parkway, Suite 140
Greenwood Village, CO 80111

Ms. Paige Olson
Review and Compliance Coordinator
South Dakota Historic Society
900 Governors Drive
Pierre, SD 57501

Mr. John Yellow Bird Steel, President
Oglala Sioux Tribe
P.O. Box 2070
Pine Ridge, SD 57770-2070

United States Department of the Interior

BUREAU OF LAND MANAGEMENT
South Dakota Field Office
310 Roundup Street
Belle Fourche, South Dakota 57717-1698
www.blm.gov/mt

In Reply Refer To:

3809 (MTC040)

TAKE PRIDE
IN AMERICA

Larry W. Camper, Director
Division of Waste Management
        And Environmental Protection
Office of Federal and State Materials
        And Environmental Management Programs
Mail Stop T-8F5
Washington, DC 20555

RE: BLM requests NRC consent as Lead Agency for the Cultural Section 106 Consultation for the Dewey Burdock In-situ Uranium Recovery Project, Custer and Fall River Counties, South Dakota

Dear Mr. Camper:

During recent NRC/BLM discussions regarding the Dewey Burdock In-situ Uranium Recovery Project, the topic of designating the NRC as lead agency for Section 106 Consultation regarding the Dewey Burdock Project was reviewed. Upon consideration, it is the position of the South Dakota Field Office that the NRC be designated as the lead Agency for the Cultural Section 106 review, and the South Dakota Field Office is seeking concurrence regarding the same.

Please contact me with any concerns you may have at the above address or at (605) 892-7001.

Sincerely,

Marian M. Atkins
South Dakota Field Manger
BLM

CC: Haimanot Yilma – NRC

        Richard Blubaugh – Powertech

UNITED STATES
PARTMENT OF THE INTERIOR
BUREAU OF LAND MANAGEMENT
South Dakota Field Office
310 Roundup Street
Belle Fourche, South Dakota 57717

OFFICIAL BUSINESS
PENALTY FOR PRIVATE USE $300

LARRY W. CAMPER
Division of Waste Management
    and Environmental Protection
Office of Federal and State Materials
    and Environmental Management Systems
Mail Stop T-8F5
Washington, DC 20555

UNITED STATES
NUCLEAR REGULATORY COMMISSION
WASHINGTON, D.C. 20555-0001

January 19, 2012

Dear Tribal Historic Preservation Officers,

The U.S. Nuclear Regulatory Commission invites you to attend a government-to-government consultation on February 14th and 15th, 2012. The meeting will be part of the ongoing consultations under Section 106 of the National Historic Preservation Act (NHPA) associated with three separate applications NRC currently has under review: the Dewey Burdock project; the Crow-Butte North Trend project; and the Crow-Butte license renewal project.

The purpose of the meeting is to hear the views of interested Tribes about the general types and descriptions of historic properties of religious and cultural significance that may be affected by the proposed projects and how these places can be identified and evaluated as part of the ongoing environmental reviews for the above listed projects.

The NRC has identified certain categories of information that will be critical to our evaluation:

1. The general types and descriptions of historic properties of religious and cultural significance to the Tribes that the Tribes know or believe to be located in the three project areas;

2. The Tribes' views and approach on how best to identify and document these potential places;

3. The kinds of potential effects the Tribes believe that these places may be subject to as a result of the proposed projects; and

4. The kinds of measures the Tribes believe might enable NRC to develop a plan to avoid or minimize effects to these places.

The meeting will be held in the Lincoln Room at the Ramkota Best Western at 2111 N. LaCrosse Street, Rapid City, South Dakota, 57701. The NRC has set aside a block of 25 rooms under "NRC Group" for your convenience. Powertech (USA) Inc. and Cameco, Inc. (the applicants) will cover travel and per diem costs for two representatives from each Tribe. In addition, the applicants will also cover, beyond those costs already noted, the travel costs of any Tribal Chair or Tribal President, who chooses to attend. If the representatives from your Tribe are interested in being reimbursed for travel costs, please contact John Schmuck, Senior Permitting Manager for Cameco Resources, at 307-316-7587.

Enclosed is a proposed agenda for the meeting. If there are additional topics that your Tribe would like added to the agenda, please contact the NRC directly. The meeting will not be open to the public because of the sensitive nature of the cultural information to be discussed. The NRC understands confidentiality is necessary in order to protect sensitive information; therefore, methods needed to protect the information will be addressed based on feedback we receive from you.

Please provide the names of the Tribal representatives who are planning to attend the meeting to Haimanot Yilma and Nathan Goodman by February 7, 2012.

2

Additionally, representatives of the applicants will attend the meeting to exchange relevant information with Tribal officials. The NRC will also allow time for the Tribes to caucus and for NRC, BLM, and Tribal representatives to engage in discussion.

If you have additional questions for the NRC staff, please don't hesitate to contact the project manager for the Dewey-Burdock project, Haimanot Yilma, via phone at 301-415-8029 or e-mail at Haimanot.Yilma@nrc.gov or the project manager for the Crow-Butte North Trend and Crow-Butte license renewal projects, Nathan Goodman, via phone at 301-415-2703 or e-mail at Nathan.Goodman@nrc.gov

Thank you very much.

Kevin Hsueh, Branch Chief
Environmental Review Branch
Environmental Protection and
  Performance Assessment Directorate
Division of Waste Management
  and Environmental Protection
Office of Federal and State Materials
  and Environmental Management Programs

Enclosures (2):

Enclosure 1: Proposed Meeting Agenda
Enclosure 2: Initial List of Meeting Attendees

DeweyBurdNonPubEm Resource

From: dianne desrosiers [dyandancer@yahoo.com]
Sent: Tuesday, January 24, 2012 3:46 PM
To: Goodman, Nathan; Yilma, Haimanot
Cc: mgraham@srifoundation.org
Subject: Re: Invitation Letter

Good afternoon,

I am writing to verify our participation in the upcoming meeting to be held in Rapid City. Jim Whitted and I are the Tribal representatives for the Sisseton Wahpeton Oyate. Thank you for your attention in this matter.

Dianne Desrosiers
Tribal Historic Preservation Officer
Sisseton Wahpeton Oyate
PO Box 907
205 Oak St. E, Suite 121
Sisseton, SD 57262
(605)698-3584 office

"Every part of this Earth is sacred to my people. We are part
of the earth and it is part of us".-Chief Seattle,1854
From: "Goodman, Nathan" <Nathan.Goodman@nrc.gov>
To: "Yilma, Haimanot" <Haimanot.Yilma@nrc.gov>; "Goodman, Nathan" <Nathan.Goodman@nrc.gov>
Sent: Thursday, January 19, 2012 3:50 PM
Subject: Invitation Letter

Dear Tribal Historic Preservation Officers,

Attached to this e-mail, you will find three enclosures. The first is an invitation letter to a Tribal Consultation meeting to take place on February 14 and 15, 2012 from our supervisor, Kevin Hsueh. The second is a proposed agenda for the Consultation meeting. And the third is a list of attendees for the same Consultation meeting. If you have any questions or problems opening any of the enclosures, please contact Haimanot or myself.

Thank you,

Nathan Goodman and Haimanot Yilma

Nathan Goodman
Project Manager
FSME/DWMEP/EPPAD/ERB
U.S. NRC
301-415-2703
Nathan.Goodman@nrc.gov

Haimanot Yilma

1

Project Manager
FSME/DWMEP/EPPAD/ERB
U.S Nuclear Regulatory Commission
Phone: 301-415-8029
email: haimanot.vilma@nrc.gov
Mail Stop : T8F05

2

UNITED STATES
NUCLEAR REGULATORY COMMISSION
WASHINGTON, D.C. 20555-0001

March 6, 2012

Dear Tribal Historic Preservation Officers:

The NRC staff has received an email indicating a conflict for many Tribes with the previously suggested Section 106 meeting dates of March 14 and 15, 2012. In the same email, alternate dates of April 17 through 19 were suggested as possible dates to hold the next Section 106 meeting. Unfortunately, while the staff can accommodate those suggested dates, several of the other anticipated meeting participants already have prior commitments during those dates that cannot be rescheduled.

In an effort to move the Section 106 consultation forward and collect your input, the staff suggests the following.

As indicated in our email invitation on Tuesday, February 28, 2012:

- The staff plans to forward the applicants' Statement of Work (SOW) for the proposed Crow Butte License Renewal, Crow Butte North Trend, and Dewey-Burdock projects by **March 9, 2012** for your review and consideration.

- The staff requests your draft SOW for the proposed Crow Butte License Renewal, Crow Butte North Trend, and Dewey-Burdock projects by **March 16, 2012**. This will allow the staff to promptly identify areas of agreement and disagreement with the applicants' proposed SOW. This will also allow the staff to identify for the applicants any areas in which they might consider revising their SOW. In brief, this will allow the consulting parties to continue working toward an acceptable SOW.

- The staff plans to review the SOWs submitted by you and the applicants, along with any additional input you or the applicants provide in the next several weeks, and prepare one comprehensive SOW for each of the proposed projects. The staff plans to circulate a comprehensive SOW for each of the projects to all consulting parties by **March 28, 2012**.

- The staff proposes having a 4 hour conference call to discuss the comprehensive SOWs during the week of April 9, 2012, or the week of April 16, 2012.

Based on the outcome of the conference call, if the parties determine another face-to-face meeting is warranted, the staff will arrange such a meeting based on the availabilities of all consulting parties.

2

If you have any questions or concerns about the above plan, please contact Mr. Nathan Goodman via email at Nathan.goodman@nrc.gov, or Ms. Haimanot Yilma via email at Haimanot.yilma@nrc.gov.

Sincerely,

Kevin Hsueh, Chief
Environmental Review Branch
Division of Waste Management
 and Environmental Protection
Office of Federal and State Materials
 and Environmental Management Programs

UNITED STATES
NUCLEAR REGULATORY COMMISSION
WASHINGTON, D.C. 20555-0001

March 9, 2012

Dear Tribal Historic Preservation Officers:

On March 6, 2012, the U.S. Nuclear Regulatory Commission (NRC) staff informed you that we would soon be forwarding the applicants' Statement of Work (SOW) for the proposed Crow Butte North Trend, Crow Butte License Renewal, and Dewey-Burdock projects for your review and consideration. These three SOWs are attached. NRC staff would appreciate any comments you may have on the SOWs before NRC issues its comprehensive SOWs on March 28, 2012.

Additionally, on March 7, 2012, NRC staff received an e-mail from Sisseton-Wahpeton's Tribal Historic Preservation Officer (THPO) stating that Ben Rhodd, a consultant of the Rosebud Tribe, would be preparing an SOW for NRC's review. This e-mail also stated that the SOW would first be given to the Tribes for their review and concurrence. As NRC staff stated in the e-mail you received on March 6, 2012, we would ask that this SOW be provided to the staff by **March 16, 2012**.

While we appreciate you sharing with us your view of having face-to-face Section 106 consultation meetings, the regulations and guidance issued under the National Historic Preservation Act (NHPA) do not limit Section 106 consultation to face-to-face meetings. Nevertheless, NRC staff agrees that face-to-face group consultation is an important part of the Section 106 consultation process. As such, NRC staff would consider the possibility of having a third face-to-face meeting with the THPOs once it receives the SOW from the Tribes and has developed one comprehensive SOW for each of the three proposed projects.

NRC staff also believes that the consulting parties can continue to make progress on developing SOWs as we await the next face-to-face meeting. NRC staff has proposed a conference call to discuss the proposed SOWs we expect to soon receive from the Tribes and the applicants. NRC staff proposed two possible weeks for conference calls in its e-mail on March 6, 2012. NRC staff proposes to host a conference call either:

- On April 10, 2012 from 2:30 – 6:30 p.m. EST;

- On April 19, 2012 from 2:30 – 6:30 p.m. EST; or

- Both dates and times if the Tribes wish to have further discussions.

This conference call will give the consulting parties an early opportunity to comment on the proposed SOWs and to bring to the staff's attention any other issues related to the NHPA.

2

If you have any questions, please don't hesitate to contact the project manager for the Dewey-Burdock project, Ms. Haimanot Yilma via phone at 301-415-8029 or e-mail at Haimanot.Yilma@nrc.gov, or the project manager for the Crow Butte North Trend and Crow Butte License Renewal projects, Mr. Nathan Goodman via phone at 301-415-2703 or e-mail at Nathan.Goodman@nrc.gov.

Sincerely,

Kevin Hsueh, Chief
Environmental Review Branch
Division of Waste Management
 and Environmental Protection
Office of Federal and State Materials
 and Environmental Management Programs

Enclosures:
1. Crow Butte North Trend SOW
2. Crow Butte License Renewal SOW
3. Dewey-Burdock SOW

March 19, 2012

Mr. Louis Maynahonah, Chairman
Apache Tribe of Oklahoma
P.O. Box 1220
Anadarko, OK 73005

SUBJECT: ONGOING SECTION 106 OF THE NATIONAL HISTORIC PRESERVATION
           ACT TRIBAL CONSULTATION LETTER FOR THE PROPOSED CROW BUTTE
           NORTH TREND, CROW BUTTE LICENSE RENEWAL, AND DEWEY-
           BURDOCK PROJECTS

Dear Chairman Maynahonah:

Enclosed please find a follow-up consultation letter to the Tribal Historic Preservation Officer
pertaining to ongoing Tribal consultation for three proposed projects: Crow Butte North Trend,
Crow Butte License Renewal, and Dewey-Burdock.

The U.S. Nuclear Regulatory Commission staff is transmitting this letter to you to keep you
informed of all Section 106 activities that are underway for the three proposed projects.

If you have any questions or concerns, please contact my staff, Ms. Haimanot Yilma via email
at Haimanot.Yilma@nrc.gov or phone at 301-415-8029 for the Dewey-Burdock project, or
Mr. Nathan Goodman via email at Nathan.Goodman@nrc.gov. or phone at 301-415-8029 for
the Crow Butte projects.

                              Sincerely,

                              /RA/

                              Larry W. Camper, Director
                              Division of Waste Management
                               and Environmental Protection
                              Office of Federal and State Materials
                               and Environmental Management Programs

cc: Mr. Lyman Guy

Enclosure:
Invitation Letter

March 26, 2012

Mr. Louis Maynahonah, Chairman
Apache Tribe of Oklahoma
P.O. Box 1220
Anadarko, OK 73005

SUBJECT: TRANSMITTAL OF TRANSCRIPTS AND ATTENDANCE LISTS FROM
SECTION 106 CONSULTATION MEETINGS

Dear Chairman Maynahonah:

Enclosed please find a copy of the transcripts and attendance lists from the Section 106
Consultation meetings held in Rapid City, South Dakota on February 14 and 15, 2012.

During the February meetings, the U.S. Nuclear Regulatory Commission (NRC) and Bureau of
Land Management (BLM) staff received the following key information:

- Tribes are concerned about confidentiality of any information they transmit to the NRC
  based on a recent undesirable experience with other consulting parties. Tribes
  expressed an interest in first developing a confidentiality agreement before submitting
  any Tribal Cultural Properties (TCP) studies to the NRC. A representative from Standing
  Rock Sioux Tribe volunteered to share a recent confidentiality agreement developed for
  another project to use as a starting point for Tribes, NRC and BLM when developing an
  agreement for the proposed Crow Butte License Renewal, Crow Butte North Trend, and
  Dewey-Burdock projects.

- Tribal Representatives requested that for future meeting invitations, the purpose be
  made clearer in order to ensure that Tribal participants have appropriate levels of
  decision-making authority.

- Tribal Representatives volunteered to develop project-specific Statements of Work
  (SOWs) to conduct TCP studies for the proposed Crow Butte License Renewal, Crow
  Butte North Trend, and Dewey-Burdock projects.

- Tribal Representatives requested another face-to-face meeting to go over the draft
  SOWs for each of the three projects. Tribal Representatives suggested March 14
  and 15, 2012 as possible meeting dates. However, due to conflicts with many
  participating Tribal Representatives, this meeting did not occur. Further discussion is
  ongoing to schedule a teleconference instead, and potentially another face-to-face
  meeting.

L. Maynahonah   2

Please note that the transcripts are not publicly available as information discussed during the February meeting is protected under the National Historic Preservation Act (NHPA)[1] and the South Dakota Codified Laws[2].

If you have any questions or concerns, please contact my staff, Ms. Haimanot Yilma via email at Haimanot.Yilma@nrc.gov or phone at 301-415-8029 for the Dewey-Burdock project, or Mr. Nathan Goodman via email at Nathan.Goodman@nrc.gov or phone at 301-415-8029 for the Crow Butte projects.

Sincerely,

/RA/

Larry W. Camper, Director
Division of Waste Management
 and Environmental Protection
Office of Federal and State Materials
 and Environmental Management Programs

Enclosures:
1. Transcript of 2/14/12 Meeting
2. Transcript of 2/15/12 Meeting
3. Attendance List from 2/14/12 Meeting
4. Attendance List from 2/15/12 Meeting

cc: Mr. Lyman Guy

---

[1] Section 304 of the National Historic Preservation Act of 1966, As amended through 2006 [16 U.S.C. 470w-3(a)] concerns the confidentiality of the location of sensitive historic resources:
(a) The head of a Federal agency or other public official receiving grant assistance pursuant to this Act, after consultation with the Secretary, shall withhold from disclosure to the public, information about the location, character, or ownership of a historic resource if the Secretary and the agency determine that disclosure may -
    (1) cause a significant invasion of privacy;
    (2) risk harm to the historic resources; or
    (3) impede the use of a traditional religious site by practitioners.

[2] The release of records pertaining to the location of archaeological sites is restricted under South Dakota Codified Laws (SDCL), specifically, SDCL § 1-20-21.2, Confidentiality of records pertaining to location of archaeological site—Exceptions.
    Any records maintained pursuant to § 1-20-21 pertaining to the location of an archaeological site shall remain confidential to protect the integrity of the archaeological site.

UNITED STATES
NUCLEAR REGULATORY COMMISSION
WASHINGTON, D.C. 20555-0001

April 5, 2012

Dear Tribal Historic Preservation Officers:

In a letter dated March 6, 2012, the U.S. Nuclear Regulatory Commission (NRC) staff informed you of our plans to forward comprehensive Statements of Work (SOWs) for the Crow Butte License Renewal, Crow Butte North Trend, and Dewey-Burdock projects to you by March 28, 2012. We planned to send the comprehensive SOWs after taking into account input from both the Applicants and the Tribes. Also in our March 6 letter, NRC staff stated its intent to forward the Applicants' SOWs to you, requested your comments on the Applicants' SOWs, and proposed a teleconference to discuss the comprehensive SOWs.

On March 7, 2012, NRC staff received an email from Sisseton-Wahpeton's Tribal Historic Preservation Officer stating that an SOW was being developed by the Tribes for NRC review.

On March 9, 2012, the NRC forwarded a copy of the Applicants' SOWs to you, and we reiterated our request to receive your SOW by March 16, 2012. The NRC also requested your comments on the Applicants' SOWs by March 28, 2012. To date, we have not received any input. For this reason, the NRC was unable to develop a comprehensive SOW for each of the three proposed projects.

In order to efficiently move the Section 106 process forward, the NRC suggests the following:

- Conduct a teleconference with all consulting parties (including the Applicants and SRIF) on **April 24, 2012 from 2:30 to 6:30 p.m. EST**. This teleconference will be a working meeting between the THPOs and NRC Staff. During this call, consulting parties would:

  ➢ Use the Applicants' SOWs as a starting point and identify elements essential for developing a comprehensive SOW for each of the three proposed projects.

  ➢ The goal is to have the consulting parties develop three comprehensive SOWs.

  ➢ Schedule dates to conduct the TCP studies.

Alternatively, if the Tribes submit draft SOWs before April 19, 2012, NRC staff can use both SOWs to initiate our discussion. The staff will highlight areas where the SOWs are different so that the consulting parties can use those differences as a starting point to work toward a consensus in those areas.

2

The Applicants' contractor (SRIF) may contact you in the near future to discuss their SOWs that the NRC forwarded to you on March 9, 2012. The intent of their call would be to solicit your feedback on their SOW.

Please let us know if you are available to participate on the April 24, 2012 teleconference by **April 13, 2012**. If you have any questions, you can contact Ms. Haimanot Yilma by phone at 301-425-8029 or via email at Haimanot.Yilma@nrc.gov or Mr. Nathan Goodman by phone at 301-415-2703 or via email at Nathan.Goodman@nrc.gov.

Sincerely,

Kevin Hsueh, Chief
Environmental Review Branch
Environmental Protection
 and Performance Assessment Directorate
Division of Waste Management
 and Environmental Protection
Office of Federal and State Materials
 and Environmental Management Programs

Rajapakse, Champa
_____

From: Goodman, Nathan
Sent: Friday, April 20, 2012 11:01 AM
To: Yilma, Haimanot; Goodman, Nathan
Subject: Upcoming teleconference on April 24, 2012
Attachments: Dewey-Burdock Draft SOW and figures .pdf; Crow Butte-NT and LR Draft SOW figures.pdf;
                Crow Butte-NT and LR Draft SOW text.pdf

Dear Tribal Historic Preservation Officers:

Haimanot and I would like to remind you of an upcoming teleconference with all consulting parties (including
the Applicants, SRIF, BLM, and EPA Region 8) on **April 24, 2012 from 2:30 to 6:30 p.m. EST**. This
teleconference will be a working meeting between the THPOs and all other consulting parties. The purpose of
this teleconference is to continue NRC's ongoing Section 106 consultations for the proposed Crow Butte
License Renewal, Crow Butte North Trend, and Dewey Burdock projects. The NRC staff is interested in
identifying areas within the three project sites that have traditional religious or cultural significance to the
Tribes, so these may be considered in our environmental reviews. During this teleconference, consulting
parties would use the Applicants' SOWs as a starting point and identify elements essential for developing a
comprehensive SOW for each of the three proposed projects and schedule dates to conduct the TCP studies.
The goal is to have the consulting parties develop three comprehensive SOWs.

Below are the proposed agenda and teleconference phone number and passcode:

Welcome   Kevin Hsueh

Introductions   All

Purpose of meeting   Michelle Ryan

Discussion of Draft SOWs   Haimanot Yilma, Nathan Goodman
    1) Crow Butte License Renewal
    2) Crow Butte North Trend
    3) Dewey-Burdock

Possible discussion topics:
    •   Purpose
    •   Scope of Work
    •   Period of Performance
    •   Reports
    •   Deliverables Schedule
    •   Level of Effort

Break –15 min (~4:30-4:45pm)

Discussion of Draft SOWs continued   Haimanot Yilma, Nathan Goodman

1

Summary of concerns and recommendations	Michelle Ryan
Next Steps	Haimanot Yilma, Nathan Goodman
Closing	Kevin Hsueh

Dial-in number: 1-800-369-1134
Passcode: 65795

We would also like to note that the call will be recorded so that a transcript of the meeting can be made available to all parties.

On March 9, 2012, the NRC forwarded a copy of the applicants' SOWs to you for your review and consideration. For your convenience, we are attaching another copy of the three SOWs to this e-mail.

Please contact us by 5:00 PM EST Monday (April 23, 2012) if you are available to participate in the April 24, 2012 teleconference. If you have any questions, you can contact either of us by phone or e-mail, which are provided below.

Thank you.

Haimanot Yilma and Nathan Goodman

Nathan Goodman
Project Manager
FSME/DWMEP/EPPAD/ERB
U.S. NRC
301-415-2703
Nathan.Goodman@nrc.gov

Haimanot Yilma
Project Manager
FSME/DWMEP/EPPAD/ERB
U.S. Nuclear Regulatory Commission
Phone: 301-415-8029
email: haimanot.yilma@nrc.gov
Mail Stop: T8F05

2

May 7, 2012

Mr. Duane Big Eagle, Chairman
Crow Creek Sioux Tribe
P.O. Box 50
Ft. Thompson, SD 57339-0050

SUBJECT: TRANSMITTAL OF APPLICANT'S DRAFT STATEMENT OF WORK
            REGARDING CROW BUTTE NORTH TREND, CROW BUTTE LICENSE
            RENEWAL, AND DEWEY-BURDOCK PROJECTS

Dear Chairman Big Eagle:

Enclosed please find a followup letter sent via email to the Tribal Historic Preservation Officers
(THPOs) forwarding draft Statement of Works (SOWs) for Identification of Properties of
Religious and Cultural Significance that the staff received from Cameco Resources Inc. and
Powertech Inc. pertaining to the proposed Crow Butte License Renewal, Crow Butte North
Trend, and Dewey-Burdock projects. The SOWs are also enclosed in this letter for your
convenience.

In the enclosed letter, the U.S. Nuclear Regulatory Commission (NRC) staff also requested for
the THPO's participation in a conference call to discuss the draft SOWs and any other issues
related to National Historic Preservation Act (NPHA) as many interested tribes had a conflict
with the previously suggested date of March 14 and 15, 2012 to host a face-to-face meeting.
The NRC staff encourages the THPOs to continue the dialog with all consulting parties via a
conference call in an effort to move the Section 106 consultation forward.

The NRC staff is transmitting this letter to you to keep you informed of all Section 106 activities
that are underway for the proposed Crow Butte License Renewal, Crow Butte North Trend, and
Dewey-Burdock projects.

D. Big Eagle   2

If you have any questions or concerns, please contact my staff, Ms. Haimanot Yilma via email at Haimanot.Yilma@nrc.gov or phone at 301-415-8029 for the Dewey-Burdock project, or Mr. Nathan Goodman via email at Nathan.Goodman@nrc.gov or phone at 301-415-8029 for the Crow Butte projects.

Sincerely,

/RA/

Larry W. Camper, Director
Division of Waste Management
  and Environmental Protection
Office of Federal and State Materials
  and Environmental Management Programs

Enclosures:
1. Followup Letter
2. Draft SOWs

cc: Wanda Wells

May 23, 2012

Mr. John Yellow Bird Steele, President
Oglala Sioux Tribe
P.O. Box 2070
Pine Ridge, SD 57770-2070

SUBJECT: TRANSMITTAL OF A LETTER SENT TO THE TRIBAL HISTORIC
            PRESERVATION OFFICERS INVITING THEM TO ATTEND A
            TELECONFERENCE REGARDING THE CROW BUTTE NORTH TREND,
            CROW BUTTE LICENSE RENEWAL, AND DEWEY-BURDOCK PROJECTS

Dear President Steele:

Enclosed please find a followup letter sent via email to the Tribal Historic Preservation Officers
(THPOs). In the letter, the Nuclear Regulatory Commission (NRC) staff invited the THPOs to
participate in a teleconference that was held on April 24, 2012. The participants in the
teleconference included the NRC staff, representatives from eight Tribes, Bureau of Land
Management, U.S. Environmental Protection Agency Region 8, and the South Dakota State
Historic Preservation Officer.

The purpose of the April 24, 2012 teleconference was to discuss the draft Statements of Work
(SOWs) for identification of historic properties of religious and cultural significance that the staff
received from Cameco Resources Inc. and Powertech (USA) Inc. The draft SOWs pertain to
the proposed Crow Butte License Renewal, Crow Butte North Trend, and Dewey-Burdock
projects. The SOWs were developed taking into account information gathered during a prior
Tribal consultation meeting held in Rapid City, South Dakota on February 14 and 15, 2012.

In the enclosed letter, the NRC staff also encouraged the THPOs to submit their own SOW for
each project prior to the scheduled teleconference so that the consulting parties can use both
the Tribes' and the applicants' documents to finalize a SOW for each project. This will allow the
parties to move forward with the identification of historic properties that may be of religious or
cultural significance to the Tribes. Based on the staff's review schedule for the three proposed
projects mentioned above, the identification of such properties will need to be completed by the
fall of 2012. To date, however, the staff has not received the Tribes' SOWs.

The NRC staff is transmitting this letter and attached correspondence to you to keep you
informed of all Section 106 activities that are underway for the proposed Crow Butte License
Renewal, Crow Butte North Trend, and Dewey-Burdock projects.

J. Yellow Bird Steele   2

If you have any questions or concerns, please contact, Ms. Haimanot Yilma via email at
Haimanot.Yilma@nrc.gov or phone at 301-415-8029 for the Dewey-Burdock project, or
Mr. Nathan Goodman via email at Nathan.Goodman@nrc.gov or phone at 301-415-8029 for the
Crow Butte projects.

Sincerely,

/RA by Bill VonTill Acting for/

Larry W. Camper, Director
Division of Waste Management
 and Environmental Protection
Office of Federal and State Materials
 and Environmental Management Programs

Enclosure:
Follow up Letter

cc: Wilmer Mesteth

June 20, 2012

Mr. Conrad Fisher
Northern Cheyenne Tribe
P.O. Box 128
Lame Deer, MT 59043-0128

SUBJECT: TRANSMITTAL OF EVALUATIVE TESTING REPORT AND ASSOCIATED MAP

Dear Mr. Fisher:

Enclosed please find an Evaluative Testing Report of 20 Sites in the proposed Dewey-Burdock
Uranium Recovery project boundary developed by the applicant (Powertech (USA) Inc.).
The associated map is also attached for your convenience.

The U.S. Nuclear Regulatory Commission (NRC) is transmitting these documents to you per
your request. Please note that some parts of the document and the Map are protected under
the National Historic Preservation Act[1] and South Dakota Codifed Laws[2].

If you have any questions or concerns, please contact my staff, Ms. Haimanot Yilma via email at
Haimanot.Yilma@nrc.gov or phone at 301-415-8029.

Sincerely,

/RA/

Kevin Hsueh, Chief
Environmental Review Brach
Division of Waste Management
  and Environmental Protection
Office of Federal and State Materials
  and Environmental Management Programs

Enclosures:
1. Supplemental ARC report
2. ARC Map

cc: Chairman Leroy Spang
     Mr. Richard Blubaugh, Powertech (USA) Inc
     Mr. Gregory R. Fesko, BLM

---

[1] Section 304 of the National Historic Preservation Act of 1966, As amended through 2006 [16 U.S.C. 470w-3(a)] concerns
the confidentiality of the location of sensitive historic resources:
(a) The head of a Federal agency or other public official receiving grant assistance pursuant to this Act, after consultation
with the Secretary, shall withhold from disclosure to the public, information about the location, character, or ownership of
a historic resource if the Secretary and the agency determine that disclosure may -
        (1) cause a significant invasion of privacy;
        (2) risk harm to the historic resources; or
        (3) impede the use of a traditional religious site by practitioners.
[2] The release of records pertaining to the location of archaeological sites is restricted under South Dakota Codified Laws
(SDCL), specifically, SDCL § 1-20-21.2, Confidentiality of records pertaining to location of archaeological site—Exceptions.
Any records maintained pursuant to § 1-20-21 pertaining to the location of an archaeological site shall remain confidential
to protect the integrity of the archaeological site.

June 26, 2012

Ms. Waste'Win Young
Standing Rock Sioux Tribe
Tribal Historic Preservation Office
P.O. Box D
Fort Yates, ND 58538-0522

SUBJECT: TRANSMITTAL OF TRANSCRIPT FROM TELECONFERENCE CONDUCTED ON APRIL 24, 2012

Dear Ms. Young:

Enclosed please find a copy of the transcript from the teleconference conducted on April 24, 2012, pertaining to the proposed Crow Butte North Trend, Crow Butte License Renewal, and Dewey-Burdock projects. The participants in the call included staff from the U.S. Nuclear Regulatory Commission (NRC), the U.S. Environmental Protection Agency (EPA) Region 8, and the Bureau of Land Management (BLM); representatives of Powertech, Cameco, and SRI Foundation (SRIF) (applicant's contractor); the South Dakota State Historic Preservation Officer (SD SHPO); and representatives of eight Tribes (Northern Cheyenne, Oglala Sioux, Rosebud Sioux, Northern Arapaho, Sisseton-Wahpeton, Standing Rock Sioux, Yankton Sioux, and Cheyenne and Arapaho).

During the teleconference, the NRC staff sought feedback from the Tribes on the applicants' proposed Statements of Work (SOWs) for conducting Tribal Cultural Properties (TCP) studies. The staff had previously sent these SOWs to the Tribes on March 9, 2012. The consulting parities discussed the following aspects of the applicants' draft SOWs:

○ Adequacy of compensation for Tribal officials conducting the fieldwork.

○ Confidentiality of information gathered by the Tribes.

○ Amount of acreage to be covered during the fieldwork. Tribes requested 100% surveys of project areas. The Tribes agreed to send two Tribal officials to visit the project areas and determine the scope and extent of the fieldwork.

○ Tribal involvement in making eligibility determinations.

○ Next steps:

- Two Tribal officials visit the project areas for initial work assessment;
- Tribes develop SOWs based on these initial visits;
- Tribes hold a teleconference to discuss their draft SOWs;
- Tribes provide copies of draft SOWs to NRC after all Tribal members agree;
- NRC distributes draft SOWs from Tribes to all other consulting parities including Tribes, applicants, and SHPO;
- NRC schedules another meeting with all consulting parties to finalize SOWs; and
- Applicants will provide dates for proposed field work.

W. Young   2

Please note that the transcript of the April 24, 2012 teleconference is not publicly available and that information discussed during the call may be protected from disclosure by the National Historic Preservation Act (NHPA) and South Dakota Codified Laws.

If you have any questions or concerns, please contact my staff Haimanot Yilma via email at Haimanot.yilma@nrc.gov or phone at 301-415-8029 for the Dewey-Burdock project or Nathan Goodman via email at Nathan.goodman@nrc.gov or phone at 301-415-8029 for the Crow Butte projects.

Sincerely,

/RA/

Kevin Hsueh, Chief
Environmental Review Branch
Environmental Protection and Performance
  Assessment Branch
Division of Waste Management
  and Environmental Protection
Office of Federal and State Materials
  and Environmental Management Programs

Enclosure:
Transcript from April 24, 2012
Teleconference

cc: Chairman Charles Murphy

June 29, 2012

Chairman Jim Shakespeare
Northern Arapaho Business Committee
P.O. Box 396
Fort Washakie, WY 82514

SUBJECT: TRANSMITTAL OF EMAIL CORRESPONDENCE PERTAINING TO
                TELECONFERENCE CONDUCTED ON APRIL 24, 2012

Dear Chairman Shakespeare:

Enclosed please find email correspondence from U.S. Nuclear Regulatory Commission (NRC) staff to Tribal Historic Preservation Officers (THPOs) requesting Tribal participation for a teleconference on April 24, 2012. The NRC staff also included three draft Statements of Work (SOWs) to the email correspondence. The purpose of the teleconference was to discuss the draft SOWs for Identification of Properties of Religious and Cultural Significance received from Cameco Resources Inc. and Powertech Inc. pertaining to the proposed Crow Butte License Renewal, Crow Butte North Trend, and Dewey-Burdock projects. The SOWs were developed taking into account information gathered during the February 2012 Tribal consultation meeting. The applicant's SOWs were first forwarded to your office on March 9, 2012 for review and comment.

The NRC staff is transmitting this letter and attached email correspondence to you to keep you informed of all Section 106 activities for the proposed Crow Butte License Renewal, Crow Butte North Trend, and Dewey-Burdock projects.

If you have any questions or concerns, please contact Ms. Haimanot Yilma via email at Haimanot.Yilma@nrc.gov or by phone at 301-415-8029 for the Dewey-Burdock project, or Mr. Nathan Goodman via email at Nathan.Goodman@nrc.gov or by phone at 301-415-8029 for the Crow Butte projects.

Sincerely,

/RA by Gregory Suber Acting for/

Larry W. Camper, Director
Division of Waste Management
 and Environmental Protection
Office of Federal and State Materials
 and Environmental Management Programs

Enclosures:
1. Followup Email
2. Draft SOWs

cc: Ms. Darlene Conrad

United States Department of the Interior

BUREAU OF LAND MANAGEMENT
South Dakota Field Office
310 Roundup Street
Belle Fourche, South Dakota 57717-1698
http://www.blm.gov/mt

In Reply Refer To

8100-R
BAS

12-MT040-15

Date: July 20, 2012

Mr. Richard E. Blubaugh
Vice President – Environmental Health & Safety Resources
Powertech (USA) Incorporated
5575 DTC Parkway, Suite 140
Greenwood Village, CO 80111

RE: Cultural Resource review of Evaluative Testing of 20 Sites in the Powertech (USA)
Inc. Dewey-Burdock Uranium Project Impact Areas: Volumes 1 and 2. For the Dewey-
Burdock Uranium Recovery Project, Fall River and Custer Counties, South Dakota.

Dear Mr. Blubaugh:

We have reviewed the appropriate volumes of the National Historic Preservation Act,
Section 106 cultural compliance reports presented by Archeology Laboratory, Augustana
College, for evaluation of cultural resource sites inside areas of potential effect for the
proposed Dewey-Burdock project area. The reports reviewed document formal
evaluation of 20 cultural resource sites inside areas proposed for the project that could
have effect. Of these 20 sites, one site is located in part on BLM administered surface
land.

Site 39FA96 was found to be significantly affected by natural erosion and therefore does
not possess adequate integrity, does not display workmanship or feeling, and it is not
associated with an important historic event. Based on the information provided in the
report the Bureau of Land Management (BLM) recommends adequate testing was
completed on site 39FA96, the site's integrity has been severely affected by deflation.
The portion on BLM administered land does not possess enough information to meet the
National Register of Historic Places criteria for an eligible archaeological site; therefore,
the BLM is in agreement with the determination for site 39FA96 on this portion, in that it
is considered not eligible for nomination to the National Register of Historic Places.
Information provided for the remaining 19 sites should to be sufficient for the lead
Federal Agency to make informed recommendations of eligibility on the historic
properties.

Mr. Richard E. Blubaugh
July 20, 2012
Page 2

Please let us know if you should need any additional information. I can be reached as
Mr. Mitch Iverson (acting South Dakota Field Manager), (605) 892-7001 or email at
Mitchell_Iverson@blm.gov or contact our archaeologist, Brenda Shierts at (605) 723-
8712 or Brenda_Shierts@blm.gov.

Sincerely,

Mitch Iverson
acting for
Marian M. Atkins
South Dakota Field Manager

cc: Paige Olson, SD SHPO
       Gary Smith, BLM MSO Historic Preservation Officer
       Mark Sant, BLM MSO Tribal Coordinator
       Mr. Greg R. Fesko, P.G., BLM MSO Solid Minerals
       Haimanot Yilma, NRC, Project Manager Environmental Review Branch

From:	Yilma, Haimanot
To:	Yilma, Haimanot
Cc:	Goodman, Nathan
Subject:	Part 1 of 2: Invitation for a Teleconference On Thursday August 9, 2012
Date:	Tuesday, August 07, 2012 5:37:00 PM
Attachments:	AreaExample_120730.pdf
	Dewey-Burdock Draft SOW Map 2.pdf
	Dewey-Burdock Draft SOW Map 1.pdf
	Revised Scope of Work Dewey-Burdock Draft 3.docx

Dear Tribal Historic Preservation Officers:

The NRC staff would like to invite you to participate in a teleconference on August 9, 2012 from 2:30 to 6:30 p.m. EST with all consulting parties (including the Applicant, SD SHPO, BLM, EPA Region 8, and SRIF, the applicant's contractor). This teleconference will be a working level meeting between the THPOs and all other consulting parties.

The purpose of this teleconference is to finalize a Statement of Work (SOW) acceptable to all consulting parties and to establish a timeframe for conducting fieldwork to identify any historic properties of religious and cultural significance to the Tribes that may be affected by the proposed Dewey-Burdock Project.

Below are the proposed agenda and teleconference phone number and passcode:

Welcome	Kevin Hsueh
Introductions	All
Purpose of the Meeting	Jean Trefethen
Discussion of Draft SOWs for the Dewey-Burdock Project	Haimanot Yilma

1) Tribes draft SOW
2) Applicant's revised draft SOW

Possible Discussion Topics:

- Purpose
- Scope of Work – *amount of land to be surveyed and coverage rate (acres per person)*
- Period of Performance – *start and duration of the fieldwork*
- Reports – *content and confidentiality*

- Level of Effort – *number of people and associated rates*
- Deliverables Schedule

Break –15 min (~4:30-4:45pm EST)

Discussion of Draft SOW continued	Haimanot Yilma
Summary of Concerns and Recommendations	Jean Trefethen
Next Steps	Jean Trefethen and Haimanot Yilma
Closing	Kevin Hsueh

Dial-in number: 1-800-779-3170
Passcode: 3569215

The teleconference will be recorded and a transcript of the meeting will be provided to all consulting parties.

For your convenience, the Dewey Burdock SOW developed by the Tribes is attached for your review. We also attach the Applicant's revised SOW for the Dewey-Burdock Project for your review. The revised Dewey Burdock SOW addresses issues raised in Tribes' draft SOW for the Dewey-Burdock Project and incorporates information gathered during the April 24, 2012 teleconference with all consulting parties. The Applicant's original SOW for the Dewey-Burdock Project was forwarded to you on March 9, 2012.

The revised SOWs for the Crow Butte North Trend, Crow Butte License Renewal, Marsland and Three Crow Projects are also attached for your review. Although these revised SOWs are not on the teleconference agenda, the Applicant will be available to answer general questions.

Please contact me by 11 AM EST Thursday August 9, 2012, if you will participate in the teleconference. If you have any questions, please contact me by phone or e-mail.

Please note that because the size of the attachments, we had to send you the materials in two parts. Part 2 of this email will follow shortly.

Thank you,

Haimanot Yilma
Project Manager
FSME/DWMEP/EPPAD/ERB
U.S Nuclear Regulatory Commission
Phone: 301-415-8029
email: haimanot.yilma@nrc.gov
Mail Stop : T8F05

From:	Yilma, Haimanot
To:	Yilma, Haimanot
Cc:	Goodman, Nathan
Subject:	Reminder: Invitation for a Teleconference today August 9, 2012 from 2:30pm to 6:30pm (EST)
Date:	Thursday, August 09, 2012 10:17:00 AM

Good Morning,

The NRC staff would like to remind you of the teleconference with all consulting parties (including the Applicant, SD SHPO, BLM, EPA Region 8, and SRIF, the applicant's contractor) scheduled for today Thursday August 9, 2012 from 2:30 to 6:30 p.m. (EST). This teleconference will be a working level meeting between the THPOs and all other consulting parties.

Below is the proposed agenda and teleconference phone number and passcode:

Dial-in number: 1-800-779-3170
Passcode: 3569215

Proposed Agenda

Welcome   Kevin Hsueh

Introductions   All

Purpose of Meeting   Jean Trefethen

Discussion of Draft SOW for   Haimanot Yilma
Dewey-Burdock Project

1) Tribes draft SOW
2) Applicant's revised draft SOW
-

Possible Discussion Topics:
-
- Purpose
- Scope of Work – *amount of land to be surveyed and coverage rate (acres per person)*
- Period of Performance – *start and duration of the fieldwork*
- Reports – *content and confidentiality*
- Level of Effort – *number of people and associated rates*
- Deliverables Schedule

Break –15 min (~4:30-4:45pm EST)

Discussion of Draft SOW (continued)   Haimanot Yilma

Summary of Concerns and Recommendations   Jean Trefethen

Next Steps   Jean Trefethen and Haimanot
                                              Yilma

Closing   Kevin Hsueh

The teleconference will be recorded and a transcript of the meeting will be provided to all consulting parties.

Thank you.

Haimanot Yilma
Project Manager
FSME/DWMEP/EPPAD/ERB
U.S Nuclear Regulatory Commission
Phone: 301-415-8029
email: haimanot.yilma@nrc.gov
Mail Stop : T8F05

From:	Yilma, Haimanot
To:	Yilma, Haimanot
Subject:	Proposed Agenda for August 21, 2012 Teleconference starting at 9:00 am (Central time)
Date:	Monday, August 20, 2012 12:15:00 PM
Attachments:	Proposed Agenda for 8-21-12 call.docx
	Dewey-Burdock Draft SOW Map 3.pdf
	Revised Scope of Work Dewey-Burdock Draft 3.docx
	Dewey-Burdock Draft SOW Map 1.pdf
	Dewey-Burdock Draft SOW Map 2.pdf

Dear Tribal Historic Preservation Officers:

During the teleconference on August 9, 2012, the consulting parties that attended the call (including representatives from the Oglala Sioux, Cheyenne River Sioux, Crow Creek Sioux, Northern Arapaho, Northern Cheyenne, Rosebud Sioux, Santee Sioux, Sisseton-Wahpeton Oyate, Standing Rock Sioux, and Yankton Sioux tribes) agreed to participate in another teleconference on Tuesday, August 21, 2012 at 9:00 am (Central Time). As requested, the teleconference is scheduled for August 21, 2012 from 9:00 am to 1:00 pm (Central time). The teleconference can be extended for an additional 4 hours if more time is required to discuss the scope estimate details highlighted below. Attached please find the agenda and call-in information.

On Tuesday August 14, 2012 the Nuclear Regulatory Commission (NRC) staff shared the following with you:

- The position of both NRC and Bureau of Land Management (BLM) staff is that the field identification survey for the Dewey-Burdock Project should focus on the proposed initial disturbance (with additional buffers). This area appears in salmon on Map 3 of the revised Statement of Work (SOW) developed by the applicant and forwarded to you on August 7, 2012 by the NRC. This area measures approximately 2,637 acres. This revised SOW is included in this email for your convenience.

- The NRC and BLM staff agree with the Tribes' recommendation that a Programmatic Agreement (PA) be developed to ensure that additional field investigations will be conducted outside this 2,637-acre buffered impact area prior to any future disturbance (such as proposed land-application areas and/or utility line locations). If a license is granted, a requirement to abide by the terms of this PA would be included as a license condition.

In the August 14, 2012 email, the NRC staff also requested that Tribal Representatives review the revised SOW prepared by the applicant for a survey of the 2,637-acre buffered impact area. The NRC staff requested that your review focus on the following important scope estimate details:

- Estimated coverage rate for field identification (# of acres per person day).
- Start date and estimated duration of the field identification effort.
- Proposed report content and confidentiality requirements (to be prepared after the field identification has been completed).
- Number of people required for the field identification, with labor classifications (e.g., surveyor, crew leader, traditional cultural expert) and associated hourly rates.
- Report deliverable schedules.

The goal for the August 21, 2012 teleconference is to discuss these scope estimates listed above and come to resolution on how to finalize the draft SOW. The NRC staff will incorporate changes discussed during the teleconference and distribute the revised SOW to all consulting parties for final review.

Dewey-Burdock is the first of three projects that will need field identification before the end of 2012 field season. For that reason, it is highly important that we schedule field identification for the Dewey-Burdock project as soon as possible.

Thank you

Haimanot Yilma
Project Manager
FSME/DWMEP/EPPAD/ERB
U.S Nuclear Regulatory Commission
Phone: 301-415-8029
email: haimanot.yilma@nrc.gov
Mail Stop : T8F05

From:	Yilma, Haimanot
To:	Yilma, Haimanot
Subject:	Reminder: Invitation for a Teleconference today August 21, 2012 from 9:00 a.m. to 1:00 p.m. (Central time)
Date:	Tuesday, August 21, 2012 8:50:00 AM

Good Morning,

The NRC staff would like to remind you of the teleconference with all consulting parties (including the Applicant, SD SHPO, BLM, EPA Region 8, and SRIF, the applicant's contractor) scheduled for today Tuesday August 21, 2012 from 9:00 a.m. to 1:00 p.m. (Central time). This teleconference will be a working level meeting between the THPOs and all other consulting parties.

Below is the proposed agenda and teleconference phone number and passcode:

Call-in Number: 800-857-9707
Passcode: 9409817

Proposed Agenda

Welcome	Kevin Hsueh
Introductions	All
Purpose of Meeting	Randy Withrow
Discussion of Draft SOWs for the Dewey-Burdock Project	Randy Withrow/ Haimanot Yilma / Lynne Sebastian

Possible Discussion Topics:

- Estimated coverage rate for field identification (# of acres per person day).
- Start date and estimated duration of the field identification effort.
- Proposed report content and confidentiality requirements (to be prepared after the field identification has been completed).
- Number of people required for the field identification, with labor classifications (e.g., surveyor, crew leader, traditional cultural expert) and associated hourly rates.
- Report deliverable schedules.

Break – 15 min (~11:30 - 11:45 am EST)

Discussion of Draft SOW (continued)	Randy Withrow/ Haimanot Yilma/ Lynne Sebastian
Summary of Concerns and Recommendations	Randy Withrow

Next Steps	Randy Withrow and Haimanot Yilma
Closing	Kevin Hsueh

The teleconference will be recorded and a transcript of the meeting will be provided to all consulting parties.

Thank You

Haimanot Yilma
Project Manager
FSME/DWMEP/EPPAD/ERB
U.S Nuclear Regulatory Commission
Phone: 301-415-8029
email: haimanot.yilma@nrc.gov
Mail Stop : T8F05

RICHARD E. BLUBAUGH
Vice President – Health, Safety
& Environmental Resources

POWERTECH (USA) INC.

August 29, 2012

Kevin Hsueh, Branch Chief
Environmental Review Branch-B
Environmental Protection and Performance Assessment Directorate
Division of Waste Management and Environmental Protection
Office of Federal and State materials and Environmental Management Programs
United States Nuclear Regulatory Commission
Washington D.C. 20555-0001

Re: August 12, 2011 letter from U.S. Nuclear Regulatory Commission (NRC) Staff to Powertech (USA) Inc. concerning information needed to complete Section 106 of the National Historic Preservation Act

Dear Mr. Hseuh:

I am writing in regard to the above-referenced letter in which NRC Staff requested that, as part of its submissions in support of its license application (docket number 040-009075) for the proposed Dewey-Burdock In Situ Leach Uranium Recovery Project (Dewey-Burdock Project), Powertech (USA) Inc. (Powertech) provide NRC Staff with information regarding potential properties of religious and cultural significance (also referred to as "traditional cultural properties") that might be affected by the proposed project.

Over the past year, Powertech has made every effort to comply with this request. First, Powertech hired third-party consultants (SRI Foundation or the Foundation) to identify and facilitate consultations with federally recognized Indian tribes (Tribes) that might ascribe religious and cultural significance to properties within the proposed project area. Once NRC Staff had informed the Tribes about its request to the applicants and explained the Foundation's role, the Foundation began the first of many contacts with the Tribes on November 4, 2011 (see Attachment 1 for a record of tribal communications). The Foundation provided information about the proposed Dewey-Burdock Project and requested Tribal input as to appropriate methods for gathering the information needed by NRC Staff. The Tribes indicated that they needed to conduct an on-the-ground field investigation within the Project area, and that they wished to discuss how to proceed with this identification effort in a face-to-face meeting with NRC.

In partnership with Cameco Resources (which had received the same request for information from NRC), Powertech sponsored a two-day face-to-face Section 106 consultation meeting on February 14 and 15, 2012, among NRC Staff, Bureau of Land Management (BLM), Environmental Protection Agency (EPA), and representatives of the following federally recognized Indian tribes:

> Cheyenne River Sioux Tribe
> Crow Creek Sioux Tribe
> Crow Tribe of Montana
> Eastern Shoshone Tribe
> Fort Peck

5575 DTC Parkway, Suite 140   Telephone: 303-790-7528   Website: www.powertechuranium.com   1
Greenwood Village, CO 80111 USA.   Facsimile: 303-790-3865   Email: info@powertechuranium.com

A–90

Northern Arapaho Tribe
Northern Cheyenne Tribe
Oglala Sioux Tribe
Rosebud Sioux Tribe
Santee Sioux Tribe of NE
Sisseton-Wahpeton Oyate
Standing Rock Sioux Tribe
Yankton Sioux Tribe

The purpose of this meeting, as established by NRC Staff, was to enable the federal agencies to hear from the Tribes what would be required in order for the Tribes to identify potential properties of religious and cultural significance to them within the Dewey-Burdock and Crow Butte/North Trend Project/license areas. No information about specific identification procedures was forthcoming during this meeting, but the Tribes in attendance proposed to provide NRC Staff with a scope of work (SOW) for the Dewey-Burdock and Crow Butte/North Trend identification efforts. The Tribes also indicated during the meeting that they would not work directly with either Powertech (USA) Inc. or its consultants.

In March of this year, Powertech, at the request of NRC Staff, developed an initial draft SOW for identification of potential properties of religious and cultural significance within the Dewey-Burdock license area. The purpose of this document was to serve as a point of departure for negotiations, along with the anticipated proposed SOW from the Tribes. NRC Staff sent Powertech's draft scope to the Tribes on March 9, 2012, and requested that the Tribes provide their promised proposed SOWs by March 16.

The Tribe's proposed SOW for the Dewey-Burdock Project was not received by NRC Staff until July 13, 2012. This SOW provided rates for items such as salaries, travel, overhead, and per diem; however, it did not provide any information on level of effort (e.g., number of field days, number of travel days, and number of crew members) which would have enabled Powertech to estimate the potential costs. NRC Staff's requests to the Tribes for clarification on level of effort issues subsequent to receipt of the Tribal SOW have been unsuccessful. On July 30, 2012, once again at the request of NRC Staff, Powertech provided a revised SOW for identification of potential properties of religious and cultural significance in the Dewey-Burdock Project area.

Powertech has participated in three conference calls sponsored by NRC and attended by BLM, EPA, and many of the Tribes on the list provided above. The first call was on April 24, 2012, the second on August 9, 2012, and the third on August 21, 2012. During the April 24th call, the Tribes requested that two tribal representatives be assisted in carrying out reconnaissance visits to both the Dewey-Burdock and Crow Butte/North Trend license areas, in order to secure information that would enable the Tribes to complete detailed proposed SOWs for these projects. Powertech accommodated this request, and the Dewey-Burdock Project reconnaissance visit took place on Saturday, May 26th. The purpose of each of these conference calls, as established by NRC Staff, was to secure input from the Tribes that would enable NRC Staff to develop a final SOW for identification of potential properties of religious and cultural significance. None of these calls succeeded in meeting this objective. In the absence of a mutually acceptable SOW, Powertech cannot contract with the Tribes or their representatives to secure the information requested by NRC to complete the identification phase of the Section 106 process.

I regret to inform you that after a year of substantial effort, Powertech is unable to provide the information on potential properties of religious and cultural significance that may be affected by the Dewey-Burdock Project as requested in your letter of August 12, 2011. Further, Powertech has concluded that additional efforts on our part are unlikely to be productive.

2

One of our primary concerns, from the beginning of this effort, has been to ensure that places of significance to the Tribes within Powertech's proposed Project area that may be affected by Project activities be identified so that Powertech can, to the extent possible, protect them from disturbance. To that end, Powertech is willing to support NRC Staff efforts to complete the Section 106 identification process by providing up to $100,000 in funding for tribal representatives to carry out fieldwork and reporting activities as agreed upon in consultations among NRC, BLM, and the tribes, provided that the fieldwork is completed this fall. Powertech also will be happy to coordinate with NRC and BLM on providing access for tribal representatives to the project area in order to carry out the agreed upon work.

Respectfully yours,

Richard Blubaugh
Vice President – Health, Safety and Environmental Resources

Enclosures
cc: R. F. Clement, Powertech
        John Mays, Powertech
        Mark Hollenbeck, Powertech
        Lynne Sebastian, SRI Foundation
        Martha Graham, SRI Foundation
        Haimanot Yilma, NRC
        Anthony Thompson, Esq.
        Christopher Pugsley, Esq.

3

From: Yilma, Halmanot
To: Yilma, Halmanot
Subject: Information Related to Section 106 Activity for Dewey-Burdock Proposed Project
Date: Thursday, August 30, 2012 2:55:33 PM

Dear Tribal Historic Preservation Officers:

The NRC staff wishes to thank all who participated in the teleconference held on August 21, 2012, to discuss the Applicant's revised Statement of Work (SOW). The consulting parties represented, included, the Oglala Sioux, Cheyenne River Sioux, Northern Cheyenne, Rosebud Sioux, Santee Sioux, Sisseton-Wahpeton Oyate, Standing Rock Sioux, Yankton Sioux tribes, EPA Region 8, BLM, NRC, Powertech Inc, Cameco Inc, SRIF (Powertech and Cameco's consultant) and Louis Berger (NRC contractor)

Participating Tribes requested an opportunity to further discuss and revise the SOW. The consulting parties agreed on the need to focus identification efforts on areas of potential ground disturbance (approximately 2637 acres). The parties also proposed developing a programmatic agreement (PA) to address any future ground disturbance.

Tribal Representatives agreed to meet with Mr. Randy Withrow (NRC contractor) and Mrs. Jean Trefethen (NRC staff), in Bismarck, North Dakota on September 5-6, 2012 to further discuss, and revise the SOW for the proposed Dewey Burdock project.

In the August 21, 2012 teleconference, tribal representatives requested adequate time be provided to record identified sites. Since majority of the sites, that might be present are confidential in nature, detailed description of the sites is not warranted. The NRC staff only requires enough information to determine eligibility pursuant to the NHPA regulations at 36 CFR 800.4 (c)(1). For these reasons, the NRC recommends the revised SOW be designed to meet, but not exceed, these information needs.

On August 29, 2012, the applicant sent a letter to the NRC that states "[in the absence of a mutually acceptable SOW, Powertech cannot contract with the Tribes or their representatives to secure the information requested by the NRC to complete identification phase of the section 106 process." (See, ML12243A158.) The applicant recognizes that "places of significance to the Tribes within Powertech's proposed Project area . . . may be affected by Project activities" and that through identification "Powertech can, to the extent possible, protect them from disturbance." The applicant is willing to provide funds up to $100,000.00 for site identification, as long as work is completed by fall 2012. The applicant will coordinate access to the project area with NRC and the Tribes.

It is the NRC staff's understanding that the working group will develop a revised SOW during the September 5-6, 2012 meeting that will ensure completion of a field survey in the fall of 2012. The NRC requires the following information:

* Estimated coverage rate for field identification (# of acres per person day).
* Start date and estimated duration of the field identification effort.

- Proposed report content and confidentiality requirements.
- Number of people required for the field identification, with labor classifications (e.g., surveyor, crew leader, traditional cultural expert) and associated hourly rates.
- Report deliverable schedules.

The NRC staff encourages the Tribal Representatives to consider the offer provided by the applicant when revising the SOW (which should include the above requested information). If Tribal Representatives are unable to provide the requested information by the end of the September 5th and 6th, 2012 meeting to support completion of a field survey in the fall of 2012, the NRC and BLM staff will develop an alternative approach for identifying historic properties, and will move the Section 106 process forward.

If you have any question regarding this email, please contact me or Mr. Withrow.
Thank you.

Haimanot Yilma
Project Manager
FSME/DWMEP/EPPAD/ERB
U.S Nuclear Regulatory Commission
Phone: 301-415-8029
email: haimanot.yilma@nrc.gov
Mail Stop : T8F05

Randy Withrow
Sr. Program Manager | Cultural Resources
The Louis Berger Group, Inc.
900 50th Street | Marion, IA 52302
Office: 319.373.3043, ext. 3035
Cell: 515-441-6497
Fax: 319.373.3045
www.louisberger.com

UNITED STATES
NUCLEAR REGULATORY COMMISSION
WASHINGTON, D.C. 20555-0001

September 18, 2012

Dear THPO:

SUBJECT: REQUEST FOR A PROPOSAL WITH COST ESTIMATE; PROPOSED DEWEY
BURDOCK IN-SITU RECOVERY PROJECT

The NRC staff wishes to thank the tribal representatives from the Crow Nation, Oglala Sioux
Tribe, Northern Cheyenne Tribe, Rosebud Sioux Tribe, Sisseton-Wahpeton Oyate, Standing
Rock Sioux Tribe, and Yankton Sioux Tribe who participated in a project meeting with Jean
Trefethen (NRC) and Randy Withrow (NRC contractor) in Bismarck, North Dakota on
September 5, 2012. This meeting was scheduled following a teleconference held on August 21,
2012, during which participating tribes requested an opportunity to revise the applicant's
proposed Statement of Work (SOW) for completing a Tribal Survey for the Dewey-Burdock
Project.

It was the U.S. Nuclear Regulatory Commission (NRC) staff's understanding that this meeting
would include an opportunity for a working group composed of NRC and tribal representatives
to develop a revised SOW for completion of a field survey in the fall of 2012. Instead, tribal
representatives provided NRC with a revised SOW (Enclosure 1) on September 3, 2012, just in
advance of the meeting. At the September 5th meeting, most of the discussion actually involved
several other topics of concern to the tribes.

Tribes requested NRC's written comment on four principal matters of concern prior to finalizing
a scope of work for a field survey limited to the area of direct effect. The tribes' first three
concerns involve general matters of compliance with the National Historic Preservation Act
(NHPA) or other laws. The NRC staff believes it has previously addressed these issues in
meetings, teleconferences and written correspondence with tribal representatives.
Nonetheless, the staff will respond to the tribal representatives' concerns below.

- Tribes are concerned that the scope of the tribal survey will be limited to the area of
  immediate direct effects (2,637 acres) and that tribes would have no assurance that
  future development outside this area would be subject to proper review prior to
  construction. Tribal representatives requested a Programmatic Agreement be
  developed for the Dewey Burdock project to address the need for phased identification
  of historic properties, including places of traditional religious and cultural significance to
  tribes.

2

Staff Response: The NRC staff agrees that a Programmatic Agreement will need to be developed to address the phased identification and evaluation of historic properties. The need for a phased approach to identification and the advantages of developing a Programmatic Agreement for the Dewey Burdock project has been discussed in previous meetings. For example, during the February 14–15, 2012, consultation meeting, the parties discussed the phased identification and evaluation of historic properties on the Dewey-Burdock site. *See* Meeting Transcripts at the Agencywide Documents Access and Management System (ADAMS) Accession Nos. ML120590330 and ML120590341. Using a phased approach to comply with the NHPA is allowed by the regulations at 36 CFR § 800.14(b). The NRC staff will continue to consult with the tribes and other consulting parities as it develops a Programmatic Agreement.

- Tribes are concerned that potential indirect effects have not yet been fully addressed and requested that the NRC and Bureau of Land Management (BLM) continue consultation with tribes, the South Dakota State Historic Preservation Office (SD SHPO) and the Advisory Council on Historic Preservation (ACHP) to define the area of potential indirect effects and then determine what level of effort is needed to identify properties and assess effects in this area.

Staff Response: The NRC staff will continue to consult with BLM, SD SHPO, and the tribes on all issues arising under Section 106 of the NHPA, including potential indirect effects. The staff will also consult with ACHP as necessary. For approximately the past year, NRC staff has been involved in discussions with the tribes over how to identify historic properties that may be affected by the proposed Dewey-Burdock Project. The staff previously sent the tribes maps identifying the area of potential effect (APE) for the entire Dewey-Burdock Project. The staff also sent the tribes maps showing areas that may be affected during the first phase of the project. These maps identify the placement of buildings, potential wellfields, land application areas, and known archaeological sites (Enclosure 2). The tribes have therefore had the resources to provide input on what areas may be affected, either directly or indirectly, during the first phase of the Dewey-Burdock project. However, to date, tribal representatives have not provided input on specific areas that may be affected during the first phase of the project.

- Tribes expressed concern about the need for confidentiality of site information associated with completion of the tribal survey and the disposition of that information. Tribes requested that the NRC endorse the confidentiality provisions included in the SOW as revised on September 3, 2012.

Staff response: The NRC staff intends to keep survey information confidential to the fullest extent allowed by law. At the same time, the staff must have sufficient information to ensure that we can make an independent recommendation as to whether properties are eligible for inclusion on the National Register of Historic Places. The staff has discussed these issues with tribal representatives previously. See February 2012, Meeting Transcripts at ADAMS Accession Nos. ML110550535, ML120590330 and ML120590341. In the "Reporting" section of the information request (Enclosure 3), the staff proposes a method of reporting fieldwork intended to address the tribes' confidentiality concerns, while at the same time meeting the staff's information needs. We ask that you provide further input on confidentiality in your response to our information request.

3

In addition to these general NHPA-related concerns, the tribes requested the following action specific to NRC's request for a cost estimate to complete the survey:

- Tribes expressed concern that the daily coverage rates (acres per person/day) requested by NRC for cost estimating purposes might be incorrectly interpreted as a precedent for other survey efforts. Tribes requested that this be waived as a requirement for the purpose of estimating survey costs.

Staff response: Since February 2012, the staff has been trying to facilitate the development of an SOW under which the applicant would contract with the tribes for a survey of the proposed Dewey-Burdock site. The initial SOW from the applicant, which the staff sent to the tribes on May 7, 2012, included coverage rates. At the end of this letter, the staff renews our request for certain information from the tribes. If the tribes object to using coverage rates to estimate survey costs, NRC invites tribes to substitute an alternative means of estimating survey cost.

As we have stated previously, the staff's schedule for completing our NHPA review is tied to our schedule for completing our review under the National Environmental Policy Act (NEPA). Because our schedule calls for issuing our final NEPA document no later than May 2013, it is imperative that we proceed with identifying any NHPA-eligible properties before the end of the 2012 field season (*i.e.,* in the fall 2012).

The staff respectfully requests that the participating tribes designate a preferred contractor to complete a cultural resources survey on their behalf and provide NRC with a written proposal with cost estimate based on the 2,637-acre area that may be disturbed during the first phase of the proposed Dewey Burdock project. For your convenience, the staff is enclosing a detailed request for information with this letter (Enclosure 3). This request repeats and consolidates the staff's prior requests for information from the tribes. *See, e.g.,* ADAMS Accession Nos. ML12143A185, ML12261A375, ML12261A429, and ML12261A476. The staff is also forwarding maps showing the location of the entire proposed project area and the proposed initial disturbed area (2,637-arce) to be surveyed. These maps were sent to you previously. *See* ADAMS Accession No. ML 12261A326.

The NRC staff requests that you submit the proposal with cost estimate stated above to Ms. Kellee Jamerson, NRC Project Manager, or Mr. Randy Withrow, NRC contractor, no later than close of business on October 1, 2012. The proposal can be submitted by email to

4

Kellee.Jamerson@nrc.gov or rwithrow@louisberger.com. Following the receipt of this information or after October 1, 2012, NRC and BLM will determine the path forward for identifying any NHPA-eligible properties before the end of the 2012 field season (*i.e.*, in the fall 2012).

Sincerely,

Kevin Hsueh, Branch Chief
Environmental Review Branch
Environmental Protection and Performance
 Assessment Directorate
Division of Waste Management
 and Environmental Protection
Office of Federal and State Materials
 and Environmental Management Programs

Enclosures:
1. Tribes Revised SOW from
   September 3, 2012
2. Powertech's Map Depicting Project
   Boundary and Proposed Known
   Disturbance (ML12261A326)
3. Detailed Request for Information
   (Request for Proposal w/Cost Estimate)

**Request for Proposal with Cost Estimate
Tribal Survey for the Proposed Dewey Burdock Project
September 18, 2012**

The U.S. Nuclear Regulatory Commission (NRC) staff requests a written proposal with cost estimate for a survey to identify places of traditional religious and cultural significance to tribes that may be affected by the first phase of the proposed Dewey Burdock Project. This request consolidates prior requests for information that the staff has made in emails, letters, teleconferences, and meetings with tribal representatives. See, e.g., ADAMS Accession Nos. ML12143A185, ML12261A375, ML12261A429, and ML12261A476.

The tribes' proposal with cost estimate should include a brief description of the work that will be completed for both field investigations and reporting. Please include the following specific information in your written proposal.

Fieldwork:

- Describe the size and composition of the survey crew (number of individuals and their titles).

- Provide a proposed start date and estimated duration of fieldwork (number of field days).

- Cost assumptions (including, for example, the estimated number of cultural features that will be recorded).

Reporting:

- Provide a schedule for completion of the following work products or deliverables.

    1. A non-confidential summary of fieldwork including a map showing where survey work was completed (this should not include specific site locations).

    2. A confidential final eligibility report that provides the location of all identified sites, a description of where each site is located in relationship to areas that will be directly impacted by planned operations, and recommendations regarding the eligibility of each site for listing in the National Register of Historic Places. The assessment of eligibility should include references to the appropriate eligibility criteria (36 CFR 60.4) and an assessment of how the site's integrity will be affected directly or indirectly by the proposed undertaking.

    3. A confidential report for use by the applicant showing the location of any eligible sites identified within the proposed Dewey Burdock license area. This report will be prepared once final determinations of eligibility have been completed and will only be shared with the applicant after tribes receive a confidentiality agreement signed by the applicant that limits use to appropriate personnel.

Enclosure 3

- 2 -

Cost Estimate:

Please provide a line-item budget that lists costs for estimated labor and related expenses for both fieldwork and reporting. For labor estimates, please include labor categories or titles, estimated number of hours for each category, and the associated hourly rates. For related expenses such as per diem or equipment rental, please include both the number of days and associated rates used to estimate total costs.

Schedule:

NRC requests that all field investigations be completed by the end of the 2012 field season (i.e., in fall 2012), and that a confidential eligibility report be completed no later than 60 days following completion of the field survey.

Access and Safety:

The applicant, Powertech (USA), will provide access to the properties, and a representative of Powertech (USA) will coordinate with Tribal preferred contractor in terms of access to land. The Powertech (USA) representative will utilize a GPS survey unit to identify all map locations selected by the Tribal preferred contractor for ground examination and will guide the Tribal personnel to the locations they select in the field. The Powertech (USA) representative will also serve as liaison with the local landowners.

Insurance:

All Tribal representatives who will be present during field work will be required to provide proof of liability insurance in the amount of $500,000 or more, or sign an indemnification statement that will hold harmless both the landowner and Powertech (USA) from any accidents that may occur in the field.

Contracting:

NRC will not contract directly with the preferred contractor selected by participating tribes. NRC will forward the proposal to the project applicant for their consideration and contracting.

October 4, 2012

Richard Blubaugh
Vice President – Health, Safety,
  and Environmental Resources
Powertech (USA) Inc.
5575 DTC Parkway, Suite 140
Greenwood Village, CO 80111

SUBJECT: TRANSMITTAL OF TRIBES' PROPOSAL AND COST ESTIMATE FOR THE
            PROPOSED DEWEY-BURDOCK ISR PROJECT

Dear Mr. Blubaugh:

On September 27, 2012, the U.S. Nuclear Regulatory Commission (NRC) received a "Proposal
with Cost Estimate for Traditional Cultural Properties Survey for Proposed Dewey Burdock
Project" from Makoche Wowapi/Mentz-Wilson Consultants, LLP (enclosed).

The NRC requests that you review the enclosed proposal and provide us with any comments by
October 10, 2012. To address the possibility that Powertech and Makoche Wowapi/Mentz-
Wilson Consultants, LLP might be unable to reach an agreement regarding the proposal, the
NRC asks that you also provide a list of alternative methods for identifying potential properties
of traditional religious and cultural importance to tribes at the proposed Dewey-Burdock site.

Please note that the cost estimate and breakdown of field crew wages in the Tribes' proposal
(pages 3 and 4) has been identified by the consultants as proprietary information and will not be
shared with all the consulting parties. In addition, the proposal with cost estimate in its entirety
is being withheld from public disclosure under 10 CFR 2.390.

Sincerely,

/RA/

Kevin Hsueh, Chief
Environmental Review Branch
Division of Waste Management
  and Environmental Protection
Office of Federal and State Materials
  and Environmental Management Programs

Enclosure:
Proposal with Cost Estimate

October 11, 2012

Chairman Cedrick Black Eagle
Crow Tribe
Baacheeitche Avenue
P.O. Box 159
Crow Agency, MT 59022

SUBJECT: TRANSMITTAL OF CORRESPONDENCE PERTAINING TO REQUEST FOR
            DETAILED INFORMATION FOR THE PROPOSED DEWEY-BURDOCK
            IN-SITU RECOVERY PROJECT

Dear Chairman Black Eagle:

Enclosed please find correspondence sent via email from the U.S. Nuclear Regulatory
Commission (NRC) staff to Tribal Historic Preservation Officers (THPOs) in response to
concerns raised during a September 5-6, 2012, meeting in Bismarck, North Dakota. The letter
requested a proposal with a cost estimate be submitted for the proposed Dewey-Burdock
project. Included with the correspondence was the Tribes' revised Statement of Work and
Powertech's maps depicting the project boundary and proposed known areas of disturbance.

The NRC staff is transmitting this letter and attached correspondence to you to keep you
informed of all Section 106 activities for the proposed Dewey-Burdock ISR project.

If you have any questions or concerns, please contact Ms. Kellee Jamerson of my staff via
email at Kellee.Jamerson@nrc.gov or by phone at 301-415-7649.

                                        Sincerely,

                                        /RA/

                                        Larry W. Camper, Director
                                        Division of Waste Management
                                         and Environmental Protection
                                        Office of Federal and State Materials
                                         and Environmental Management Programs

Enclosure:
Letter w/Detailed Request for
 Information (Request for Proposal
 w/Cost Estimate)

cc: Hubert Two Leggings, THPO

UNITED STATES
NUCLEAR REGULATORY COMMISSION
WASHINGTON, D.C. 20555-0001

October 12, 2012

Dear Tribal Historic Preservation Officer:

SUBJECT: TRANSMITTAL OF TRIBES' PROPOSAL WITH COST ESTIMATE FOR THE
PROPOSED DEWEY-BURDOCK ISR PROJECT

On September 27, 2012, the U.S. Nuclear Regulatory Commission (NRC) received a "Proposal
with Cost Estimate for Traditional Cultural Properties Survey for the Proposed Dewey Burdock
Project" from Makoche Wowapi/Mentz-Wilson Consultants, LLP.

The NRC is aware of significant differences in the proposal submitted by Makoche
Wowapi/Mentz-Wilson Consultants, LLP and the proposal[1] submitted by Powertech. The NRC
anticipates that resolving these differences will not support completion of a field survey in the fall
of 2012 for the Dewey-Burdock In-Situ Recovery (ISR) Project and for this reason it seeks
alternatives.

The NRC recognizes that there are additional methods for identifying potential properties of
traditional religious and cultural importance to tribes at the proposed Dewey-Burdock site.
Alternatives include opening the site to interested tribal specialists over a period of several
weeks with payments to be made to individual tribes, or seeking ethnohistorical and
ethnographic information from tribal specialists in interviews at tribal headquarters.

The NRC requests that you provide us with your ideas on alternative methods for identifying
potential properties by close of business Friday, October 19, 2012.

Also, enclosed is Powertech's "Reply to October 4, 2012 Letter and Statement of Work (SOW),"
dated October 9, 2012.

Please note that the cost estimate and breakdown of field crew wages in the Tribes' proposal
(pages 3 and 4) has been identified by the consultants as proprietary information and will not be
shared with all the consulting parties. In addition, the proposal with cost estimate in its entirety
are being withheld from public disclosure under 10 CFR 2.390.

Sincerely,

Kevin Hsueh, Chief
Environmental Review Branch
Division of Waste Management
    and Environmental Protection
Office of Federal and State Materials
    and Environmental Management Programs

Enclosures:
    1. Proposal with Cost Estimate
    2. Powertech letter dated 10/9/12 (ML12285A425)

---

[1] On August 7, 2012, the NRC forwarded Powertech's revised statement of work (SOW) dated July 30, 2012
(ML12261A333). The NRC received a letter dated August 29, 2012 from Powertech in response to an August 12,
2011 request concerning information needed to complete Section 106 (ML12243A156).

APPENDIX B

ALTERNATE CONCENTRATION LIMITS

ALTERNATE CONCENTRATION LIMITS

In-situ recovery (ISR) facilities operate by first extracting uranium from specific areas called wellfields. After uranium recovery has ended, the groundwater in the wellfield contains constituents that the lixiviant mobilized. Licensees shall commence aquifer restoration in each wellfield soon after the uranium recovery operations end (NRC, 2009). Aquifer restoration criteria for the site-specific baseline constituents are determined either for each individual well or as a wellfield average.

U.S. Nuclear Regulatory Commission (NRC) licensees are required to return water quality parameters to the standards in 10 CFR Part 40, Appendix A, Criterion 5B(5). As stated in the regulations: "5B(5)—At the point of compliance, the concentration of a hazardous constituent must not exceed—(a) The Commission approved background concentration of that constituent in the groundwater; (b) The respective value given in the table in paragraph 5C if the constituent is listed in the table and if the background level of the constituent is below the value listed; or (c) An alternate concentration limit (ACL) is established by the Commission."

For an ACL to be considered by the NRC, a licensee must submit a license amendment application to request an ACL. In this ACL license amendment request, the licensee must provide the basis for any proposed limits, including consideration of practicable corrective actions that limits are as low as reasonably achievable (ALARA), and information on the factors the Commission must consider. NRC will establish a site-specific ACL for a hazardous constituent as provided in Criterion 5B(5) if NRC finds the proposed limit ALARA, after considering practicable corrective actions, and determining that the constituent will not pose a substantial present or potential hazard to human health or the environment as long as the ACL is not exceeded.

To determine if the ACL does not pose a potential hazard to human health or the environment, NRC performs three risk assessments (NRC, 2003a). The first is a hazard assessment which evaluates the radiological dose and toxicity of the constituents in question and the risk to human health and environment. The second is an exposure assessment to examine the existing distribution of hazardous constituents, as well as potential sources for future releases and the potential consequences associated with the human and environmental exposure to the hazardous constituents. The last assessment is a corrective action assessment, which evaluates (i) all applicant proposed corrective actions; (ii) the technical feasibility of each proposed corrective actions; (iii) the costs and benefits associated with each proposed corrective action; and (iv) the preferred corrective action to achieve the hazardous constituent concentration, which is protective of human health and the environment.

To perform these assessments, the NRC staff uses a rigorous review process. Licensees must provide a comprehensive ACL amendment that addresses groundwater and surface water quality and expected impacts on human health and the environment. Such information required in an amendment request pursuant to 10 CFR Part 40, Appendix A, Criterion 5B(6) includes the following factors:

- Potential adverse effects on groundwater quality, considering the following:

  — The physical and chemical characteristics of the waste in the licensed site including its potential for migration

1   — The hydrogeologic characteristics of the facility and surrounding land
2
3   — The quantity of groundwater and the direction of groundwater flow
4
5   — The proximity and withdrawal rates of groundwater users
6
7   — The current and future uses of groundwater in the area
8
9   — The existing quality of groundwater, including other sources of contamination and
10   their cumulative impact on the groundwater quality
11
12   — The potential for health risks caused by human exposure to waste constituents
13
14   — The potential damage to wildlife, crops, vegetation, and physical structures
15   caused by exposure to waste constituents
16
17   — The persistence and permanence of the potential adverse effects
18
19 • Potential adverse effects on hydraulically connected surface water quality, considering
20   the following:
21
22   — The volume and physical and chemical characteristics of the waste in the
23   licensed site
24
25   — The hydrogeologic characteristics of the facility and surrounding land
26
27   — The quantity and quality of groundwater, and the direction of groundwater flow
28
29   — The patterns of rainfall in the region
30
31   — The proximity of the licensed site to surface waters
32
33   — The current and future uses of surface waters in the area and any water quality
34   standards established for those surface waters
35
36   — The existing quality of surface water including other sources of contamination
37   and the cumulative impact on surface water quality
38
39   — The potential for health risks caused by human exposure to waste constituents
40
41   — The potential damage to wildlife, crops, vegetation, and physical structures
42   caused by exposure to waste constituents
43
44   — The persistence and permanence of the potential adverse effects
45
46 Although state "class of use" standards are not recognized in NRC's regulations as restoration
47 standards, these standards may be considered as one factor in evaluating ACL requests for ISR
48 facilities located in South Dakota. Furthermore, in considering ACL requests, particular

1   importance is placed on protecting underground sources of drinking water (USDWs). The use
2   of modeling and additional groundwater monitoring may be necessary to show that ACLs in ISR
3   wellfields would not adversely impact USDWs. It must be demonstrated that the licensee it has
4   attempted to restore hazardous constituents in groundwater to background or a maximum
5   contaminant level—whichever level is higher.
6
7   Before an ISR licensee is allowed to extract uranium, the U.S. Environmental Protection Agency
8   (EPA) under 40 CFR 146.4 and in accordance with the Safe Drinking Water Act must issue an
9   aquifer exemption covering the portion of the aquifer in which the uranium-bearing rock is
10 located. EPA cannot exempt the portion of the aquifer unless it is found that "it does not
11 currently serve as a source of drinking water" and "cannot now and will not in the future serve as
12 a source of drinking water." Due to these criteria, only impacts outside of the exempted aquifer
13 are evaluated. In most cases, the water in aquifers adjacent to the uranium ore zones does not
14 meet drinking water standards. The staff will not approve an ACL if it will impact any adjacent
15 USDWs. Therefore, the impact of granting an ACL request is SMALL.
16
17 Further guidance for the review of ACLs for ISR facilities is being developed in a revision of
18 NUREG–1569 (NRC, 2003a). Existing guidance for the review of ACLs for conventional mills is
19 in NUREG–1620, "Standard Review Plan for the Review of a Reclamation Plan for Mill Tailings
20 Sites Under Title II of the Uranium Mill Tailings Radiation Control Act of 1978" (NRC, 2003b).
21
22 **References**
23
24 10 CFR Part 40. Appendix A. *Code of Federal Regulations,* Title 10, *Energy*, Part 40,
25 Appendix A. "Criteria Relating to the Operations of Uranium Mills and to the Disposition of
26 Tailings or Wastes Produced by the Extraction or Concentration of Source Material from Ores
27 Processed Primarily from their Source Material Content." Washington, DC: U.S. Government
28 Printing Office.
29
30 40 CFR Part 146. *Code of Federal Regulations,* Title 40, *Protection of Environment,* Part 146.
31 "Underground Injection Control Program: Criteria and Standards." Washington, DC:
32 U.S. Government Printing Office.
33
34 NRC (U.S. Nuclear Regulatory Commission). NUREG-1910, "Generic Enviromental Impact
35 Statement for In-Situ Leach Uranium Milling Facilities." ML091480244, ML091480188.
36 Washington, DC. NRC. May 2009.
37
38 NRC. NUREG–1569, "Standard Review Plan for *In-Situ* Leach Uranium Extraction License
39 Applications." Final Report. Washington, DC: NRC. June 2003a.
40
41 NRC. NUREG–1620, "Standard Review Plan for the Review of a Reclamation Plan for Mill
42 Tailings Sites Under Title II of the Uranium Mill Tailings Radiation Control Act of 1978."
43 Final Report. Washington, DC: NRC. June 2003b.

APPENDIX C
NONRADIOLOGICAL AIR EMISSIONS ESTIMATES

NONRADIOLOGICAL AIR EMISSIONS ESTIMATES

C1   Introduction

This appendix provides detailed nonradiological air emissions information associated with the proposed action. The information in the appendix consolidates and supplements information from several sources (Powertech, 2009, 2010a–c, 2012 and Inter-Mountain Labs, 2012), which is then summarized in the SEIS.

While NRC is responsible for assessing the potential environmental impacts from the proposed action pursuant to the National Environmental Policy Act of 1969 as amended, NRC does not have the authority to develop or enforce regulations to control nonradiological air emissions from equipment licensees use. For the proposed Dewey-Burdock ISR Project, this authority rests with the South Dakota Department of Environment and Natural Resources (SDDENR). To ensure the air quality of South Dakota is adequately protected, in addition to addressing all NRC regulatory requirements that address radiological emissions, NRC applicants and licensees must also comply with all applicable state and federal air quality regulatory compliance and permitting requirements.

NRC staff acknowledges that SDDENR has not yet conducted the formal air quality permitting for the proposed Dewey-Burdock ISR Project (see Table 1.6-1). In the absence of a formal determination and permitting by the SDDENR, NRC staff will characterize the magnitude of air effluents from the proposed project in part by comparing (i) the emission levels to Prevention of Significant Deterioration and Title V thresholds and (ii) the modeled concentrations to regulatory standards such as NAAQS. This characterization is meant to provide a context for understanding the magnitude of the proposed project's air effluents. The NRC description in this SEIS does not document or represent the formal determination by the SDDENR. As such, the SDDENR determination and permitting may vary with the NRC description.

C2   Non-Greenhouse Combustion Exhaust Emissions

The non-greenhouse combustion exhaust emissions discussion is divided into three sections. Section C.2.1 addresses the emissions inventory that describes the amount or mass of pollutants generated by the proposed action. Section C.2.2 discusses the combustion exhaust emissions from drill rigs. Section C.2.3 addresses the air dispersion modeling that predicts pollutant concentrations based on the emissions inventory.

C2.1   Emission Inventory

The non-greenhouse combustion emissions inventory addresses both stationary and mobile sources associated with the proposed action. Stationary source emissions are limited to the operation phase and are presented in Table C–1 (for ease of reading, all tables are located at the end of this appendix). Mobile source emissions, which occur in each of the four phases of the proposed action, are presented in Table C–2. These two tables identify some individual sources and provide the associated emission levels. In addition, the mobile sources were categorized into one of two source classifications: construction and drilling field equipment or other mobile sources (i.e., light-duty trucks and vehicles) excluding commuters. The construction and drilling field equipment source classification was further categorized into three emission vehicle types: deep well drill rigs, other drill rigs, and other construction and drilling field equipment. The deep well drill rigs are used for drilling the Class V deep injection disposal

1   wells. The other drill rigs are used for drilling the delineation, monitoring, production, and
2   injection wells. The other construction and drilling field equipment classification includes
3   sources such as bulldozers, graders, scrappers, cranes, forklifts, and backhoes. Table C–3
4   contains the detailed information used to calculate the mobile sources emission levels. The
5   stationary and mobile emission levels are summarized in Tables 2.1-1 and 2.1-2.
6
7   The applicant revised the initial mobile combustion emission inventory to in part incorporate
8   mitigation measures and improve the accuracy of the emissions expected from the ISR
9   activities. The revised emission inventory is the one used in the SEIS text and documented in
10 Table C–2. In association with the revised inventory, the applicant committed to the following
11 actions (Powertech, 2012):
12
13 •   Lowering the drill rig engine horsepower from 550 horsepower to 300 horsepower,
14   except for the deep well drill rig
15
16 •   Use of Tier 1, or higher, drill rig engines and Tier 3, or higher, for construction
17   equipment engines
18
18 The revised emissions inventory is calculated using emission factors based on these
19 commitments which resulted in lower annual pollution levels relative to the initial inventory.
20 Emission factors are values used to relate the levels of activities to the amounts of pollution
21 produced. In this case the emission factor relates the amount of fuel consumed by the
22 equipment to the mass of pollutants generated. The initial inventory used mostly uncontrolled
23 emission factors (i.e., emission factors based on older engines with greater emission in contrast
24 to newer engines that meet stricter emission standards). The various tiers refers to a phased
25 program of standards that the Federal Government mandated that requires newly manufactured
26 engines to generate lower pollutant emission levels. Higher tier numbers mean stricter emission
27 standards and lower the pollutant levels. Table C–4 describes the effectiveness (i.e., the
28 percent that the emissions are reduced) of the different tier levels based on the associated
29 emission factors. The revised inventory also incorporated equipment load factors (i.e., the
30 fraction of available power utilized). The initial inventory assumed 100 percent duty at
31 maximum horsepower. The revised inventory applied load factors ranging from 25 percent to
32 59 percent depending on the type of equipment and application. The specifics are available in
33 the Powertech Dewey-Burdock Project Emissions Inventory (Powertech, 2012). Reducing the
34 load factors result in lower emission levels and lower pollutant concentrations. The applicant
35 identified other mitigation they would implement (see SEIS Table 6.2-1). However these other
36 mitigations were not incorporated in the calculation of the revised emissions inventory.
37
38 ISR phases may occur simultaneously. To account for overlapping phases, a total emission
39 estimate was calculated by adding together the annual emissions for all four phases. This total
40 or peak year estimate accounts for when all four phases occur simultaneously and represents
41 the highest amount of emissions the proposed action would generate in any one project year.
42 Table 2.1.2 contains the peak year emissions for the mobile sources. The stationary phase did
43 not require a peak year calculation because the emissions are limited to only the operation
44 phase (see Table 2.1-1). Table 2.1-3 contains the peak year estimate for when the stationary
45 and mobile source emissions are combined. The only phase being performed in project year
46 one is construction. The construction phase in project year one consists of two main activities
47 (i) facilities construction and (ii) well field construction. Facilities construction will be completed
48 at the end of project year one. The construction phase associated with the remaining life of the
49 project is limited to well field construction. Therefore, the peak year emission calculations which

1 account for overlapping phase use the construction emission levels associated with the well
2 field only.
3
4 ## C2.2   NAAQS Pollutant Emissions from Drilling Activities
5
6 Information in Table C–4 reveals that the construction phase generates the most NAAQS
7 pollutant emissions relative to the other phases and, within the construction phase, the drill rigs
8 generate the majority of the NAAQS pollutant emissions (PM_{10}, SO_2, NO_x, and CO). The drill
9 rigs are used to dig the various wells associated with ISR. Five types of wells are proposed for
10 this project: delineation wells, monitoring wells, production wells, injection wells, and Class V
11 deep disposal wells. The type of drill rig required for the job can vary based on the type of well.
12 The first four well types require rigs that can drill wells to a depth of less than 305 m, [1,000 ft].
13 The Class V deep disposal well requires drilling equipment suitable to reach depths of about
14 914 m [3,000 ft]. The emission estimates include the drilling of eight Class V deep disposal
15 wells over the life of the project. In project year one, four Class V deep disposal wells would be
16 drilled. After project year one, the emission estimates assume that no more than one Class V
17 deep disposal well will be drilled in any single project year. For the pollutants in Table C–4, the
18 percentage of emissions from the construction phase compared to the other phases ranged
19 from 68 to 79 percent depending on the particular pollutant. The percentage of emissions from
20 the drill rigs (excluding the deep well drill rig) compared to all of the construction phase
21 emissions ranged from 61 to 81 percent depending on the pollutant. The percentage of
22 emissions from the deep well drill rig compared to all of the construction phase emissions
23 ranged from 0.2 to 0.3 percent depending on the pollutant. The deep well drill rig emission
24 contribution is relatively small because the proposed project only requires the drilling of up to
25 eight Class V wells.
26
27 ### C.2.3   Air Quality Modeling
28
29 Expressing the proposed project's emissions in concentrations can help characterize the
30 magnitude of the emission levels because standards such as NAAQS and Prevention of
31 Significant Deterioration are also expressed in concentrations. The AERMOD dispersion model
32 was used to predict pollutant concentrations based on the emission mass flow rates from the
33 initial air emissions inventory at 47 various locations in and around the proposed site
34 (Powertech, 2010a). These concentrations were provided for each of the four phases of the
35 proposed action: construction (Table C–5), operations (Table C–6), aquifer restoration
36 (Table C–7) and decommissioning (Table C–8). The applicant revised the air emissions
37 inventory. However revised air dispersion modeling results were not provided with the revised
38 inventory. The applicant has committed to perform air dispersion modeling using the revised
39 emission inventory prior to the preparation of the final SEIS (Powertech, 2012). The final SEIS
40 analyses would be based on this updated modeling. SEIS Section 4.7.1 describes the scope of
41 this update.
42
43 The modeling results (i.e., pollution concentrations) based on the initial emission inventory were
44 used to generate pollution concentrations for the updated emission inventory. Multiplication
45 factors will be used for each pollutant to calculate the concentration for the revised inventory.
46 A multiplication factor is a value when multiplied by the pollution concentration from the initial
47 inventory yields the pollution concentration associated with the revised inventory. Multiplication
48 factors are generated by calculating the percent difference between the revised emission levels
49 and the initial emission levels associated with a particular set of modeling results (pollutant
50 concentrations). Table C–9 contains the information associated with the generation of the
51 multiplication factors for calculating the peak year concentration values (i.e., the values for when

1   all four phase occur simultaneously) for the revised inventory. The emission values for the
2   revised inventory presented in Table C–9 came from Table C–3. The emission values for the
3   initial inventory presented in Table C–9 are for the construction phase (Powertech, 2010a).
4   Table C–10 contains the information associated with the calculation of the peak year
5   concentration values for the revised inventory. The concentrations for the initial inventory
6   presented in Table C–10 came from the applicant (see Table ER_RAI AQ8.1 from Powertech,
7   2010a) and the multiplication factors came from Table C–9. These pollutant concentrations
8   results for the revised emission inventory from the combustion emissions from stationary and
9   mobile sources for the peak year of the proposed action are summarized in Table 2.1-4.
10
11 The peak year concentrations are important because they account for when all four phases
12 occur simultaneously and represent the highest amount of emissions the proposed action would
13 generate in any one project year. However, the SEIS analyses also examine emissions
14 associated with individual phases. Pollutant concentrations associated with each phase
15 during the peak year can be calculated by knowing the relative contribution from each phase.
16 Table C–11 contains the percent of emissions by phase for various NAAQS pollutants from
17 combustion emissions from stationary and mobile sources when all phases occur
18 simultaneously. A slight adjustment is needed to address the construction phase emissions in
19 project year one. As described in Section C.2.1, the only phase conducted in project year one
20 is construction and these emissions (presented in SEIS Table 2.1-2) include both facility and
21 wellfield construction. In the subsequent project years when the phases can overlap, the
22 construction phase only entails wellfield construction. Based on the information in Table 2.1-2,
23 the project year one construction NAAQS pollutant emissions would be no more than about
24 13 percent greater than the construction emissions in the remaining project years.
25
26 ## C3   Greenhouse Gas Emissions
27
28 Greenhouse gas emissions were provided for each of the four phases of the proposed action:
29 construction (Table C–12), operations (Table C–13), aquifer restoration (Table C–14), and
30 decommissioning (Table C–15). Each table identifies the various activities associated with
31 each phase as well as the various emission sources that compose each activity. In addition,
32 each emission source in these tables was categorized into one of the following two sources:
33 stationary or mobile. Emission information in Tables C-12 through C-15 were added by source
34 classification for each phase and summarized in Table 2.1-5, which contains a third category of
35 emissions: electrical consumption.
36
37 ## C4   Fugitive Dust Emissions
38
39 The fugitive dust emissions discussion is divided into two sections. Section C.4.1 addresses the
40 emissions inventory that describes the amount or mass of pollutants generated by the proposed
41 action. Section C.4.2 addresses the air dispersion modeling (i.e., the concentration of the
42 particulate matter in the air)
43
44 ## C4.1   Emission Inventory
45
46 Fugitive dust emissions are provided for vehicle travel on unpaved roads and wind erosion to
47 disturbed land. The applicant revised the initial fugitive dust emission inventory to in part
48 incorporate mitigation measures and improve the accuracy of the emissions expected from the
49 ISR activities. The applicant initially committed to mitigate fugitive dust emission by watering
50 unpaved roads (Powertech 2010a). However, the initial inventory did not account for this when

calculating the emission levels. The revised inventory credits water spray for a 50 percent reduction of all fugitive emissions generated from unpaved roads. The 50 percent reduction for water spray is a conservative, industry accepted value for this particular mitigation (Powertech, 2012). The applicant identified other mitigation they would implement. However, these other mitigations were not incorporated in the calculation of the revised inventory. Changes made to the calculation of the revised inventory intended to improve the accuracy are as follows:

- The number of passenger vehicles and light-duty trucks used in the calculation was reduced to conform more closely with the construction and operation plan.

- The silt content value used in the fugitive dust calculations was lowered to 8.5 percent. This value is typical of western surface mines and unpaved industrial roads (EPA, 1996).

Silt content is one of the variables used to calculate the emission factor for travel on unpaved roads. Emission factors are values used to relate the levels of activities to the amounts of pollution produced. In this case the emission factor relates the number of miles a vehicle travels to the mass of fugitive dust generated. Tables C–16 to C–20 contain the detailed information used to calculate the fugitive dust emissions for vehicle travel on unpaved roads for the various phases.

The revised fugitive emission inventory included wind erosion. Dust generated by wind blowing over land that has been disturbed is an example of wind erosion. The amount of fugitive emissions from wind erosion is a function of the amount of disturbed land. The two liquid waste disposal options, deep well disposal and land application, did vary in the amount of land disturbed. The deep disposal well option disturbed up to around 39.5 hectares [97.5 acres] of land while the land application option disturbed up to around 116.6 hectares [288.2 acres] (Inter-Mountain Labs, 20912). An emission factor was used relate the amount of total suspended particles generated annually to the amount of land disturbed [i.e., 0.345 metric tons [0.38 short tons] of total suspended particles for each acre disturbed (Powertech, 2012)]. Total suspended particles include particles larger than PM_{10}. Here, 30 percent of total suspend particles is comprised of PM_{10} and 15 percent of PM_{10} is comprised of $PM_{2.5}$ (Powertech, 2012). Table C-16 compares the onsite fugitive emission mass flow rate estimates for the two liquid waste disposal options. The emission estimates in this table includes both fugitive sources (i.e., travel on upaved roads and wind erosion) and provides estimates for all phases as well as the peak year when all phases occur simultaneously.

C4.2 Air Quality Modeling

Fugitive dust emissions were not included in the modeling based on the initial emission inventory described in Section C.2.3. The applicant committed to perform air dispersion modeling using the revised emission inventory prior to the preparation of the final SEIS (Powertech, 2012). The final SEIS analyses would be based on this updated modeling. SEIS Section 4.7.1 describes the scope of this update. To help characterize the Dewey-Burdock fugitive emissions, modeling from a similar project will be used. The similar project is the Atlantic Rim Natural Gas Development Project. This project examined the drilling and

1 development of 2,000 new natural gas wells in Carbon County, Wyoming. The similarities
2 between the Dewey-Burdock and Atlantic Rim projects are as follows:
3
4 • Both projects are similar in scope (i.e., they generate wellfields with a large number wells)
5
6 • Both projects have similar sources of fugitive dust: travel on unpaved roads and
7   wind erosion
8
9 • Both projects produce similar levels of onsite fugitive dust
10
11 • Both projects are located about the same distance away from the nearest Class I area
12
13 The Atlantic Rim project analysis included modeling of onsite maximum concentrations that
14 could occur from all sources operating simultaneously in the field (i.e., the peak year value).
15 The annual mass flow rates used in the Atlantic Rim modeling analyses are at 708.1 metric tons
16 [780.4 short tons] for PM_{10} and 154.8 metric tons [170.6 short tons] for $PM_{2.5}$ (TRC
17 Environmental Corporation, 2006).
18
19 The modeling results (i.e., pollution concentrations) from the Atlantic Rim project are used to
20 generate pollution concentrations for the proposed project. Multiplication factors will be used to
21 calculate the fugitive emission concentrations for the revised inventory. Multiplication factors
22 are generated by calculating the percent difference between the particulate matter annual mass
23 flow rates of the Atlantic Rim project and Dewey-Burdock proposed project. The peak year
24 onsite emission level estimates for travel on unpaved roads for the proposed project are at
25 481.8 metric tons [531.1 short tons] for PM_{10} and 48.2 metric tons [53.1 short tons] for $PM_{2.5}$.
26 The multiplication factors for PM_{10} and $PM_{2.5}$ are 0.68 and 0.31, respectively. Based on the
27 concentration results for the Atlantic Rim project in Table F1.5.1 (TRC Environmental
28 Corporation, 2006), the Dewey-Burdock onsite peak year fugitive dust concentrations
29 (24-hour mean) would be 23.3 $\mu g/m^3$ for PM_{10} and 1.2 $\mu g/m^3$ for $PM_{2.5}$. Table C–17 contains the
30 percentage of fugitive dust emissions from travel on unpaved roads that each of the four phases
31 contribute to the peak year total when all phases occur simultaneously. Table C–18 contains
32 the onsite fugitive dust emission concentrations from travel on unpaved roads for each of the
33 phases which were calculated using the percentages in Table C–17.
34

C5   References

36
37 40 CFR Part 86. *Code of Federal Regulations*, Title 40, *Protection of the Environment*, Part 86.
38 "Control of Emissions from New and In-Use Highway Vehicles and Engines." Washington, DC:
39 U.S. Government Printing Office.
40
41 EPA. Letter (March 3) from C. Rushin, Acting Regional Administrator, U.S. Environmental
42 Protection Agency, to M. Lesar, Chief, U.S. Nuclear Regulatory Commission. "Subject:
43 NUREG–1910, Supplements 1, 2, and 3 Draft SEIS for Three Wyoming Uranium ISR
44 Projects: Lost Creek ISR Project CEQ# 20090425; Moore Ranch ISR Project CEQ# 20090421;
45 and Nichols Ranch ISR Project CEQ# 20090423." ML100680712. Washington, DC:
46 U.S. Environmental Protection Agency. 2010.
47
48 EPA. "Exhaust and Crankcase Emission Factors for Nonroad Engine
49 Modeling—Compression-Ignition." EPA420–P–04–009, NR–009c. Washington DC:
50 U.S. Environmental Protection Agency. April 2004.

1   EPA. "Compilation of Air Pollutant Emission Factors, Volume 1: Stationary and Point
2   Area Sources." AP–42, Fifth Edition, Chapters 3.3, 3.4, and 13.2. Washington, DC:
3   U.S. Environmental Protection Agency. October 1996.
4
5   Inter-Mountain Labs. "Follow-Up on Dewey-Burdock Questions." Email (August 2) from
6   R. Smith to Bradley Werling (Southwest Research Institute®). ML12216A215. San Antonio,
7   Texas: Southwest Research Institute®. 2012.
8
9   NRC. NUREG–1910 Supplement 1, "Environmental Impact Statement for the Moore Ranch
10   ISR Project in Campbell County, Wyoming: Supplement to the Generic Environmental Impact
11   Statement for In-Situ Leach Uranium Milling Facilities." ML102290470. Washington, DC:
12   U.S. Nuclear Regulatory Commission. August 2010.
13
14   NRC. NUREG–1910, "Generic Environmental Impact Statement for In-Situ Leach Uranium
15   Milling Facilities." ML091480244, ML091480188. Washington, DC: U.S. Nuclear Regulatory
16   Commission. May 2009.
17
18   Powertech. "Dewey-Burdock Project Emissions Inventory Revisions." Email (July 31) from
19   R. Blubaugh to Bradley Werling (Southwest Research Institute®). ML12216A220. San Antonio,
20   Texas: Southwest Research Institute®. 2012.
21
22   Powertech. "Response to the U.S. Nuclear Regulatory Commission's (NRC) Request for
23   Additional Information for the Dewey-Burdock Uranium Project Environmental Report.
24   Submitted August 11, 2009." ML102380516. Greenwood Village, Colorado: Powertech
25   (USA) Inc. 2010a.
26
27   Powertech. Letter (November 4) from R. Blubaugh, Vice President-Environmental Health and
28   Safety Resources, Powertech, to R. Burrows, Project Manager, Office of Federal and Sate
29   Materials and Environmental Management Programs. "Subject: Powertech (USA), Inc.'s
30   Responses to the U.S. Nuclear Regulatory Commission (NRC) Staff's Verbal and Email
31   Requests for Clarification of Selected Issues Related to the Dewey-Burdock Uranium Project.
32   Environmental Review Docket No. 40-9075, TAC No. J 00533." ML110820582. Greenwood,
33   Colorado: Powertech (USA) Inc. 2010b.
34
35   Powertech. Letter (November 17) from R. Blubaugh, Vice President-Environmental Health and
36   Safety Resources, Powertech, to R. Burrows, Project Manager, Office of Federal and State
37   Materials and Environmental Management Programs. "Subject: Powertech (USA), Inc.'s
38   Responses to the U.S. Nuclear Regulatory Commission (NRC) Staff's Verbal Request for
39   Clarification of Response Regarding Inclusion of Emissions from Drilling Disposal Wells;
40   Dewey-Burdock Uranium Project. Environmental Review, Docket No. 40-9075, TAC
41   No. J 00533." ML103220208. Greenwood, Colorado: Powertech (USA) Inc. 2010c.
42
43   Powertech. "Dewey-Burdock Project, Application for NRC Uranium Recovery License Fall
44   River and Custer Counties, South Dakota—Environmental Report." ML092870160.
45   Greenwood Village, Colorado: Powertech (USA) Inc. 2009.
46
47   TRC Environmental Corporation. "Air Quality Technical Support Document, Atlantic Rim
48   Natural Gas Project and the Seminoe Road Gas Development Project, Wyoming." Laramie,
49   Wyoming: TRC Environmental Corporation. July 2006.

Table C–1. Nonradiological Combustion Emission Estimates Mass Flow Rates (Short Tons* Per Year) From Stationary Sources† for Various Phases of the Proposed Action

Activity	Emission Source	Quantity	Capacity	Operating Hours				Emission Factors						Emissions					
				hr/d	d/wk	mo/yr	hr/yr	PM$_{10}$ lb/hp-hr	SO$_x$ lb/hp-hr	NO$_x$ lb/hp-hr	CO lb/hp-hr	TOC lb/hp-hr	Aids‡ lb/hp-hr	PM$_{10}$ t/yr	SO$_x$ t/yr	NO$_x$ t/yr	CO t/yr	TOC t/yr	Aids t/yr
Central Processing Plant	Propane Heating	1	18 gal/hr	24	7	6	4,368	0.7	0.0002	13.0	7.5	1.000	NA	0.03	7.7E-06	0.5	0.29	0.04	NA
	Thermal Fluid Heater—propane	2	16 gal/hr	24	7	12	8,736	0.7	0.0002	13.0	7.5	1.000	NA	0.10	2.8E-05	1.9	1.07	0.14	NA
	Emergency Backup Generator—propane	1	12 gal/hr	0.25	1	12	13	0.7	0.0002	13.0	7.5	1.000	NA	5.6E-05	1.6E-08	1.0E-03	6.0E-04	8.0E-05	NA
	Fire Suppression System—diesel pump	1	100	0.25	1	12	13	0.002	0.002	0.031000	0.007	0.003	0.0005	0.001	0.001	0.02	0.004	0.002	0.00
Satellite Facility	Propane Heating	1	4 gal/hr	24	7	6	4,368	0.7	0.0002	13.0	7.5	1.000	NA	0.01	1.6E-06	0.1	0.06	0.01	NA
	Emergency Backup Generator—propane	1	6 gal/hr	0.25	1	12	13	0.7	0.0002	13.0	7.5	1.000	NA	2.8E-05	8.0E-09	5.2E-04	3.0E-04	4.0E-05	NA
	Fire Suppression System—diesel pump	1	100	0.25	1	12	13	0.002	0.002	0.031	0.007	0.003	0.0005	1.4E-03	1.3E-03	0.02	4.3E-03	1.6E-03	0.00
Office Building	Propane Heating	1	1 gal/hr	24	7	6	4,368	0.7	0.0002	13.0	7.5	1.000	NA	0.002	4.9E-07	0.03	0.02	0.002	NA
Maintenance and Warehouse Building	Propane Heating	1	3 gal/hr	24	7	6	4,368	0.7	0.0002	13.0	7.5	1.000	NA	0.004	1.2E-06	0.08	0.04	0.01	NA

Source: Modified from Powertech (2010a)
*Source and appendix mass expressed in short tons only (dual units used in SEIS text with metric being primary).
†All stationary source emissions are associated with the operation phase.
‡Aids = Aldehydes

Table C–2. Nonradiological Combustion Emission Mass Flow Rate Estimates (Short Tons* Per Year) From Mobile Sources for Various Phases of the Proposed Action

Pollutant	Emission Vehicle	Phase†						Source Classification
		Con - F+W	Con -F only	Con - W only	Operation	AR	Decom	
Particulate Matter PM₁₀‡	Drill Rig	2.65	0	2.65	0	0	0	C&DFE (§)
	Deep Well Drill Rig	0.03	0.02	0.01	0	0	0	C&DFE
	Other C&DFE (¶)	1.15	0.38	0.77	0.33	0	0.51	C&DFE
	Light Duty Vehicles	0.37	0.12	0.25	0.56	0.1	0.13	Other Mobile (#)
Sulfur Dioxide	Drill Rig	6.13	0	6.13	0	0	0	C&DFE
	Deep Well Drill Rig	0.08	0.06	0.02	0	0	0	C&DFE
	Other C&DFE	4.8	1.12	3.68	1.58	0	2.11	C&DFE
	Light Duty Vehicles	0.3	0.1	0.2	0.47	0.07	0.11	Other Mobile
Nitrogen Oxides	Drill Rig	45.22	0	45.22	0	0	0	C&DFE
	Deep Well Drill Rig	0.57	0.43	0.14	0	0	0	C&DFE
	Other C&DFE	20.55	6.22	14.33	6.17	0	10.72	C&DFE
	Light Duty Vehicles	5.55	1.78	3.77	8.84	1.26	1.96	Other Mobile
Carbon Monoxide	Drill Rig	56.03	0	56.03	0	0	0	C&DFE
	Deep Well Drill Rig	0.7	0.52	0.18	0	0	0	C&DFE
	Other C&DFE	13.88	3.39	10.49	4.43	0	6.04	C&DFE
	Light Duty Vehicles	3.52	1.13	2.39	5.59	0.8	1.24	Other Mobile
Total Hydrocarbon	Drill Rig	6.39	0	6.39	0	0	0	C&DFE
	Deep Well Drill Rig	0.08	0.06	0.02	0	0	0	C&DFE
	Other C&DFE	5.82	1.37	4.45	1.9	0	2.55	C&DFE
	Light Duty Vehicles	11.11	3.56	7.55	17.67	2.52	3.93	Other Mobile
Formaldehyde	Drill Rig	1.38	0	1.38	0	0	0	C&DFE
	Deep Well Drill Rig	0.2	0	0	0	0	0	C&DFE
	Other C&DFE	0.91	0.28	0.83	0.36	0	0.48	C&DFE
	Light Duty Vehicles	0.24	0.07	0.17	0.39	0.06	0.08	Other Mobile

Source: Modified from Powertech (2012)
*Source document and appendix table mass expressed in short tons only (dual units used in SEIS text with metric being primary).
†Phases = the following abbreviations: Con = construction, F = facilities, W = well fields, AR = aquifer restoration, and Decom = decommissioning
‡PM₂.₅ detailed emission inventory not available
§C&DFE = Construction and Drilling Field Equipment. Examples include drill rigs, bulldozers, graders, and scrappers.
¶Other C&DFE represents construction and drilling field equipment other than drill rigs
#Other mobile = additional mobile sources (i.e., light-duty trucks and vehicles) other than C&DFE and commuter traffic

Table C-3: Tailpipe Emission Factors and Equipment Hours

Basis For Equipment Tailpipe Emissions	Vehicle Parameters			Tailpipe Emission Factors (lb/hp-hr)								Total Equipment Hours per Year by Phase				
	Horse-power	Load Factor	Fuel	THC	NOx	CO	SO2	CO2	PM10	PM2.5	Formalde-hydes	Constr Facilities/WF†	Constr WF Only	Ops‡	Aquifer Rest§	Decom‖
Scraper	462	40%	Diesel	0.00247	0.00661	0.00573	0.00205	1.15000	0.00033	0.00032	0.00046	1299				2601
Bulldozer	410	40%	Diesel	0.00247	0.00661	0.00573	0.00205	1.15000	0.00033	0.00032	0.00046	433				867
Compactor	315	40%	Diesel	0.00247	0.00661	0.00573	0.00205	1.15000	0.00033	0.00032	0.00046	433				867
Motor Grader	297	40%	Diesel	0.00247	0.00661	0.00573	0.00205	1.15000	0.00033	0.00032	0.00046	1196	763	416		867
Water Truck (1,500 gal)	325	40%	Diesel	0.00247	0.00661	0.00573	0.00205	1.15000	0.00033	0.00032	0.00046	15600	13520	1040		867
Fueling Truck	325	40%	Diesel	0.00247	0.00661	0.00573	0.00205	1.15000	0.00033	0.00032	0.00046	130	520	520		520
Heavy Duty Diesel Truck	325	40%	Diesel	0.00247	0.00661	0.00573	0.00205	1.15000	0.00033	0.00032	0.00046	2080		520		1388
Logging Truck	325	25%	Diesel	0.00247	0.00661	0.00573	0.00205	1.15000	0.00033	0.00032	0.00046	8320	8320	2080		
Electrical Pole Truck	325	25%	Diesel	0.00247	0.00661	0.00573	0.00205	1.15000	0.00033	0.00032	0.00046	3466	3466			
Truck Mounted Drill Rig, Tier 1	300	59%	Diesel	0.00214	0.01512	0.01873	0.00205	1.15000	0.00089	0.00086	0.00046	33800	33800			
Deep Well Drill Rig, Tier 1	425	59%	Diesel	0.00214	0.01512	0.01873	0.00205	1.15000	0.00089	0.00086	0.00046	300	75			
Trackhoe	268	40%	Diesel	0.00247	0.00661	0.00573	0.00205	1.15000	0.00033	0.00032	0.00046	3120	3120			1300
Backhoe	93	40%	Diesel	0.00247	0.00661	0.00573	0.00205	1.15000	0.00033	0.00032	0.00046	5200	5200			1300
Loader	351	40%	Diesel	0.00247	0.00661	0.00573	0.00205	1.15000	0.00033	0.00032	0.00046					650
Tractor	530	40%	Diesel	0.00247	0.03100	0.00573	0.00205	1.15000	0.00033	0.00032	0.00046			160		650
Resin-hauling Semi-Truck	430	40%	Diesel	0.00247	0.00661	0.00573	0.00205	1.15000	0.00033	0.00032	0.00046			1040		
Pump Pulling Truck	325	40%	Diesel	0.00247	0.00661	0.00573	0.00205	1.15000	0.00033	0.00032	0.00046			6240		2130
Product Transport Truck	430	40%	Diesel	0.00247	0.00661	0.00573	0.00205	1.15000	0.00033	0.00032	0.00046			208		
Crane	516	25%	Diesel	0.00247	0.03100	0.00668	0.00205	1.15000	0.00220	0.00213	0.00046	694	7720			693
Forklift	100	25%	Diesel	0.00247	0.03100	0.00668	0.00205	1.15000	0.00220	0.00213	0.00046	9360		2132		2079
Manlift	50	25%	Diesel	0.00247	0.03100	0.00668	0.00205	1.15000	0.00220	0.00213	0.00046	4160		208		2772
Cementer	90	59%	LPG	0.00009	0.00118	0.00668	0.00000	1.13636	0.00006	0.00006	0.00000	8320	8320			2130
Welding Equipment	47	40%	LPG	0.00009	0.00118	0.00668	0.00000	1.13636	0.00006	0.00006	0.00000	9880				
HDPE Fusion Equipment	83	40%	LPG	0.00009	0.00118	0.00668	0.00000	1.13636	0.00006	0.00006	0.00000	6240				
Light Duty Pickup	265	25%	Gas¶	0.02200	0.01100	0.00696	0.00059	1.08000	0.00072	0.00070	0.00049	10000	6500	18736	2912	4000

Table C-3: Tailpipe Emission Factors and Equipment Hours (continued)

Basis For Equipment Tailpipe Emissions	Vehicle Parameters			Tailpipe Emission Factors (lb/hp-hr)							Total Equipment Hours per Year by Phase				
	Horse-power	Load Factor	Fuel	THC	NO$_x$	CO	SO$_2$	CO$_2$	PM$_{10}$	PM$_{2.5}$	Formalde-hydes	Constr[a] Facilities/ WF[†]	Constr WF Only	Ops[‡]	Aquifer Rest§	Decom‖
Light Duty Pickup Vehicle	150	25%	Gas	0.02200	0.01100	0.00696	0.00059	1.08000	0.00072	0.00070	0.00049	9263	6825	9720	975	2438

Emission Factor Sources:

1. EPA, AP-42 Table 3.3-1, Emission Factors for Uncontrolled Gasoline and Diesel Industrial Engines (THC, SO2, CO2, Aldehydes)
2. EPA, Exhaust and Crankcase Emission Factors for Non-Road Engine Modeling - Compression Ignition, April 2004 (PM2.5)
3. EPA, Control of Emissions of Air Pollution from Non-Road Diesel Engines; Final Rule, Subpart 89.112, October 1998 (all Tiers: NOx, CO, PM10; THC for Tier 1)
4. EPA, AP-42 Table 1.5-1, Emission Factors for LPG Combustion

*Construction
† Wellfield
‡Operations
§Restoration (aquifer)
‖Decomissioning
¶Gasoline

Table C-4. Effect of Using Updated Emissions Factors That Account for Pollution Controls for 300–600 Horsepower Engines

Pollutant	Tier 0 Emission Factor g/hp-hr*	Tier 1 Emission Factor g/hp-hr	Tier 1 Percent Emissions Reduced From Tier 0 Levels†	Tier 2 Emission Factor g/hp-hr	Tier 2 Percent Emissions Reduced From Tier 0 Levels‡	Tier 3 Emission Factor g/hp-hr	Tier 3 Percent Emissions Reduced From Tier 0 Levels§	Tier 4 Emission Factor g/hp-hr	Tier 4 Percent Emissions Reduced From Tier 0 Levels‖
Nitrogen Oxides	8.38	6.0153	28	4.3351	48	2.5	70	0.276	97
Carbon Monoxid¶	2.7	1.3060	52	0.8425	69	0.8425	69	0.084	94
Particulate Matter PM_{10}#	0.402	0.2008	50	0.1316	67	0.15	63	0.0092	98

Source: Modified from EPA (2004)

*Table only expressed emission factors in units of g/hp-hr. Dual units were not calculated because the value of interest is the percent emissions, which is unitless.

†Calculated using the following equation: [1-(Tier 1 emission factor/Tier 0 emission factor)]*100

‡Calculated using the following equation: [1-(Tier 2 emission factor/Tier 0 emission factor)]*100

§Calculated using the following equation: [1-(Tier 3 emission factor/Tier 0 emission factor)]*100

‖Calculated using the following equation: [1-(Tier 4 emission factor/Tier 0 emission factor)]*100

¶For carbon monoxide, the tier 2 and tier 3 emission standards are the same and the tier 2 and tier 3 emission factors used in the modeling are also the same values.

#For PM_{10}, the tier 2 and tier 3 emission standards are the same. However, the tier 2 emission factor which is based on actual certification data is actually lower than the tier 3 emission factor which is based on the emission standard.

Table C–5. Nonradiological Combustion Concentration Estimates From 47 Locations in and Around the Dewey-Burdock Site From Multiple Sources for the Construction Phase

Location*	SO_2 Maximum 24-Hour Mean ppm	SO_2 Annual Mean ppm	NO_x Annual Mean ppb	CO Maximum 8-Hour Mean ppm	CO Maximum 1-Hour Mean ppm	PM_{10} Maximum 24-Hour Mean µg/m³†	TOC Annual Mean µg/m³	Aldehydes Annual Mean ppm
CPP	0.00982	0.00011	0.02	0.039	0.309	24.4	1.36	0.021
SF-NE	0.0000682	0.0000041	0.0074	0.00035	0.002	0.2	0.05	0.00081
SF-E	0.0159	0.00017	0.031	0.06	0.483	39.5	2.15	0.034
SF-SE	0.0000446	0.0000017	0.003	0.00014	0.0009	0.11	0.02	0.00032
SF-S	0.00542	0.00006	0.11	0.021	0.172	13.4	0.74	0.012
SF-SW	0.0000655	0.0000034	0.0061	0.00034	0.002	0.2	0.04	0.00067
SF-W	0.0121	0.00013	0.24	0.049	0.392	30.1	1.65	0.026
SF-NW	0.011	0.00012	0.22	0.042	0.338	27.2	1.49	0.023
SF-N	0.0000738	0.000004	0.0072	0.00036	0.003	0.2	0.05	0.00079
CPP-N	0.000112	0.0000071	0.013	0.0005	0.004	0.3	0.09	0.0014
CPP-NE	0.000175	0.0000031	0.0055	0.00064	0.005	0.4	0.04	0.0006
CPP-E	0.0138	0.00015	0.27	0.051	0.411	34.3	1.88	0.03
CPP-SE	0.0000639	0.00000092	0.0016	0.00029	0.002	0.2	0.01	0.00018
CPP-S	0.000233	0.0000042	0.0076	0.00086	0.007	0.6	0.05	0.00083
CPP-SW	0.000189	0.0000035	0.0063	0.0007	0.006	0.5	0.04	0.00068
CPP-W	0.00778	0.000086	0.15	0.031	0.25	19.3	1.06	0.017
CPP-NW	0.000047	0.0000026	0.0047	0.00015	0.001	0.12	0.03	0.00051
B.C. Ranch	0.00908	0.0001	0.18	0.035	0.28	22.5	1.24	0.019
Burdock School	0.000149	0.0000025	0.0045	0.00055	0.004	0.4	0.03	0.00049
Daniels Ranch	0.0119	0.00018	0.33	0.045	0.359	29.5	2.25	0.036
LA-2	0.0000481	0.0000016	0.0029	0.0002	0.0013	0.12	0.02	0.00032
SF	0.0132	0.00014	0.26	0.052	0.42	32.8	1.79	0.028
Heck Ranch	0.000127	0.000003	0.0053	0.00047	0.004	0.3	0.04	0.00058
Mining Unit 5	0.00345	0.000049	0.089	0.011	0.089	8.6	0.61	0.0097
SF-SSW	0.0000473	0.0000018	0.0032	0.00017	0.0012	0.12	0.02	0.00035
SF-WSW	0.00498	0.000055	0.099	0.016	0.126	12.3	0.68	0.011
SF-WNW	0.0118	0.00013	0.23	0.046	0.37	29.2	1.6	0.025
SF-NNW	0.00386	0.000043	0.078	0.013	0.1	9.6	0.54	0.0085

Table C–5. Nonradiological Combustion Concentration Estimates From 47 Locations in and Around the Dewey-Burdock Site From Multiple Sources for the Construction Phase (continued)

	SO$_2$		NO$_x$	CO		PM$_{10}$	TOC	Aldehydes
	Maximum 24-Hour Mean	Annual Mean	Annual Mean	Maximum 8-Hour Mean	Maximum 1-Hour Mean	Maximum 24-Hour Mean	Annual Mean	Annual Mean
Location*	ppm	ppm	ppb	ppm	ppm	µg/m^3†	µg/m^3	ppm
CPP-NNW	0.000334	0.0000091	0.016	0.0014	0.011	0.8	0.11	0.0018
CCP-NNE	0.000103	0.0000061	0.011	0.0003	0.0014	0.3	0.08	0.0012
CPP-ENE	0.0239	0.000062	1.1	0.074	0.579	59.3	7.7	0.12
CPP-ESE	0.0000938	0.0000021	0.0037	0.00035	0.002	0.2	0.03	0.00041
CPP-SSE	0.0000743	0.0000012	0.0021	0.00027	0.002	0.2	0.01	0.00023
CPP-SSW	0.0181	0.00045	0.8	0.065	0.455	45	5.55	0.088
CCP-WSW	0.00439	0.000049	0.087	0.014	0.115	10.9	0.6	0.0095
CPP-WNW	0.0000549	0.000003	0.0054	0.00028	0.002	0.1	0.04	0.00059
Puttman Ranch	0.0000347	0.00000041	0.00074	0.00013	0.00076	0.09	0.01	0.00008
Background	0.0000256	0.00000048	0.00086	0.000073	0.00043	0.06	0.01	0.000094
Englebert Ranch	0.0000527	0.00000044	0.00078	0.00026	0.0015	0.13	0.01	0.000086
LA-1	0.0118	0.00013	0.23	0.047	0.372	29.2	1.6	0.025
Edgemont	0.00017	0.0000047	0.0084	0.00084	0.007	0.4	0.06	0.00091
Spencer Ranch	0.0000379	0.0000017	0.003	0.00013	0.00092	0.09	0.02	0.00033
Mining Unit 2	0.0142	0.00016	0.28	0.055	0.44	35.3	1.92	0.03

Source: Modified from Powertech (2010a)
*Locations are specified in Figure 2.1-13
†To convert µg/m^3 to oz/yd^3, multiply by 2.74 × 10^{-8}

Table C–6. Nonradiological Combustion Concentration Estimates From 47 Locations in and Around the Dewey-Burdock Site From Multiple Sources for the Operations Phase

	SO$_2$		NO$_x$	CO		PM$_{10}$	TOC	Aldehydes
	Maximum 24-Hour Mean	Annual Mean	Annual Mean	Maximum 8-Hour Mean	Maximum 1-Hour Mean	Maximum 24-Hour Mean	Annual Mean	Annual Mean
Location*	ppm	ppm	ppb	ppm	ppm	µg/m^3†	µg/m^3	ppm
CPP	0.00315	0.000087	0.37	0.031	0.074	7.8	2.14	0.028
SF-NE	0.0000862	0.0000061	0.025	0.00079	0.005	0.2	0.14	0.0019
SF-E	0.00632	0.00017	0.63	0.056	0.207	16	3.71	0.051
SF-SE	0.0000334	0.0000025	0.0093	0.00023	0.0017	0.09	0.06	0.00076
SF-S	0.00174	0.000047	0.2	0.017	0.042	4.3	1.14	0.015
SF-SW	0.0045900	0.000048	0.076	0.019	0.15	12.5	0.58	0.011
SF-W	0.0047	0.00012	0.46	0.043	0.142	11.8	2.71	0.037
SF-NW	0.00364	0.0001	0.41	0.035	0.087	9.1	2.38	0.032
SF-N	0.0000841	0.0000059	0.024	0.00077	0.005	0.2	0.14	0.0019
CPP-N	0.00925	0.000096	0.15	0.033	0.265	25.1	1.16	0.021
CPP-NE	0.0000931	0.0000052	0.021	0.00085	0.005	0.2	0.12	0.0016
CPP-E	0.00446	0.00021	0.64	0.041	0.111	12	3.99	0.059
CPP-SE	0.0000749	0.0000025	0.006	0.00034	0.002	0.2	0.04	0.00063
CPP-S	0.00592	0.00013	0.19	0.022	0.173	16.1	1.48	0.028
CPP-SW	0.000125	0.0000062	0.021	0.00075	0.006	0.3	0.13	0.0018
CPP-W	0.00253	0.000066	0.28	0.025	0.061	6.3	1.64	0.022
CPP-NW	0.00574	0.000058	0.085	0.023	0.183	15.6	0.66	0.012
B.C. Ranch	0.00301	0.000083	0.34	0.029	0.071	7.5	1.98	0.026
Burdock School	0.00022	0.0000058	0.017	0.00097	0.008	0.6	0.11	0.0016
Daniels Ranch	0.0171	0.00048	1.1	0.059	0.431	46.3	7.28	0.12
LA-2	0.0000743	0.0000028	0.011	0.00068	0.0036	0.18	0.06	0.00085
SF	0.00514	0.00013	0.51	0.047	0.158	13	2.99	0.041
Heck Ranch	0.0000553	0.0000036	0.015	0.00051	0.004	0.1	0.09	0.0012
Mining Unit 5	0.0224	0.00049	0.77	0.079	0.572	60.8	5.81	0.11

Table C–6. Nonradiological Combustion Concentration Estimates From 47 Locations in and Around the Dewey-Burdock Site From Multiple Sources for the Operations Phase (continued)

	SO$_2$		NO$_x$	CO		PM$_{10}$	TOC	Aldehydes
	Maximum 24-Hour Mean	Annual Mean	Annual Mean	Maximum 8- Hour Mean	Maximum 1-Hour Mean	Maximum 24-Hour Mean	Annual Mean	Annual Mean
Location*	ppm	ppm	ppb	ppm	ppm	µg/m^3†	µg/m^3	ppm
SF-NNE	0.0000851	0.000006	0.024	0.00078	0.005	0.2	0.14	0.0019
SF-ENE	0.000257	0.000019	0.079	0.0021	0.01	0.6	0.46	0.0061
SF-ESE	0.00654	0.000066	0.1	0.026	0.205	17.79	0.77	0.014
SF-SSE	0.00139	0.000041	0.17	0.013	0.031	3.4	1.01	0.013
SF-SSW	0.0000279	0.0000027	0.0098	0.00026	0.0019	0.07	0.06	0.00081
SF-WSW	0.00609	0.000096	0.26	0.029	0.197	16.4	1.66	0.025
SF-WNW	0.00406	0.00011	0.44	0.038	0.101	10.1	2.56	0.034
SF-NNW	0.000979	0.000039	0.17	0.0091	0.025	2.4	0.96	0.013
CPP-NNW	0.00829	0.000089	0.15	0.03	0.24	22.5	1.11	0.02
CCP-NNE	0.0106	0.00011	0.16	0.037	0.3	28.8	1.25	0.023
CPP-ENE	0.00731	0.00056	2.3	0.04	0.187	18.5	13.3	0.18
CPP-ESE	0.00661	0.000066	0.092	0.029	0.231	18	0.73	0.014
CPP-SSE	0.000255	0.0000064	0.013	0.0012	0.008	0.7	0.09	0.0015
CPP-SSW	0.00543	0.00038	1.6	0.031	0.14	13.7	9.04	0.12
CCP-WSW	0.00123	0.000042	0.18	0.012	0.027	3.1	1.02	0.013
CPP-WNW	0.000027	0.0000044	0.018	0.00056	0.003	0.2	0.11	0.0014
Puttman Ranch	0.000027	0.00000035	0.0011	0.00015	0.00078	0.07	0.01	0.000097
Background	0.0000194	0.00000083	0.0029	0.0002	0.0012	0.05	0.02	0.00024
Englebert Ranch	0.0000542	0.00000073	0.0024	0.0003	0.0015	0.14	0.01	0.00021
LA-1	0.00509	0.00013	0.46	0.044	0.18	13.4	2.77	0.038
Edgemont	0.000102	0.0000067	0.024	0.00098	0.005	0.3	0.15	0.002
Spencer Ranch	0.0000332	0.0000026	0.0097	0.00025	0.0019	0.09	0.06	0.00078
Mining Unit 2	0.00576	0.00015	0.56	0.051	0.193	14.6	3.31	0.046

Source: Modified from Powertech (2010a)
*Locations are specified in Figure 2.1-13
†To convert µg/m^3 to oz/yd^3, multiply by 2.74 ×10^{-8}

Table C–7. Nonradiological Combustion Concentration Estimates From 47 Locations in and Around the Dewey-Burdock Site From Multiple Sources for the Aquifer Restoration Phase

	SO$_2$		NO$_x$	CO		PM$_{10}$	TOC	Aldehydes
	Maximum 24-Hour Mean	Annual Mean	Annual Mean	Maximum 8- Hour Mean	Maximum 1-Hour Mean	Maximum 24-Hour Mean	Annual Mean	Annual Mean
Location*	ppm	ppm	ppb	ppm	ppm	µg/m^3†	µg/m^3	ppm
CPP	0.0022	0.000058	0.18	0.0088	0.021	1.562	1.232	0.0074
SF-NE	0.00000469	0.000000052	0.00016	0.000018	0	0.003	0.001	0.0000055
SF-E	0.00000621	0.000000085	0.00027	0.000024	0	0.004	0.002	0.000011
SF-SE	0.0000178	0.0000011	0.0034	0.000067	0.0005	0.013	0.024	0.00014
SF-S	0.00122	0.000032	0.098	0.0048	0.012	0.862	0.677	0.0041
SF-SW	0.00000839	0.00000019	0.00059	0.000027	0	0.006	0.004	0.000025
SF-W	0.00000756	0.000000085	0.00027	0.000024	0	0.006	0.002	0.000011
SF-NW	0.00000617	0.000000077	0.00023	0.000019	0	0.004	0.002	0.000011
SF-N	0.00000472	0.000000052	0.00016	0.000018	0	0.003	0.001	0.0000055
CPP-N	0.00000874	0.00000012	0.00039	0.000034	0	0.006	0.003	0.000017
CPP-NE	0.000019	0.000001	0.0032	0.000061	0	0.014	0.022	0.00013
CPP-E	0.003	0.000083	0.26	0.012	0.27	2.126	1.771	0.011
CPP-SE	0.0000114	0.00000032	0.00097	0.000037	0	0.008	0.007	0.000042
CPP-S	0.00000782	0.00000015	0.00047	0.000025	0	0.006	0.003	0.000019
CPP-SW	0.00000522	0.00000039	0.0012	0.000016	0	0.004	0.008	0.00005
CPP-W	0.00177	0.000046	0.14	0.0071	0.17	1.254	0.972	0.0058

Table C–7. Nonradiological Combustion Concentration Estimates From 47 Locations in and Around the Dewey-Burdock Site From Multiple Sources for the Aquifer Restoration Phase (continued)

Location*	SO₂		NOₓ	CO		PM₁₀	TOC	Aldehydes
	Maximum 24-Hour Mean	Annual Mean	Annual Mean	Maximum 8-Hour Mean	Maximum 1-Hour Mean	Maximum 24-Hour Mean	Annual Mean	Annual Mean
	ppm	ppm	ppb	ppm	ppm	µg/m³†	µg/m³	ppm
CPP-NW	0.0000124	0.00000019	0.00057	0.000039	0	0.009	0.004	0.000025
B.C. Ranch	0.00000677	0.000000077	0.00023	0.000021	0	0.005	0.002	0.0000083
Burdock School	0.00000491	0.00000044	0.0013	0.000016	0	0.003	0.009	0.000055
Daniels Ranch	0.00261	0.000072	0.22	0.01	0.024	1.848	1.522	0.0091
LA-2	0.0000147	0.00000099	0.003	0.000051	0.00039	0.01	0.021	0.00012
SF	0.00000726	0.000000085	0.00027	0.000023	0	0.005	0.002	0.000011
Heck Ranch	0.0000387	0.0000022	0.0068	0.00014	0.001	0.027	0.047	0.00028
Mining Unit 5	0.000613	0.000025	0.075	0.0023	0.006	0.435	0.521	0.0031
SF-NNE	0.00000474	0.000000052	0.00016	0.000018	0	0.003	0.001	0.0000055
SF-ENE	0.00000519	0.00000006	0.00019	0.00002	0	0.004	0.001	0.0000083
SF-ESE	0.0000117	0.00000017	0.00051	0.000037	0	0.008	0.004	0.000022
SF-SSE	0.000973	0.000028	0.087	0.0038	0.009	0.69	0.599	0.0036
SF-SSW	0.000019	0.0000011	0.0035	0.000071	0.00051	0.013	0.024	0.00015
SF-WSW	0.00000883	0.00000011	0.00035	0.000028	0	0.006	0.002	0.000014
SF-WNW	0.00000659	0.000000081	0.00024	0.000021	0	0.005	0.002	0.000011
SF-NNW	0.00000485	0.000000064	0.0002	0.000019	0	0.003	0.001	0.0000083
CPP-NNW	0.00000859	0.00000013	0.00039	0.000029	0	0.006	0.003	0.000017
CCP-NNE	0.00000916	0.00000012	0.00038	0.000035	0.00019	0.006	0.003	0.000017
CPP-ENE	0.0000276	0.0000012	0.0036	0.000088	0.001	0.02	0.025	0.00015
CPP-ESE	0.0000176	0.0000014	0.0043	0.000066	0	0.013	0.029	0.00018
CPP-SSE	0.00000943	0.00000021	0.00065	0.00003	0	0.007	0.004	0.000028
CPP-SSW	0.00000754	0.00000016	0.00048	0.000023	0	0.005	0.003	0.000019
CCP-WSW	0.000861	0.000028	0.085	0.0033	0.008	0.61	0.587	0.0035
CPP-WNW	0.0000424	0.0000026	0.008	0.00016	0.001	0.03	0.056	0.00033
Puttman Ranch	0.00000189	0.000000028	0.00009	0.0000071	0.000026	0.001	0.001	0.0000028
Background	0.00000433	0.00000015	0.00045	0.000013	0.000093	0.003	0.003	0.000019
Englebert Ranch	0.0000163	0.00000017	0.00052	0.000061	0.00035	0.012	0.004	0.000022
LA-1	0.00000793	0.000000093	0.00028	0.000025	0	0.006	0.002	0.000011
Edgemont	0.0000604	0.0000035	0.011	0.00028	0.001	0.043	0.074	0.00044
Spencer Ranch	0.000019	0.0000012	0.0037	0.000071	0.00053	0.013	0.025	0.00015
Mining Unit 2	0.00000676	0.000000089	0.00027	0.000021	0	0.005	0.002	0.000011

Source: Modified from Powertech (2010a)
*Locations are specified in Figure 2.1-13.
†To convert µg/m³ to oz/yd³, multiply by 2.74 × 10⁻⁸.

Table C–8. Nonradiological Combustion Concentration Estimates From 47 Locations In and Around the Dewey-Burdock Site From Multiple Sources for the Decommissioning Phase

Location*	SO₂		NOₓ	CO		PM₁₀	TOC	Aldehydes
	Maximum 24-Hour Mean	Annual Mean	Annual Mean	Maximum 8-Hour Mean	Maximum 1-Hour Mean	Maximum 24-Hour Mean	Annual Mean	Annual Mean
	ppm	ppm	ppb	ppm	ppm	µg/m³†	µg/m³	ppm
CPP	0.00614	0.000058	0.18	0.012	0.093	9.14	1.232	0.0074
SF-NE	0.0000426	0.000000052	0.00016	0.00011	0.001	0.06	0.001	0.0000055
SF-E	0.00995	0.000000085	0.00027	0.018	0.145	14.81	0.002	0.000011
SF-SE	0.0000279	0.0000011	0.0034	0.000042	0.00027	0.04	0.024	0.00014
SF-S	0.00339	0.000032	0.098	0.0064	0.052	5.04	0.677	0.0041
SF-SW	0.0000409	0.00000019	0.00059	0.0001	0.001	0.06	0.004	0.000025
SF-W	0.00757	0.000000085	0.00027	0.015	0.117	11.27	0.002	0.000011
SF-NW	0.00684	0.000000077	0.00023	0.013	0.101	10.19	0.002	0.000011
SF-N	0.0000461	0.000000052	0.00016	0.00011	0.001	0.07	0.001	0.0000055
CPP-N	0.0000702	0.00000012	0.00039	0.00015	0.001	0.1	0.003	0.000017
CPP-NE	0.000109	0.000001	0.0032	0.00019	0.002	0.16	0.022	0.00013

Table C–8. Nonradiological Combustion Concentration Estimates From 47 Locations In and Around the Dewey-Burdock Site From Multiple Sources for the Decommissioning Phase (continued)

Location*	SO₂ Maximum 24-Hour Mean	SO₂ Annual Mean	NOₓ Annual Mean	CO Maximum 8-Hour Mean	CO Maximum 1-Hour Mean	PM₁₀ Maximum 24-Hour Mean	TOC Annual Mean	Aldehydes Annual Mean
	ppm	ppm	ppb	ppm	ppm	µg/m³†	µg/m³	ppm
CPP-E	0.00865	0.000083	0.26	0.015	0.123	12.88	1.771	0.011
CPP-SE	0.00004	0.00000032	0.00097	0.000088	0.001	0.06	0.007	0.000042
CPP-S	0.000146	0.00000015	0.00047	0.00026	0.002	0.22	0.003	0.000019
CPP-SW	0.000118	0.00000039	0.0012	0.00021	0.002	0.16	0.008	0.00005
CPP-W	0.00487	0.000046	0.14	0.0094	0.075	7.24	0.972	0.0058
CPP-NW	0.0000298	0.00000019	0.00057	0.000046	0	0.04	0.004	0.000025
B.C. Ranch	0.00568	0.000000077	0.00023	0.01	0.084	8.45	0.002	0.0000083
Burdock School	0.0000933	0.00000044	0.0013	0.00017	0.001	0.14	0.009	0.000055
Daniels Ranch	0.00743	0.000072	0.22	0.013	0.108	11.06	1.522	0.0091
LA-2	0.00003	0.00000099	0.003	0.000059	0.00038	0.04	0.021	0.00012
SF	0.00826	0.000000085	0.00027	0.016	0.126	12.3	0.002	0.000011
Heck Ranch	0.0000797	0.0000022	0.0068	0.00014	0.001	0.12	0.047	0.00028
Mining Unit 5	0.00215	0.000025	0.075	0.0034	0.027	3.21	0.521	0.0031
SF-NNE	0.0000446	0.000000052	0.00016	0.00011	0.001	0.07	0.001	0.0000055
SF-ENE	0.000726	0.00000006	0.00019	0.0013	0.01	1.08	0.001	0.0000083
SF-ESE	0.0000324	0.00000017	0.00051	0.00008	0.001	0.05	0.004	0.000022
SF-SSE	0.00293	0.000028	0.087	0.005	0.04	4.36	0.599	0.0036
SF-SSW	0.0000296	0.0000011	0.0035	0.00005	0.00035	0.04	0.024	0.00015
SF-WSW	0.00311	0.00000011	0.00035	0.0047	0.038	4.63	0.002	0.000014
SF-WNW	0.00735	0.000000081	0.00024	0.014	0.111	10.94	0.002	0.000011
SF-NNW	0.00241	0.000000064	0.0002	0.0038	0.03	3.59	0.001	0.0000083
CPP-NNW	0.000209	0.00000013	0.00039	0.00042	0.003	0.31	0.003	0.000017
CCP-NNE	0.0000645	0.00000012	0.00038	0.00009	0.00043	0.1	0.003	0.000017
CPP-ENE	0.0149	0.0000012	0.0036	0.022	0.174	22.25	0.025	0.00015
CPP-ESE	0.0000586	0.0000014	0.0043	0.0001	0.001	0.09	0.029	0.00018
CPP-SSE	0.0000464	0.00000021	0.00065	0.000082	0.001	0.07	0.004	0.000028
CPP-SSW	0.0113	0.00000016	0.00048	0.019	0.136	16.86	0.003	0.000019
CCP-WSW	0.00274	0.000028	0.085	0.0043	0.034	4.08	0.587	0.0035
CPP-WNW	0.0000343	0.0000026	0.008	0.000085	0.001	0.05	0.056	0.00033
Puttman Ranch	0.0000217	0.000000028	0.00009	0.000039	0.00023	0.03	0.001	0.0000028
Background	0.000016	0.00000015	0.00045	0.000022	0.00013	0.02	0.003	0.000019
Englebert Ranch	0.000033	0.00000017	0.00052	0.000077	0.00046	0.05	0.004	0.000022
LA-1	0.00735	0.000000093	0.00028	0.014	0.112	10.93	0.002	0.000011
Edgemont	0.000106	0.0000035	0.011	0.00025	0.002	0.16	0.074	0.00044
Spencer Ranch	0.0000237	0.0000012	0.0037	0.000039	0.00027	0.04	0.025	0.00015
Mining Unit 2	0.00889	0.000000089	0.00027	0.016	0.132	13.23	0.002	0.000011

Source: Modified from Powertech (2010a)
*Locations are specified in Figure 2.1-13
†To convert µg/m³ to oz/yd³, multiply by 2.74×10^{-8}

Table C–9. Quantitative Difference Between the Revised and Initial Nonradiological Combustion Emission Mass Flow Rate Estimates (Short Tons* per Year) Including Both Stationary and Mobile Sources.

Pollutant	Revised Inventory	Initial Inventory	Multiplication Factor (Revised/Initial)	Percent (Revise/Initial) ×100
Particulate Matter PM₁₀†	5.4	39	0.138	13.8
Sulfur Dioxide	14.4	36	0.40	40
Nitrogen Oxides	95.0	77	1.234	123.4
Carbon Monoxide	88.7	143	0.62	62

Source: Revised inventory modified from Powertech (2012) and initial inventory modified from Powertech (2010a)
*Source document and appendix table mass expressed in short tons only (dual units used in SEIS text with metric being primary)
†PM₂.₅ detailed emission inventory not available

Table C–10. Calculation for the NAAQS Pollutant Peak Year Concentrations for the Revised Inventory Utilizing the Air Dispersion Modeling Results from the Initial Inventory

Pollutant	Time	Concentration from Dispersion Modeling of Initial Inventory*		Multiplication Factor‡	Calculated Concentration for Revised Inventory	
		Expressed in µg/m³†	Expressed in ppb		Expressed in µg/m³	Expressed in ppb
Particulate Matter PM$_{10}$§	24-hour mean	59.3	na	0.138	8.2	na
Sulfur Dioxide	24-hour mean	59.3	23.9	0.40	23.7	9.6
Sulfur Dioxide	Annual mean	1.5	0.62	0.40	0.6	0.25
Nitrogen Oxides	Annual mean	3.8	1.1	1.23	4.7	1.3
Carbon Monoxide	8-hour mean	418.8	74	0.62	260.0	45.9
Carbon Monoxide	1-hour mean	3,286	579	0.62	2037	359

Source: Revised inventory modified from Powertech (2012) and initial inventory modified from Powertech (2010a).
*Most concentrations in this table are expressed in two different units to accommodate for comparisons to the NAAQS and Prevention of Significant Deterioration regulations which can differ in units used to express the concentrations
‡Multiplication factor = the value when multiplied by the concentration from the dispersion modeling of the initial inventory yields the peak year concentrations of the revised inventory. The multiplication factor comes from the percent difference calculated in Table C–9
†To convert µg/m³ to oz/yd³, multiply by 2.74 x 10^{-8}
§PM$_{2.5}$ emission inventory not available for inclusion in air dispersion modeling
¶Not applicable. See footnote *

Table C–11. Percentage of Emissions by Phase for Various NAAQS Pollutants from Combustion Emissions From Stationary and Mobile Sources When All Phases Occur Simultaneously (i.e., a Peak Year*)

Pollutant	Phase			
	Construction (Well Field Only)	Operation	Aquifer Restoration	Decommissioning
Particulate Matter PM$_{10}$	69.0	19.0	1.8	10.1
Sulfur Dioxide	70.2	13.8	0.7	15.3
Nitrogen Oxides	66.7	18.7	1.3	13.3
Carbon Monoxide	77.9	13.0	0.9	8.2

Source: Modified from Powertech (2012)
*Peak year accounts for when all four phase occur simultaneously and represents the highest amount of emission the proposed action would generate in any one project year.

Table C–12. Nonradiological Combustion Emission Estimates (Mass* Per Year) for Greenhouse Gases for the Construction Phase of the Proposed Action

Activity	Emission Vehicle	Pollutant			Source Classification
		CO$_2$	CH$_4$	N$_2$O	
Earthworks Construction	Scraper	345	0.01	0.01	Mobile
	Bulldozer	102	0.01	0.01	Mobile
	Compactor	0	0	0	Mobile
	Motor Grader	74	0.01	0.01	Mobile
	Heavy-Duty Water Truck	389	0.02	0.03	Mobile
	Fueling Truck	24	0	0	Mobile
	Light-Duty Pickup	74	0.05	0.06	Mobile
Facilities Construction	Crane	206	0.01	0.01	Mobile
	Welding Equipment	225	0.02	0.03	Mobile
	Forklift	119	0.02	0.03	Mobile
	Man Lift	120	0.02	0.03	Mobile
	Heavy-Duty Diesel Truck	389	0.02	0.03	Mobile
	Light Duty Truck	744	0.59	0.19	Mobile

Table C–12. Nonradiological Combustion Emission Estimates (Mass* Per Year) for Greenhouse Gases for the Construction Phase of the Proposed Action (continued)

Activity	Emission Vehicle	Pollutant			Source Classification
		CO_2	CH_4	N_2O	
Wellfield/Electric Construction	High-Density Polyethylene (HDPE) Fusion Equipment	298	0.07	0.09	Mobile
	Trackhoe	481	0.07	0.09	Mobile
	Backhoe	167	0.07	0.09	Mobile
	Welding Equipment	42	0.03	0.04	Mobile
	Electrical Pole Truck	648	0.04	0.05	Mobile
	Motor Grader	130	0.02	0.02	Mobile
	Forklift	179	0.07	0.09	Mobile
	Light-Duty Truck	2,679	3.53	1.15	Mobile
Drilling	Truck-Mounted Rotary Drill Rig	10,689	0.1	0.07	Mobile
	Heavy-Duty Water Truck	2,527	1.2	0.38	Mobile
	Backhoe	111	0.05	0.06	Mobile
	Forklift	239	0.05	0.06	Mobile
	Cementer (gas)	404	0.05	0.06	Mobile
	Logging Truck	1,555	0.05	0.06	Mobile
	Light-Duty Truck	1,116	0.59	0.19	Mobile

Source: Modified from Powertech (2010a)
*Source and appendix mass expressed in short tons only (dual units used in SEIS text with metric being primary)

Table C–13. Nonradiological Combustion Emission Estimates (Mass* Per Year) for Greenhouse Gases for the Operations Phase of the Proposed Action

Activity	Emission Source	Pollutant			Source Classification
		CO_2	CH_4	N_2O	
Central Processing Plant	Propane Heating	483	0.04	na†	Stationary
	Thermal Fluid Heater	1,780	0.14	na	Stationary
	Emergency Backup Generator	1	0.00008	na	Stationary
	Fire Suppression System	0.7	0	0	Stationary
Satellite Facility	Propane Heating	102	0.01	na	Stationary
	Emergency Backup Generator	0.5	0.00004	na	Stationary
	Fire Suppression System	0.7	0.0003	0	Stationary
Office Building	Propane Heating	31	0.013	na	Stationary
Maintenance + Warehouse Building	Propane Heating	72	0.01	na	Stationary
Central Processing Plant Operations	Man Lift	6	0	0.01	Mobile
	Welding Equipment	17	0.01	0.02	Mobile
	Forklift (Warehouse)	56	0	0.01	Mobile
	Forklift (Packaging)	24	0	0.01	Mobile
	Light-Duty Truck	1,786	1.8	0.58	Mobile
	Light-Duty Vehicles	708	2.5	0.81	Mobile
Satellite Facility and Warehouse Facility Operations	Resin Hauling Semi Truck	257	0	0.03	Mobile
	Pump-Pulling Truck	1,166	0	0.04	Mobile
	Motor Grader	71	0.01	0.01	Mobile
	Logging Truck	389	0.05	0.06	Mobile
	Light-Duty Truck	2,500	9.9	3.23	Mobile
	Light-Duty Vehicles	253	1.8	0.58	Mobile
Product Transport	Diesel Semi with Trailer	51	0	0.01	Mobile

Source: Modified from Powertech (2010a)
*Source and appendix mass expressed in short tons only (dual units used in SEIS text with metric being primary).
†Not applicable.

Table C–14. Nonradiological Combustion Emission Estimates (Mass* Per Year) for Greenhouse Gases for the Aquifer Restoration Phase of the Proposed Action

Activity	Emission Vehicle	Pollutant			Source Classification
		CO_2	CH_4	N_2O	
Restoration Operations	Cementer (gas)	17	0.39	0.13	Mobile
	Light-Duty Truck	446	1.8	0.58	Mobile
	Light-Duty Vehicle	126	1.8	0.58	Mobile

Source: Modified from Powertech (2010a)
*Source and appendix mass expressed in short tons only (dual units used in SEIS text with metric being primary).

Table C–15. Nonradiological Combustion Emission Estimates (Mass* Per Year) for Greenhouse Gases for the Decommissioning Phase of the Proposed Action

Activity	Emission Vehicle	Pollutant			Source Classification
		CO_2	CH_4	N_2O	
Earthwork	Scraper	691	0.02	0.02	Mobile
	Motor Grader	148	0.02	0.02	Mobile
	Compactor	148	0.02	0.02	Mobile
	Bulldozer	204	0.02	0.02	Mobile
	Excavator	200	0.01	0.02	Mobile
	Backhoe	70	0.01	0.02	Mobile
	Loader	131	0.01	0.02	Mobile
	Tractor	198	0.01	0.02	Mobile
	Fueling Truck	97	0.01	0.01	Mobile
	Light-Duty Truck	186	0.73	0.24	Mobile
Demolition	Crane	206	0.02	0.02	Mobile
	Welding/Cutting Equipment	75	0.02	0.02	Mobile
	Man Lift	80	0.02	0.02	Mobile
	Forklift	119	0.02	0.02	Mobile
	Heavy-Duty Truck (Diesel)	259	0.01	0.01	Mobile
	Light-Duty Truck	496	0.78	0.26	Mobile
	Light-Duty Vehicle	421	1.18	0.38	Mobile

Source: Modified from Powertech (2010a)
*Source and appendix mass expressed in short tons only (dual units used in SEIS text with metric being primary)

Table C–16. Facilities and Initial Well Field Construction Phase—Year 1 Only

Equipment Item	Quantity	Hours	Speed (mph)	Weight (tons)	Lb/VMT	VMT	Lb/Hr	Control Efficiency	PM^{10} Tons/yr	Emission Factor Reference§
Scraper	3	433	15	30	3.10	19,485		50%	15.10	
Bulldozer	1	433					0.70	0%	0.15	AP-42 Section 13.2.2
Compactor	1	433	5	5	1.38	2,165		50%	0.75	AP-42 Section 13.2.2
Motor Grader	1	1,196	10		3.06	11,960		50%	9.15	AP-42 Table 11.9-1

Table C–16. Facilities and Initial Well Field Construction Phase—Year 1 Only (continued)

Equipment Item	Quantity	Hours	Speed (mph)	Weight (tons)	Lb/VMT	VMT	Lb/Hr	Control Efficiency	PM^{10} Tons/yr	Emission Factor Reference§
Water Truck (1,500 gallon)	15	1,040	15	16	2.34	234,000		50%	136.65	AP-42 Section 13.2.2
Fueling Truck	1	130	15	10	1.89	1,950		50%	0.92	AP-42 Section 13.2.2
Heavy Duty Diesel Truck	2	260	15	20	2.58	7,800		50%	5.04	AP-42 Section 13.2.2
Logging Truck	4	2,080	15	10	1.89	124,800		50%	58.99	AP-42 Section 13.2.2
Electrical Pole Truck	2	1,733	15	10	1.89	51,990		50%	24.57	AP-42 Section 13.2.2
Truck Mounted Drill Rig*	13	2,600	20				0.07	0%	1.10	AP-42 Table 11.9-4
Deep Well Drill Rig*	1	300	75				0.02	0%	0.00	AP-42 Table 11.9-4
Trackhoe†	1	3,120					1.66	0%	2.59	AP-42 Table 11.9-4
Backhoe†	1	5,200					1.33	0%	3.46	AP-42 Table 11.9-4
Forklift	4	2,340	5	1	0.67	46,800		50%	7.85	AP-42 Section 13.2.2
Manlift	4	1,1040	2	10	1.89	8,320		50%	3.93	AP-42 Section 13.2.2
Light Duty Pickup (onsite use)	5	2,000	15	3	1.10	150,000		50%	41.24	AP-42 Section 13.2.2
Onsite Passenger Vehicle‡	57	4	25	250	0.63	57,000		50%	9.03	AP-42 Section 13.2.2
Total Onsite PM$_{10}$ Emissions (tons/year)									**320.53**	
Offsite Passenger Vehicle‡	57	22	40	250	0.80	313,500		0%	125.70	AP-42 Section 13.2.2
Heavy Duty Diesel Truck	2	780	25	20	2.58	39,000		0%	50.36	AP-42 Section 13.2.2
Total Offsite PM$_{10}$ Emissions (tons/year)									**176.06**	

Source: Modified from Powertech, 2010a
Notes: measurements in this table are not converted to metric units (dual units used in SEIS text with metric being primary)
   mph = miles per hour
   lb/VMT = pounds per vehicle miles traveled
   lb/hr = pounds per hour
   PM$_{10}$ = particles with a diameter of 10 micrometers or less
*For drill rigs, "Speed (mph)" = average per hours per hole drilled
†For trackhoe and backhoe, assumed 1.56 and 1.25-cubic yard buckets and specific gravity of 1.6
‡For passenger vehicle, "Hours" column = round-trip miles; "Weight (tons)" column = trips/year
§ Constants for PM$_{10}$ Calculations:
   AP-42 Industrial Unpaved Roads: k = 1.5
   AP-42 Industrial Unpaved Roads: a = 0.9
   AP-42 Industrial Unpaved Roads: b = 0.45
   AP-42 Public Unpaved Roads: k = 1.8
   AP-42 Public Unpaved Roads: a = 1.0
   AP-42 Public Unpaved Roads: c = 0.2
   AP-42 Public Unpaved Roads: d = 0.5
   AP-42 Public Unpaved Roads: C = 0.00047
   Average silt content (%): s = 8.5
   Average moisture content (%): M = 10.4
Where separate factors were not given, PM$_{10}$ was assumed to be 30% of total suspended particulates (TSP) (AP-42 Section 13.2.2, at 12% silt, K_{PM10}/K_{TSP} = 1.5/4.9 = 0.306

Table C–17. Well Field Construction Phase

Equipment Item	Quantity	Hours	Speed (mph)	Weight (tons)	Lb/VMT	VMT	Lb/Hr	Control Efficiency	PM$_{10}$ Tons/yr	Emission Factor Reference§
Motor Grader	1	347	10		3.06	3,470		50%	2.65	AP-42 Table 11.9-1
Water Truck (1,500 gallon)	13	1,040	15	16	2.34	202,800		50%	118.43	AP-42 Section 13.2.2
Heavy Duty Diesel Truck	1	133	15	20	2.58	1,995		50%	1.29	AP-42 Section 13.2.2
Logging Truck	4	2,080	15	10	1.89	124,800		50%	58.99	AP-42 Section 13.2.2
Electrical Pole Truck	2	1,733	15	10	1.89	51,990		50%	24.57	AP-42 Section 13.2.2
Truck Mounted Drill Rig*	13	2,600	20				0.07	0%	1.10	AP-42 Table 11.9-4
Deep Well Drill Rig*	1	75	75				0.2	0%	0.00	AP-42 Table 11.9-4
Trackhoe†	1	3,120					1.66	0%	2.59	AP-42 Table 11.9-4
Backhoe†	2	2,600					1.33	0%	3.46	AP-42 Table 11.9-4
Forklift	4	1,820	5	1	0.67	36,400		50%	6.10	AP-42 Section 13.2.2
Light Duty Pickup (onsite use)	13	500	15	3	1.10	97,500		50%	26.81	AP-42 Section 13.2.2
Onsite Passenger Vehicle‡	42	4	25	250	0.63	42,000		50%	6.66	AP-42 Section 13.2.2
Total Onsite PM$_{10}$ Emissions (tons/year)									**252.66**	
Offsite Passenger Vehicle‡	42	22	40	250	0.80	231,000		0%	92.62	AP-42 Section 13.2.2
Heavy Duty Diesel Truck	1	387	25	20	2.58	9,675		0%	12.46	AP-42 Section 13.2.2
Total Offsite PM$_{10}$ Emissions (tons/year)									**105.11**	

Source: Modified from Powertech, 2010a

Notes: measurements in this table are not converted to metric units (dual units used in SEIS text with metric being primary)

mph = miles per hour

lb/VMT = pounds per vehicle miles traveled

lb/hr = pounds per hour

PM$_{10}$ = particles with a diameter of 10 micrometers or less

*For drill rigs, "Speed (mph)" = average per hours per hole drilled

†For trackhoe and backhoe, assumed 1.56 and 1.25-cubic yard buckets and specific gravity of 1.6

‡For passenger vehicle, "Hours" column = round-trip miles; "Weight (tons)" column = trips/year

§ Constants for PM$_{10}$ Calculations:

AP-42 Industrial Unpaved Roads: k = 1.5

AP-42 Industrial Unpaved Roads: a = 0.9

AP-42 Industrial Unpaved Roads: b = 0.45

AP-42 Public Unpaved Roads: k = 1.8

AP-42 Public Unpaved Roads: a = 1.0

AP-42 Public Unpaved Roads: c = 0.2

AP-42 Public Unpaved Roads: d = 0.5

AP-42 Public Unpaved Roads: C = 0.00047

Average silt content (%): s = 8.5

Average moisture content (%): M = 10.4

Where separate factors were not given, PM$_{10}$ was assumed to be 30% of total suspended particulates (TSP) (AP-42 Section 13.2.2, at 12% silt, K_{PM10}/K_{TSP} = 1.5/4.9 = 0.306

Table C–18. Operation Phase

Equipment Item	Quantity	Hours	Speed (mph)	Weight (tons)	lb/VMT	VMT	Lb/Hr	Control Efficiency	PM10 Tons/yr	Emission Factor Reference†
Motor Grader	1	416	10		3.06	4,160		50%	3.18	AP-42 Table 11.9-1
Logging Truck	1	2,080	15	10	1.89	31,200		50%	14.75	AP-42 Section 13.2.2
Resin-haul Semi Truck	1	1,040	15	20	2.58	15,600		50%	10.07	AP-42 Section 13.2.2
Water Truck (1,500 gallon)	1	1,040	15	16	2.34	15,600		50%	9.11	AP-42 Section 13.2.2
Heavy Duty Diesel Truck	1	133	15	20	2.58	1,995		50%	1.29	AP-42 Section 13.2.2
Pump Pulling Truck	4	1,560	15	10	1.89	93,600		50%	44.24	AP-42 Table 11.9-4
Forklift	1	2,132	5	1	0.67	10,660		50%	1.79	AP-42 Section 13.2.2
Manlift	1	208	2	10	1.89	416		50%	0.20	AP-42 Section 13.2.2
Product Transport Truck	1	27	15	40	0.49	405		0%	0.10	AP-42 Section 13.2.2
Light Duty Pickup (onsite use)	12	1,561	15	3	1.10	281,040		50%	77.27	AP-42 Section 13.2.2
Onsite Passenger Vehicle*	60	4	25	250	0.63	60,000		50%	9.51	AP-42 Section 13.2.2
Total Onsite PM10 Emissions (tons/year)									**171.50**	
Offsite Passenger Vehicle*	60	22	40	250	0.80	330,000		0%	132.31	AP-42 Section 13.2.2
Product Transport Truck	1	181	25	40	0.63	4,525		0%	1.43	AP-42 Section 13.2.2
Heavy Duty Diesel Truck	1	387	25	20	2.58	9,675		0%	12.49	AP-42 Section 13.2.2
Total Offsite PM10 Emissions (tons/year)									**146.24**	

Source: Modified from Powertech, 2010a
Notes: measurements in this table are not converted to metric units (dual units used in SEIS text with metric being primary)
    mph = miles per hour
    lb/VMT = pounds per vehicle miles traveled
    lb/hr = pounds per hour
    PM10 = particles with a diameter of 10 micrometers or less
*For passenger vehicle, "Hours" column = round-trip miles; "Weight (tons)" column = trips/year
†Constants for PM10 Calculations:
    AP-42 Industrial Unpaved Roads: k = 1.5
    AP-42 Industrial Unpaved Roads: a = 0.9
    AP-42 Industrial Unpaved Roads: b = 0.45
    AP-42 Public Unpaved Roads: k = 1.8
    AP-42 Public Unpaved Roads: a = 1.0
    AP-42 Public Unpaved Roads: c = 0.2
    AP-42 Public Unpaved Roads: d = 0.5
    AP-42 Public Unpaved Roads: C = 0.00047
    Average silt content (%): s = 8.5
    Average moisture content (%): M = 10.4

Table C–19. Aquifer Restoration Phase

Equipment Item	Quantity	Hours	Speed (mph)	Weight (tons)	lb/VMT	VMT	Lb/Hr	Control Efficiency	PM10 Tons/yr	Emission Factor Reference†
Light Duty Pickup (onsite use)	1	2,912	15	3	1.10	43,680		50%	12.01	AP-42 Section 13.2.2
Onsite Passenger Vehicle*	6	4	25	250	0.63	6,000		50%	0.95	AP-42 Section 13.2.2
Total Onsite PM10 Emissions (tons/year)									**12.96**	

Table C–19. Aquifer Restoration Phase (continued)

Equipment Item	Quantity	Hours	Speed (mph)	Weight (tons)	lb/VMT	VMT	Lb/Hr	Control Efficiency	PM10 Tons/yr	Emission Factor Reference†
Offsite Passenger Vehicle*	6	22	40	250	0.80	33,000		0%	13.23	AP-42 Section 13.2.2
Total Offsite PM₁₀ Emissions (tons/year)									**13.23**	

Source: Modified from Powertech, 2010a
Notes: measurements in this table are not converted to metric units (dual units used in SEIS text with metric being primary)
  mph = miles per hour
  lb/VMT = pounds per vehicle miles traveled
  lb/hr = pounds per hour
  PM_{10} = particles with a diameter of 10 micrometers or less
*For passenger vehicle, "Hours" column = round-trip miles; "Weight (tons)" column = trips/year
†Constants for PM_{10} Calculations:
  AP-42 Industrial Unpaved Roads: k = 1.5
  AP-42 Industrial Unpaved Roads: a = 0.9
  AP-42 Industrial Unpaved Roads: b = 0.45
  AP-42 Public Unpaved Roads: k = 1.8
  AP-42 Public Unpaved Roads: a = 1.0
  AP-42 Public Unpaved Roads: c = 0.2
  AP-42 Public Unpaved Roads: d = 0.5
  AP-42 Public Unpaved Roads: C = 0.00047
  Average silt content (%): s = 8.5
  Average moisture content (%): M = 10.4

Table C–20. Decommissioning Phase

Equipment Item	Quantity	Hours	Speed (mph)	Weight (tons)	lb/VMT	VMT	Lb/Hr	Control Efficiency	PM10 Tons/yr	Emission Factor Reference‡
Scraper	3	867	15	30	3.10	39,015		50%	30.23	AP-42 Section 13.2.2
Bulldozer	1	867					0.70	0%	0.00	AP-42 Table 11.9-1
Compactor	1	867	5	5	1.38	4,335		50%	1.50	AP-42 Section 13.2.2
Motor Grader	1	867	10		3.06	8,670	30.60	50%	6.63	AP-42 Table 11.9-1
Water Truck (1,500 gallon)	1	867	15	16	2.34	13,005		50%	7.59	AP-42 Section 13.2.2
Fueling Truck	1	520	15	10	1.89	7,800		50%	3.69	AP-42 Section 13.2.2
Loader	1	65					17.76	0%	0.00	AP-42 Table 11.9-4
Heavy Duty Diesel Truck	4	87	15	20	2.58	5,205		50%	3.36	AP-42 Section 13.2.2
Pump Pulling Truck	1	2,130	15	10	1.89	31,950		50%	15.10	AP-42 Table 11.9-4
Trackhoe*	2	650					1.66	0%	0.00	AP-42 Table 11.9-4
Backhoe*	2	650					1.33	0%	0.00	AP-42 Table 11.9-4

Table C–20. Decommissioning Phase (continued)

Equipment Item	Quantity	Hours	Speed (mph)	Weight (tons)	lb/VMT	VMT	Lb/Hr	Control Efficiency	PM10 Tons/yr	Emission Factor Reference‡
Forklift	3	693	5	1	0.67	10,395		50%	1.74	AP-42 Section 13.2.2
Manlift	4	693	2	10	1.89	5,544		50%	2.62	AP-42 Section 13.2.2
Tractor	1	650	5	1.38	3,250			0%	2.25	AP-42 Section 13.2.2
Light Duty Pickup (onsite use)	2	2,000	15	3	1.10	60,000		50%	16.50	AP-42 Section 13.2.2
Onsite Passenger Vehicle†	15	4	25	250	0.63	15,000		50%	2.38	AP-42 Section 13.2.2
Total Onsite PM₁₀ Emissions (tons/year)									**93.59**	
Offsite Passenger Vehicle†	15	22	40	250	0.80	82,500		0%	33.08	AP-42 Section 13.2.2
Heavy Duty Diesel Truck	4	260	25	20	2.58	26,025		0%	33.61	AP-42 Section 13.2.2
Total Offsite PM₁₀ Emissions (tons/year)									**66.69**	

Source: Modified from Powertech, 2010a

Notes: measurements in this table are not converted to metric units (dual units used in SEIS text with metric being primary)

mph = miles per hour

lb/VMT = pounds per vehicle miles traveled

lb/hr = pounds per hour

PM_{10} = particles with a diameter of 10 micrometers or less

*For trackhoe and backhoe, assumed 1.56 and 1.25-cubic yard buckets and specific gravity of 1.6

†For passenger vehicle, "Hours" column = round-trip miles; "Weight (tons)" column = trips/year

‡Constants for PM_{10} Calculations:

AP-42 Industrial Unpaved Roads: k = 1.5

AP-42 Industrial Unpaved Roads: a = 0.9

AP-42 Industrial Unpaved Roads: b = 0.45

AP-42 Public Unpaved Roads: k = 1.8

AP-42 Public Unpaved Roads: a = 1.0

AP-42 Public Unpaved Roads: c = 0.2

AP-42 Public Unpaved Roads: d = 0.5

AP-42 Public Unpaved Roads: C = 0.00047

Average silt content (%): s = 8.5

Average moisture content (%): M = 10.4

Where separate factors were not given, PM_{10} was assumed to be 30% of total suspended particulates (TSP)

(AP-42 Section 13.2.2, at 12% silt, K_{PM10}/K_{TSP} = 1.5/4.9 = 0.306

APPENDIX D

DOCUMENTS RELATED TO HISTORIC AND CULTURAL RESOURCES

Rapid City 026642.

4—1008-R.

The United States of America,

To all to whom these presents shall come, Greeting:

WHEREAS, a Certificate of the Register of the Land Office at **Rapid City, South Dakota,**

has been deposited in the General Land Office, whereby it appears that, pursuant to the Act of Congress of May 20, 1862,

"To Secure Homesteads to Actual Settlers on the Public Domain," and the acts supplemental thereto, the claim of

Emaline Richardson

has been established and duly consummated, in conformity to law, for the **west half of the southeast quarter**

of Section ten and the west half of the northeast quarter of Section fifteen

in Township seven south of Range one east of the Black Hills Meridian,

South Dakota, containing one hundred sixty acres,

according to the Official Plat of the Survey of the said Land, returned to the GENERAL LAND OFFICE by the Surveyor-General:

NOW KNOW YE, That there is, therefore, granted by the UNITED STATES unto the said claimant   the tract of Land above described:
TO HAVE AND TO HOLD the said tract of Land, with the appurtenances thereof, unto the said claimant   and to the heirs and assigns of
the said claimant   forever; subject to any vested and accrued water rights for mining, agricultural, manufacturing, or other purposes, and
rights to ditches and reservoirs used in connection with such water rights, as may be recognized and acknowledged by the local customs, laws,
and decisions of courts; and there is reserved from the lands hereby granted, a right of way thereon for ditches or canals constructed by the
authority of the United States.

IN TESTIMONY WHEREOF, I, **Woodrow Wilson**

President of the United States of America, have caused these letters to be made

Patent; and the seal of the General Land Office to be hereunto affixed.

GIVEN under my hand, at the City of Washington, the   **TWENTY-THIRD**

(SEAL)   day of   **JANUARY**   in the year of our Lord one thousand

nine hundred and   **FIFTEEN**   and of the Independence of the

United States the one hundred and   **THIRTY-NINTH.**

By the President: *Woodrow Wilson*

By   *M. O. LeRoy*   Secretary,

R. C. Lamar
Recorder of the General Land Office.

RECORD OF PATENTS: Patent Number ___455381___   4—6177

United States Department of the Interior

BUREAU OF LAND MANAGEMENT
South Dakota Field Office
310 Roundup Street
Belle Fourche, South Dakota 57717-1698
http://www.blm.gov/mt

TAKE PRIDE
IN AMERICA

In Reply Refer To:

8100-R
BAS

12-MT040-15

Date: July 20, 2012

Mr. Richard E. Blubaugh
Vice President – Environmental Health & Safety Resources
Powertech (USA) Incorporated
5575 DTC Parkway, Suite 140
Greenwood Village, CO 80111

RE: Cultural Resource review of Evaluative Testing of 20 Sites in the Powertech (USA)
Inc. Dewey-Burdock Uranium Project Impact Areas: Volumes 1 and 2. For the Dewey-
Burdock Uranium Recovery Project, Fall River and Custer Counties, South Dakota.

Dear Mr. Blubaugh:

We have reviewed the appropriate volumes of the National Historic Preservation Act,
Section 106 cultural compliance reports presented by Archeology Laboratory, Augustana
College, for evaluation of cultural resource sites inside areas of potential effect for the
proposed Dewey-Burdock project area. The reports reviewed document formal
evaluation of 20 cultural resource sites inside areas proposed for the project that could
have effect. Of these 20 sites, one site is located in part on BLM administered surface
land.

Site 39FA96 was found to be significantly affected by natural erosion and therefore does
not possess adequate integrity, does not display workmanship or feeling, and it is not
associated with an important historic event. Based on the information provided in the
report the Bureau of Land Management (BLM) recommends adequate testing was
completed on site 39FA96, the site's integrity has been severely affected by deflation.
The portion on BLM administered land does not possess enough information to meet the
National Register of Historic Places criteria for an eligible archaeological site; therefore,
the BLM is in agreement with the determination for site 39FA96 on this portion, in that it
is considered not eligible for nomination to the National Register of Historic Places.
Information provided for the remaining 19 sites should to be sufficient for the lead
Federal Agency to make informed recommendations of eligibility on the historic
properties.

Mr. Richard E. Blubaugh
July 20, 2012
Page 2

Please let us know if you should need any additional information. I can be reached as
Mr. Mitch Iverson (acting South Dakota Field Manager), (605) 892-7001 or email at
Mitchell_Iverson@blm.gov or contact our archaeologist, Brenda Shierts at (605) 723-
8712 or Brenda_Shierts@blm.gov.

Sincerely,

Marian M. Atkins
South Dakota Field Manager

cc: Paige Olson, SD SHPO
        Gary Smith, BLM MSO Historic Preservation Officer
        Mark Sant, BLM MSO Tribal Coordinator
        Mr. Greg R. Fesko, P.G., BLM MSO Solid Minerals
        Haimanot Yilma, NRC, Project Manager Environmental Review Branch

SOUTH DAKOTA STATE HISTORIC PRESERVATION OFFICE

HISTORIC SITES SURVEY STRUCTURE FORM   06-01-2012

SOUTH DAKOTA
STATE HISTORICAL SOCIETY

SHPOID	SiteID	StructureID
CU00000050	9202	13997

SITE INFORMATION

*Survey Date:	6/20/1988 12:00:00 AM	*Quarter1:	SE
*Surveyor:	Unknown	*Quarter2:	SE
*Property Address:	Unknown	*Township:	6S
*County:	cu	*Range:	1E
*City:	Dewey	*Section:	31
		Acres:	12.000
		Quadname:	Twenty-one Divide

Legal Description:

Location Description: approx. 3 miles south of Dewey

Owner Code1: P	Owner Name:	Andes, Clark
Owner Code2:	Owner Address:	PO Box 560
Owner Code3:	Owner City:	Black Hawk
	Owner State:	SD
	Owner Zip:	57718

HISTORIC SIGNIFICANCE

*DOE:	NR Eligible	Register Name:	Young, Edna, and Ernest, Ranch
*DOE Date:	7/5/1990 12:00:00 AM	Multiple Property Name	Ranches of Southwestern Custer County
Nomination Status:	NR listed	SignificanceLevel1:	Local
Listed Date:	7/5/1990 12:00:00 AM	SignificanceLevel2:	Local
Ref Num:	90000949	NR Criteria 1:	A
Period:	1912-40	NR Criteria 2:	
Category:	District	NR Criteria 3:	
Historic District Rating:		NR Criteria 4:	

SOUTH DAKOTA STATE HISTORIC PRESERVATION OFFICE

HISTORIC SITES SURVEY STRUCTURE FORM   06-01-2012

SOUTH DAKOTA
STATE HISTORICAL SOCIETY

Significance Notes : Significant in the area of Exploration/Settlement, because it represents the development of the legal homestead rancher in the southwestern corner of Custer County, SD. In particular, this ranch represents homesteading on more than one claim.

STRUCTURE DETAILS

'Structure Name: Bakewell Ranch

Other Name: Andes, Clark

Date Of Construction: 1912   Significant Person: Young, Ernie

Cultural Affiliation:

Type: Walls: Stone

Style: Stories:

Roof Shape: Gambrel   Foundataion:

Roof Material: Metal   'UTM Zone: 14

Occupied: 'UTM Easting: 92020.0106

Accessible: 'UTM Northing: 4827170.3755

Structural System: Restricted: N

Altered/Moved Notes:

Interior Notes:

* = REQUIRED FIELD   Page 2 of 4

SOUTH DAKOTA STATE HISTORIC PRESERVATION OFFICE

HISTORIC SITES SURVEY STRUCTURE FORM   06-01-2012

SOUTH DAKOTA
STATE HISTORICAL SOCIETY
Department of Tourism and State Development

Physical Notes: House (contributing) 1912. A sandstone masonry house, the material quarried from the nearby "racetrack" of the Black Hills formation. The gable ends are sheathed with white asphalt shingles, two shed roofed dormers, wood frame summer kitchen is attached to the NW, fenestrations irregular, a replica of an early shed roof porch was recent added. Bunkhouse (contributing) Stucco-covered wood frame, with a tin-clad gable roof.

Small House (contributing) One story, rectangular pen building, on a wood sill, sided in clapboard and tar paper, gable roof covered in tar paper. Privy (contributing) Wood frame, wood sill capped by a board-covered gable roof.

Granary (contributing) Rectangular pen wood frame, on stone piers, capped by a tin-covered shed roof.

Granary (noncontributing) A modern round metal bin with a conical metal roof.

Sheep wagon (contributing) Small frame building, with a segmental arch roof clad with tar paper. Once mounted on wheels, it was moved to pasture ranges to provide domestic shelter for persons who tended grazing sheep herds.

Garage (contributing) Wood frame building clad with stucco, resting on a stone foundation, capped by a tar paper-covered gable roof.

Small Garage (contributing) One- car garage, wood frame covered with stucco and capped by a tin-clad segmental arch roof.

Livestock Building (contributing) Wood frame, gable roof.

Oil Storage Building (contributing) Wood frame, gable roof.

Log Outbuilding (contributing) Rectangular outbuilding constructed of rough hewn logs with double vertical corner notches, chinked with plaster and cement, gable roof covered with sod.

Three Sheds (contributing) Rectangular pen, wood frame with shed roofs. Used fro working livestock. Rectangular pen,

Other Notes: The Young Ranch illustrates the common western SD practice of creating a larger, more economically productive ranch by various family members homesteading several adjacent parcels. In 1908, Edna Petty Young recived a patent for a parcel in the SW quarter of section 31. Four years later, her husband Ernest Young received a patent to the adjoinging NE quarter. Throughout the 1920s, the two homesteaders continued to prove up contiguous and discontinguous claims in Custer County and in Niobrara County, WY. As late as 1942, Ida Kirby Young (Ernest's mother) received a patent to land in Wyoming. However, all of the permanent domestic and agricutlural operations of the ranch took place on the nominated ranch site in Section 31. Originally from nearby Pringle, SD, the Youngs were principally cattle and sheep ranchers. Although with generally little success, they attempted to irrigate some of their land with overflow from Beaver Creek. In later years, Ernest Young also raised Appaloosa horses. Edna and Ernest Young's one daughter, Lena, earned a teaching certificate at the normal school in Spearfish, SD. Upon graduation, she returned to her parent's ranch and taught in the local schools. In 1965, the Young family sold the ranch to Robert and Lois Bakewell, who operated it on a nonresident basis for 15 years. Between 1980 and 1985, John Holmes leased the property. In 1985, Clark Andis purchased the property from the Bakewells.

SOUTH DAKOTA STATE HISTORIC PRESERVATION OFFICE

HISTORIC SITES SURVEY STRUCTURE FORM   06-01-2012

SOUTH DAKOTA
STATE HISTORICAL SOCIETY
Department of Tourism and State Development

Link to National Register Nomination:

http://pdfhost.focus.nps.gov/docs//NRHP//Text//90000949.pdf

NRC FORM 335
(12-2010)
NRCMD 3.7

U.S. NUCLEAR REGULATORY COMMISSION

BIBLIOGRAPHIC DATA SHEET

(See instructions on the reverse)

1. REPORT NUMBER
(Assigned by NRC, Add Vol., Supp., Rev., and Addendum Numbers, if any.)

NUREG-1910, Supplement 4, Volume 2

2. TITLE AND SUBTITLE

Environmental Impact Statement for the Dewey-Burdock In-Situ Recovery Project in Fall River and Custer Counties, South Dakota

Supplement to the Generic Environmental Impact Statement for In-Situ Leach Uranium Milling Facilities

3. DATE REPORT PUBLISHED

MONTH	YEAR
November	2012

4. FIN OR GRANT NUMBER

5. AUTHOR(S)

6. TYPE OF REPORT

Technical

7. PERIOD COVERED (Inclusive Dates)

8. PERFORMING ORGANIZATION - NAME AND ADDRESS (If NRC, provide Division, Office or Region, U. S. Nuclear Regulatory Commission, and mailing address; if contractor, provide name and mailing address.)

Division of Waste Management and Environmental Protection
Office of Federal and State Materials and Environmental Management Programs
U.S. Nuclear Regulatory Commission
Washington, DC 20555-001

9. SPONSORING ORGANIZATION - NAME AND ADDRESS (If NRC, type "Same as above", if contractor, provide NRC Division, Office or Region, U. S. Nuclear Regulatory Commission, and mailing address.)

Same as above

10. SUPPLEMENTARY NOTES

11. ABSTRACT (200 words or less)

By letter dated August 10, 2009, Powertech (USA), Inc. (Powertech, the applicant) submitted a source material license application to the U.S. Nuclear Regulatory Commission (NRC) for the Dewey-Burdock in-situ recovery (ISR) Project. Powertech is proposing to construct, operate, conduct aquifer restoration, and decommission an ISR facility at the Dewey-Burdock ISR Project site, located in Fall River and Custer Counties, South Dakota. The NRC staff evaluated site-specific data and information to assess whether the applicant-proposed activities were consistent with activities considered in NUREG-1910, "Generic Environmental Impact Statement for In-Situ Leach Uranium Milling Facilities" (GEIS) and determined which GEIS data and analyses could be incorporated by reference and what resource areas required site-specific review. The draft SEIS describes the environment potentially affected by the proposed site activities, describes the potential environmental impacts, and describes Powertech's environmental monitoring program and proposed mitigation measures. The NRC staff will respond to public comments received on the draft SEIS in the final SEIS.

12. KEY WORDS/DESCRIPTORS (List words or phrases that will assist researchers in locating the report.)

Uranium Recovery
In-Situ Recovery Process
Uranium
Environmental Impact Statement
Supplemental Environmental Impact Statement

13. AVAILABILITY STATEMENT

unlimited

14. SECURITY CLASSIFICATION

(This Page)

unclassified

(This Report)

unclassified

15. NUMBER OF PAGES

16. PRICE

NUREG-1910
Supplement 4, Vol. 2
Draft

Environmental Impact Statement for the Dewey-Burdock Project in
Custer and Fall River Counties, South Dakota

November 2012

www.ingramcontent.com/pod-product-compliance
Lightning Source LLC
Chambersburg PA
CBHW080234180526
45167CB00006B/2268